Nanostructured Magnetic Materials and their Applications

T0137681

NATO Science Series

A Series presenting the results of scientific meetings supported under the NATO Science Programme.

The Series is published by IOS Press, Amsterdam, and Kluwer Academic Publishers in conjunction with the NATO Scientific Affairs Division

Sub-Series

I.   **Life and Behavioural Sciences**   IOS Press
II. **Mathematics, Physics and Chemistry**   Kluwer Academic Publishers
III. **Computer and Systems Science**   IOS Press
IV. **Earth and Environmental Sciences**   Kluwer Academic Publishers
V.   **Science and Technology Policy**   IOS Press

The NATO Science Series continues the series of books published formerly as the NATO ASI Series.

The NATO Science Programme offers support for collaboration in civil science between scientists of countries of the Euro-Atlantic Partnership Council. The types of scientific meeting generally supported are "Advanced Study Institutes" and "Advanced Research Workshops", although other types of meeting are supported from time to time. The NATO Science Series collects together the results of these meetings. The meetings are co-organized bij scientists from NATO countries and scientists from NATO's Partner countries – countries of the CIS and Central and Eastern Europe.

Advanced Study Institutes are high-level tutorial courses offering in-depth study of latest advances in a field.
Advanced Research Workshops are expert meetings aimed at critical assessment of a field, and identification of directions for future action.

As a consequence of the restructuring of the NATO Science Programme in 1999, the NATO Science Series has been re-organised and there are currently Five Sub-series as noted above. Please consult the following web sites for information on previous volumes published in the Series, as well as details of earlier Sub-series.

http://www.nato.int/science
http://www.wkap.nl
http://www.iospress.nl
http://www.wtv-books.de/nato-pco.htm

Series II: Mathematics, Physics and Chemistry – Vol. 143

Nanostructured Magnetic Materials and their Applications

edited by

Bekir Aktaş

Gebze Institute of Technology,
Gebze, Kocaeli, Turkey

Lenar Tagirov

Kazan State University,
Kazan, Russia

and

Faik Mikailov

Institute of Physics of NAS,
Baku, Azerbaijan

Kluwer Academic Publishers

Dordrecht / Boston / London

Published in cooperation with NATO Scientific Affairs Division

Proceedings of the NATO Advanced Research Workshop on
Nanostructured Magnetic Materials and their Applications
Istanbul, Turkey
1–4 July 2003

A C.I.P. Catalogue record for this book is available from the Library of Congress.

ISBN 1-4020-2004-X (PB)
ISBN 1-4020-2003-1 (HB)
ISBN 1-4020-2200-X (e-book)

Published by Kluwer Academic Publishers,
P.O. Box 17, 3300 AA Dordrecht, The Netherlands.

Sold and distributed in North, Central and South America
by Kluwer Academic Publishers,
101 Philip Drive, Norwell, MA 02061, U.S.A.

In all other countries, sold and distributed
by Kluwer Academic Publishers,
P.O. Box 322, 3300 AH Dordrecht, The Netherlands.

Printed on acid-free paper

Contents

Contributing Authors

A. Al-Jibouri
Nordiko Ltd., Havant, Hampshire, PO9 2NL, UK

B. Aktaş
Gebze Institute of Technology, P.O. 141,
41400, Kocaeli, Turkey

O.D. Asenchik
Gomel State Technical University, October av. 48,
Gomel, 246746, Belarus

I. Avgin
Electrical and Electronics Engineering Department,
Ege University, Bornova, Izmir 35100 Turkey

B. Bakar
Universiteit Antwerpen Departement Natuurkunde
Groenenborgerlaan, 171 B2020 Antwerpen Belgie

M. Bal
University of Massachusetts, Department of Physics,
Amherst, MA 01003, USA

N. Berdunov
SFI Laboratories, Trinity College,
Dublin 2, Ireland

F. S. Bergeret
Ruhr-University Bochum,
D-44780 Bochum, Germany

H. Braak
Institut für Festkörperforschung, Forschungszentrum Jülich GmbH,
D-52425 Jülich, Germany d.buergler@fz-juelich.de

G. Brown
School for Computational Science and Information Technology,
Florida State University,   Tallahassee, FL, USA

H. Brückl
University of Bielefeld, Department of Physics, Nanodevice group,
P.O. Box 100 131, 33501 Bielefeld, Germany

M. Brzeska
University of Bielefeld, Department of Physics, Nanodevice group,
P.O. Box 100 131, 33501 Bielefeld, Germany

D. E. Bürgler
Institut für Festkörperforschung, Forschungszentrum Jülich GmbH,
D-52425 Jülich, Germany

M. Buchmeier
Institut für Festkörperforschung, Forschungszentrum Jülich GmbH,
D-52425 Jülich, Germany d.buergler@fz-juelich.de

S. Budak
Fatih University, Faculty of Art and Science,
Physics Department, 34900, Istanbul,Turkey

S.F. Ceballos
SFI Laboratories, Trinity College,
Dublin 2, Ireland

H. Cheng
Laboratorio de Física de Sistemas Pequeños y Nanotecnología
Consejo Superior de Investigaciones Científicas
Serrano 144, Madrid 28006, Spain

J.P. Clerc
Ecole Polytechnique, Universitaire de Marseille,
Technopole de Chateau Gombert, 13453, Marseille, France

D. Dolgy,
Russian Research Center "Kurchatov Institute",
Kurchatov sq. 1, Moscow 123182, Russia

A. Domantovsky,
Russian Research Center "Kurchatov Institute",
Kurchatov sq. 1, Moscow 123182, Russia

K.B. Efetov
Ruhr-University Bochum,
D-44780 Bochum, Germany

M. Z. Fattakhov
Zavoisky Physical-Technical Institute,
Russian Academy of Sciences, 420029 Kazan

N. García
Laboratorio de Física de Sistemas Pequeños y Nanotecnología
Consejo Superior de Investigaciones Científicas
Serrano 144, Madrid 28006, Spain

R.R. Gareev
Institut für Festkörperforschung, Forschungszentrum Jülich GmbH,
D-52425 Jülich, Germany

I. A. Garifullin
Zavoisky Physical-Technical Institute,
Russian Academy of Sciences, 420029 Kazan

N.N. Garifyanov
Zavoisky Physical-Technical Institute,
Russian Academy of Sciences, 420029 Kazan

A. Gedanken
Department of Chemistry, Bar-Ilan University,
52900, Ramat-Gan, Israel

G. Gorodetsky
Department of Physics, Ben-Gurion University of the Negev,
P.O. Box 653, 84105, Beer-Sheva, Israel

A. Granovsky
P Faculty of Physics, Lomonosov Moscow State University,
119992 Moscow, Russia

P. Grünberg
Institut für Festkörperforschung, Forschungszentrum Jülich GmbH,
D-52425 Jülich, Germany d.buergler@fz-juelich.de

C. Guerrero
Laboratorio de Física de Sistemas Pequeños y Nanotecnología
Consejo Superior de Investigaciones Científicas
Serrano 144, Madrid 28006, Spain

A. Gupta
IBM T.J. Watson R.C., Yorktown Heights,
New York 10598, USA

B. Gurovich
Russian Research Center "Kurchatov Institute",
Kurchatov sq. 1, Moscow 123182, Russia

T. He
Department of Chemical and Materials Engineering,
University of Cincinnati, Cincinnati, OH 45221-0012, USA

B. Heinrich
Simon Fraser University, Burnaby,
BC, V5A 1S6, Canada

A. Hütten
University of Bielefeld, Department of Physics, Nanodevice group,
P.O. Box 100 131, 33501 Bielefeld, Germany

M. Inoue
Toyohashi University of Technology,
Toyohashi 441-8580, Japan

J.S. Jiang
Materials Science Division, Argonne National Laboratory,
Argonne, Illinois USA

S. Kämmerer
University of Bielefeld, Department of Physics, Nanodevice group,
P.O. Box 100 131, 33501 Bielefeld, Germany

I.B. Khaibullin
Kazan Physical-Technical Institute, Sibirsky Trakt 10/7,
420029 Kazan, Russia

R.I. Khaibullin
Kazan Physical-Technical Institute, Sibirsky Trakt 10/7,
420029 Kazan, Russia

S. Ya. Khlebnikov
Zavoisky Physical-Technical Institute,
Russian Academy of Sciences, 420029 Kazan

D.K. Kim
Royal Institute of Technology, Materials Chemistry Division,
Stockholm, Sweden

H. Koop
University of Bielefeld, Department of Physics, Nanodevice group,
P.O. Box 100 131, 33501 Bielefeld, Germany

A. Kozlov
P Faculty of Physics, Lomonosov Moscow State University,
119992 Moscow, Russia

Y. Köseoğlu
Fatih University, Physics Department,
Istanbul, Turkey

U. Kreibig
Physikalisches Institut IA der RWTH, Sommerfeldstrasse 14,
52056 Aachen, Germany

V. Krivoruchko
Donetsk Physics & Technology Institute,
Donetsk-114, 83114 Ukraine

E. Kuleshova,
Russian Research Center "Kurchatov Institute",
Kurchatov sq. 1, Moscow 123182, Russia

A. Layadi
Département de Physique, Université Ferhat Abbas,
Sétif 19000, Algeria

V. Leiner
Institut Laue Langevin,
Grenoble, France

L.F. Lemmens
Universiteit Antwerpen Departement Natuurkunde
Groenenborgerlaan 171 B2020 Antwerpen Belgie

J. Lian
Department of Nuclear Engineering and Radiological Science,
University of Michigan, Ann Arbor, MI 48109, USA

Z. Lu
CRIST, School of Computing, Communication and Electronics,
Univerisyt of Plymouth, Plymouth, Devon, PL4 8AA, UK

G. Mariotto
SFI Laboratories, Trinity College,
Dublin 2, Ireland

K. Maslakov,
Russian Research Center "Kurchatov Institute",
Kurchatov sq. 1, Moscow 123182, Russia

E. Meilikhov,
Russian Research Center "Kurchatov Institute",
Kurchatov sq. 1, Moscow 123182, Russia

D. Meyners
University of Bielefeld, Department of Physics, Nanodevice group,
P.O. Box 100 131, 33501 Bielefeld, Germany

M. Muhammed
Royal Institute of Technology, Materials Chemistry Division,
Stockholm, Sweden

W. Nawrocki
Poznan University of Technology, ul. Piotrowo 3A,
60-965 Poznan, Poland

N.D. Nikolic
Laboratorio de Física de Sistemas Pequeños y Nanotecnología
Consejo Superior de Investigaciones Científicas
Serrano 144, Madrid 28006, Spain

M.A. Novotny
Mississippi State University,
Mississippi State, MS, USA

C. Okay
Faculty of Art and Science, Physics Department,
Marmara University, P.K. 81040, Göztepe/Istanbul, Turkey

M. Özdemir
Faculty of Art and Science, Physics Department,
Marmara University, Istanbul Turkey

K. Özdoğan
Gebze Institute of Technology, P.K.141,
41400, Kocaeli, Turkey

Y. Öztürk
Electrical and Electronics Engineering Department,
Ege University, Bornova, Izmir 35100 Turkey

G. Pan
CRIST, School of Computing, Communication and Electronics,
Univerisyt of Plymouth, Plymouth, Devon, PL4 8AA, UK

G. Pang
Department of Chemistry, Bar-Ilan University,
52900, Ramat-Gan, Israel

A.C. Papageorgopoulos
Laboratorio de Física de Sistemas Pequeños y Nanotecnología
Consejo Superior de Investigaciones Científicas
Serrano 144, Madrid 28006, Spain

R. Pecenka
Physikalisches Institut IA der RWTH, Sommerfeldstrasse 14,
52056 Aachen, Germany

L.L. Pohlmann
Institut für Festkörperforschung, Forschungszentrum Jülich GmbH,
D-52425 Jülich, Germany d.buergler@fz-juelich.de

K. Prikhodko,
Russian Research Center "Kurchatov Institute",
Kurchatov sq. 1, Moscow 123182, Russia

B.Z. Rameev
Kazan Physical-Technical Institute, Sibirsky Trakt 10/7,
420029 Kazan, Russia

A. Reinholdt
Physikalisches Institut IA der RWTH, Sommerfeldstrasse 14,
52056 Aachen, Germany

G. Reiss
University of Bielefeld, Department of Physics, Nanodevice group,
P.O. Box 100 131, 33501 Bielefeld, Germany

P.A. Rikvold
Center for Materials Research and Technology and Department of
Physics, Florida State University, Tallahassee, FL, US

E. Rozenberg
Department of Physics, Ben-Gurion University of the Negev,
P.O. Box 653, 84105, Beer-Sheva, Israel

V. Rylkov,
Russian Research Center "Kurchatov Institute", Kurchatov sq. 1,
Moscow 123182, Russia

J. Schmalhorst
University of Bielefeld, Department of Physics, Nanodevice group,
P.O. Box 100 131, 33501 Bielefeld, Germany

T. Schmitte
Institut für Experimentalphysik/Festkörperphysik,
Ruhr-Universitaet, D 44780 Bochum, Germany

R. Schreiber
Institut für Festkörperforschung, Forschungszentrum Jülich GmbH,
D-52425 Jülich, Germany d.buergler@fz-juelich.de

A.I. Shames
Department of Physics, Ben-Gurion University of the Negev,
P.O. Box 653, 84105, Beer-Sheva, Israel

D. Shi
Department of Chemical and Materials Engineering, University of
Cincinnati, Cincinnati, OH 45221-0012, USA

I.V. Shvets
SFI Laboratories, Trinity College,
Dublin 2, Ireland

A. Smirnov
P. L. Kapitza Institute for Physical Problems RAS,
117334 Moscow, Russia

E.G. Starodubtsev
Gomel State Technical University, October av. 48,
Gomel, 246746, Belarus

A.L. Stepanov
Kazan Physical-Technical Institute, Sibirsky Trakt 10/7,
420029 Kazan, Russia

S. M. Stinnett
Mississippi State University., Mississippi State, MS, USA

L.R. Tagirov
Kazan State University, 420008 Kazan, Russian Federation

S. Tarapov
Institute of Radiophysics and Electronics NAS of Ukraine,
12 Ac Proskura St., 61085, Kharkov, Ukraine

K. Theis-Bröhl
Institut für Experimentalphysik/Festkörperphysik,
Ruhr-Universität Bochum, 44780 Bochum, Germany

A. Thomas
MIT, Francis Bitter Magnet Lab., NW 14-2128,
170 Albany St., 02139 Cambridge, MA, USA

D.A. Tikhonov
Zavoisky Physical-Technical Institute,
Russian Academy of Sciences, 420029 Kazan

M.T. Tuominen
University of Massachusetts,
Department of Physics,
Amherst, MA 01003, USA

R. Urban
Simon Fraser University, Burnaby,
BC, V5A 1S6, Canada

B.P. Vodopyanov
Kazan Physico-technical Institute of RAS,
420029 Kazan, Russian Federation

F. Volkov
Ruhr-University Bochum,
D-44780 Bochum, Germany

H. Wang
Laboratorio de Física de Sistemas Pequeños y Nanotecnología
Consejo Superior de Investigaciones Científicas
Serrano 144, Madrid 28006, Spain

L.M. Wang
Department of Nuclear Engineering and Radiological Science,
University of Michigan, Ann Arbor, MI 48109, USA

M. Wawrzyniak
Poznan University of Technology,
ul. Piotrowo 3A, 60-965 Poznan, Poland

K. Westerholt
Institut für Experimentalphysik/Festkörperphysik,
Ruhr-Universität Bochum, 44780 Bochum, Germany

A. Westphalen
Institut für Experimentalphysik/Festkörperphysik,
Ruhr-Universitaet, D 44780 Bochum, Germany

G. Woltersdorf
Simon Fraser University, Burnaby,
BC, V5A 1S6, Canada

A. Yakubovsky
Russian Research Center "Kurchatov Institute",
Kurchatov sq. 1, Moscow 123182, Russia

O. Yalçın
Department of Physics, Gaziosmanpaşa University,
60110, Tokat Turkey

R. Yilgin
Gebze Institute of Technology, P.K.141,
41400, Kocaeli, Turkey

F. Yıldız
Gebze Institute of Technology, P.K.141,
41400, Kocaeli, Turkey

A. Yurasov
P Faculty of Physics, Lomonosov Moscow State University,
119992 Moscow, Russia

H. Zabel
Institut für Experimentalphysik/Festkörperphysik,
Ruhr-Universität Bochum, 44780 Bochum, Germany

V.A. Zhikharev
Kazan Physical-Technical Institute of RAS,
420029 Kazan, Russia

M. Ziese
Department of Superconductivity and Magnetism,
University of Leipzig, 04103 Leipzig, Germany

Preface

The interests to research on nanoscale materials have been steadily increasing. The nano-structured magnetic materials exhibit new and interesting physical properties, which cannot be found in the bulk. Many of these unique properties have high potential for technical applications in magneto-sensors, bio-sensors, magneto-electronics, data storage, magnetic heads of computer hard disks, single-electron devises, microwave electronic devices, etc. The scientific research concentrates on the device design, synthesis and characterization of nanostructured materials.

This book contains a collection of papers that were presented at NATO Advanced Research Workshop on Nanostructured Magnetic Materials and their Application (NMMA-2003) held in Istanbul (Turkey) on July 1-4, 2003. The contributions have concentrated on magnetic properties of nanoscale magnetic materials, especially, on fabrication, characterization, physics behind the unique properties these structures and device design.

We would like to thank all participants for their invaluable contributions. We acknowledge great efforts of our colleagues, Dr. Engin Başaran, Dr. Bulat Rameev and others who made major contribution to the organization of the meeting. Special thanks to Prof. Alinur Büyükaksoy, Rector of Gebze Institute of Technology, who supported every stages of preparation and holding of the meeting, and to the Scientific and the Technical Council of Turkey (TÜBİTAK) for the support of this workshop.

Bekir Aktaş, Lenar Tagirov, Faik Mikailov
The Editors

Acknowledgments

The Editors are grateful to NATO Scientific Affairs Division for supporting the Advanced Research Workshop on Nanostructured Magnetic Materials and also for the preparation of this Proceeding Book.

1

PART 1: SYNTHESIS AND CHARACTERIZATION OF NANOSTRUCTURED MATERIALS

PLASMA COATING OF MAGNETIC NANOPARTICLES FOR BIO-PROBE APPLICATIONS

D. Shi [a], T. He [a], J. Lian [b], L. M. Wang [b]

[a] *Department of Chemical and Materials Engineering, University of Cincinnati, Cincinnati, OH 45221-0012, USA*

[b] *Department of Nuclear Engineering and Radiological Science, University of Michigan, Ann Arbor, MI 48109, USA*

Abstract: For bio-probe applications, a bio-film has to be attached to the nanoparticle surfaces to react with virus. However, the surfaces of many nanoparticles are not ideal for these particular bio-films. To provide an adhesive interface between the bio-film and the nanoparticle substrate, an ultrathin polymer film such as polystyrene and acrylic acid was deposited on several nanoparticles and carbon nanotubes using a plasma polymerization treatment. High Resolution Transmission Electron Microscopy (HRTEM) experiments showed that an extremely thin polymer film was uniformly deposited on the surfaces of these nanoparticles and nanotubes. The deposited films were characterized by depressive energy spectrum (EDS) and infrared spectrum (FTIR). The deposition mechanism is discussed.

Key words: bio-film, surface coating,

1. INTRODUCTION

Nanoparticles are used in many applications because of their desirable bulk properties [1-6]. Unfortunately, the surface of the particle is often not ideal for the particular application. The ability to deposit well-controlled coatings on particles would offer a wide range of technological opportunities based on changes to both the physical and chemical properties of the particles. Atomic layer controlled coatings on particles, for example, would allow particles to retain their bulk properties but yield more desirable surface properties. These ultrathin coatings could act to activate, passivate or

B. Aktaş et al. (eds.), Nanostructured Magnetic Materials and their Applications, 3–11.

functionalize the particle to achieve both desirable bulk and surface properties. In some of the applications, the nanostructure involves an ultrathin film on the nanoparticle surface that can be tailored into multilayers. Both the substrate nanoparticle and the ultrathin film serve certain functionalities for specific applications. For instance, in bio-probes, the substrate must be magnetic for virus separation in a solution. An antigen bio-film has to be deposited on the magnetic nanoparticles. However, before the attachment of the bio-film, a polymer film needs to be deposited which provides the adhesive and coherent bonding between the bio-film and the nanoparticle substrate. The deposited polymer film must be extremely thin and uniform. Plasma polymerization is a unique method that we have developed to coat thin and uniform films on nanoparticles.

Magnetic Labeling, designed to quantitatively detect target biomolecules using bound magnetic beads, is a relatively new field [7-12]. In this technology, the system measures the small magnetic field retained by magnetic particles after they are removed from a magnetizing field. The sensitivity of this technique is reported on the order of 10 pg of 10 nm ferric oxide labels. The method utilizes magnetic particles as reporters and detects the induced magnetic field generated by the paramagnetic particles bound to the analyte. The commercial instrument based on this concept can detect biomolecules captured on the paramagnetic beads using electrochemiluminescent signals generated by Ruthenium chelate labels and provides higher sensitivity and broader dynamic range than most of the conventional detection methods.

However, all conventional magnetic particles are in the size range of 0.5 – 10 μm. Applications of paramagnetic nanoparticles (<500 nm) for biological separation and detection have been rarely reported. Since nanoparticles can provide higher sensitivity and greater surface area/volume ratios of the reporter, the diagnostic sensitivity is increased when particle size is no longer a limiting factor in analyte recognition. High surface area to volume ratios is critical when functionalizing particles for the attachment of biological ligands. To attach the bio-film onto the substrates, an adhesive polymer film is needed as the first layer on these magnetic nanoparticles.

In this experiment we have attempted to coat a thin film of polymer onto various nanoparticles including magnetic $NiFe_2O_4$ nanoparticles, carbon nanotubes, and ZnO. The purpose of the experiment was to investigate the optimum plasma coating conditions for deposition of thin and uniform polymer films on a variety of nanoparticles. After coating, the experiments were focused on the characterization of the deposited thin films. High Resolution Transmission Electron Microscopy (HRTEM) was used to study the extremely thin film on the nanoparticle surfaces. FTIR and EDS experiments were carried out to analyze the composition and the structure of the coated thin films.

3.   EXPERIMENTAL DETAILS

The schematic diagram of the plasma reactor for thin film deposition of nanoparticles is shown in Fig. 1. The vacuum chamber of plasma reactor consisted of a Pyrex glass column about 80 cm in height and 6 cm in internal diameter [13,14]. The nanoparticles were vigorously stirred at the bottom of the tube and thus the surface of nanoparticles can be continuously renewed and exposed to the plasma for thin film deposition during the plasma polymerization process. A magnetic bar was used to stir the powders. The gases and monomers were introduced from the gas inlet during the plasma cleaning treatment or plasma polymerization. The system pressure was measured by a pressure gauge. A discharge by RF power of 13.56 MHz was used for the plasma film deposition.

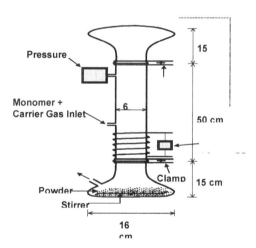

Figure 1. Schematic diagram of the plasma reactor for thin polymer film coating of the nano-particles

Before the plasma treatment, the basic pressure was pumped down to less than 2 Pa and then the plasma gases or monomer vapors were introduced into the reactor chamber. The operating pressure was adjusted by the gas/monomer mass flow rate. The base pressure was less than 1 Pa.

Polystyrene and acrylic acid were used as monomers for plasma polymerization. During the plasma polymerization process, the input power was 10 W and the system pressure was 25 Pa. The plasma treatment time was 240 min. Per batch 40 grams of powder were treated.

After the plasma treatment, the nanoparticles were examined by using high-resolution transmission electron microscopy (HRTEM). The high-resolution TEM experiments were performed on a JEM 4000EX TEM.

4.   RESULTS AND DISCUSSION

The original and plasma-coated nanoparticles were dispersed onto the poly-carbon film supported by Cu-grids for TEM operated at 400 kV. Fig. 2a and 2b are the bright field images of the coated ZnO nanoparticles at different magnifications. Fig. 2a shows the low-magnification of a ZnO particle coated with acrylic acid (AA) film. As can be seen, the coating is uniform on the entire surface of the particle. In our TEM observation we found that, although these particles have different diameters, the film remains the same thickness indicating a uniform distribution of active radicals in the plasma chamber. Fig. 2b is an image of a higher magnification. The uniformity of the AA thin film can be clearly seen in this photograph. The film thickness is about 5 nm. From these images, the film was identified as typical amorphous structure by HREM observation over different particles. The lattice image of ZnO forms a clear contrast with the amorphous AA film at the interfaces.

Using the same method, polystyrene thin films were also deposited on carbon nanotubes. An ultrathin film amorphous layer can be clearly seen covering both the inner and outer surfaces of the nanotubes after plasma treatment (Fig. 3). The thin film is uniform on both surfaces, however, with a larger thickness on the outer wall (7 nm) than on the inner wall (1~3 nm) surface (Fig. 3b). The thickness of ultrathin film is approximately 2~7 nm completely surrounding the nanotube surfaces.

Figure 2. (a) Bright-field TEM image of the acrylic film coated ZnO nanoparticles at low magnification. (b) HRTEM image showing the AA coated ZnO nanoparticle surfaces at higher magnification

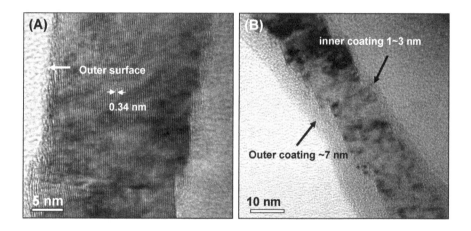

Figure 3. HRTEM images showing the polystyrene coated carbon nanotube surfaces at higher magnification

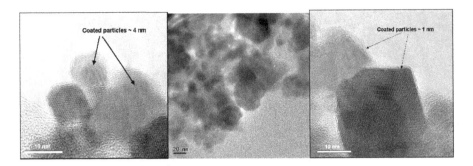

Figure 4. (a) TEM images of (a) uncoated $NiFe_2O_4$ nanoparticles; (b), (c) polystyrene coated $NiFe_2O_4$ nanoparticles

For bio-probe applications, we chose magnetic nanoparticles $NiFe_2O_4$ as shown in Figure 4. These nanoparticles are irregular in shape and have the dimensions on the order of 10-50 nm (Fig. 4a). Fig. 4b and 4c show the

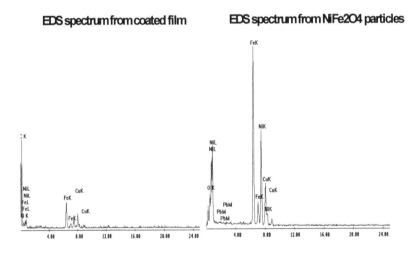

Figure 5. EDS spectrum of coated $NiFe_2O_4$ nanoparticles

TEM images of coated $NiFe_2O_4$ nanoparticles. The coated polystyrene can be seen from these figures. The coating layer is about 2-3 nm thick. With these polymer films, the bio-film can be possibly attached for bio-probe applications. Energy disperssive spectrum (EDS) was also performed on the coated and uncoated $NiFe_2O_4$ nanoparticles (Fig. 5). As can be seen, the uncoated spectrum has strong Fe and Ni peaks showing the naked surface of the nanoparticles (Fig. 5a). In coated spectrum, however, the Fe and Ni peaks are significantly suppressed. The spectrum shows intensive carbon peaks indicatinging the polystyrene film on the surfaces.

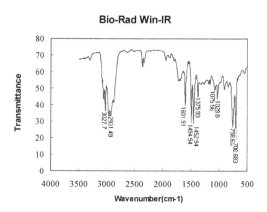

Figure 6. FTIR spectrum of coated NiFe$_2$O$_4$ nanoparticles.

To further characterize the coated polymer films on NiFe$_2$O$_4$, FTIR was performed on these coated nanoparticles (Fig. 6). As shown in Fig. 6, there are strong peaks around 1300-1600 (cm^{-1}) that are result of the special absorption of benzene rings. Near 700 (cm^{-1}), there are two other peaks due to the substitution of benzene rings. These are clear indications of coated polystyrene on the NiFe$_2$O$_4$ nanoparticle surfaces.

Nanoparticles are known to have extremely high surface areas [15-19]. As a result of the high surface area, the surface energy can reach the order of 100 kJ/mol for a variety of materials. During film deposition, the polymer is introduced as a vapor and the collision frequency increases with the gas pressure. Due to high surface energy of the nanoparticles, condensation of the polymer vapor on the nanoparticles naturally lowers the surface energy by forming an extremely thin film. On a large two-dimensional flat substrate, however, the deposition of the film may cluster severely and form small islands initially, due to its surface tension. A smooth film will require continued deposition at an appreciable thickness. In our deposition process [20-23], both energy terms (surface energy of the nanoparticle and the surface tension of the polymer) were balanced by controlling the plasma coating parameters including electron density, temperature, and energy density. The gas pressure must be moderate for a low collision rate on the nanoparticle surfaces. In addition, polymerization should take place relatively fast after the condensation on the particle surfaces. These will ensure a uniform coating on the order of several nanometers for all particle sizes.

5.   SUMMARY

In summary, we have deposited ultrathin films of polystyrene and acrylic acid on the nanoparticles of $NiFe_2O_4$, ZnO, and carbon nanotubes by means of a plasma polymerization treatment. The polymer layer is not only uniform on all particle sizes, but also deposited in an extremely thin thickness of several nanometers. Such ultrathin film deposition characteristics are essential in establishing multi-layer nanostructures, particularly for attachment of various bio-films in bio-probe applications. Our future work will focus on the interface study in terms of structure, adhesion behavior, and related properties. Improved adhesive coatings will also be selected for coating of the bio-films.

ACKNOWLEDGMENT

The TEM analyses were conducted at the Electron Microbeam Analysis Laboratory at the University of Michigan, Ann Arbor, Michigan.

REFERENCES

1. R. W. Siegel, *Nanostructured Materials.* **3**, 1 (1993).
2. G. C. Hadjipanayis and R. W. Siegel, *Nanophase materials, Synthesis-properties applications* (Kluwer Press, Dordrecht, 1994).
3. G. M. Whitesides, J. P. Mathias, and C. T. Seto, *Science*, **254**, 1312 (1991).
4. C. D. Stucky, and J. E. MacDougall, *Science*, **247**, 669 (1990).
5. H. Gleiter, *Nanostructured Materials*, **6**, 3 (1995).
6. *Nanotechnology*, A. T. Wolde, ED. (STT Netherlands Study Center for Technology Trends, The Hague, The Netherlands, 1998).
7. G. Blasse and B.C. Grabmaier, Luminescent Materials, Springer-Verlag, 1994.
8. Y. Mita in Phosphor Handbook, Edited by S. Shionoya and W. M. Yen, CRC Press, 1999.
9. F. E. Auzel, *Proc. IEEE*, 61, 758-786 (1973).
10. R.H. Page, K.I. Schaffers, P.I. Waide, J.B. Tassano, S. A. Payne, W.F. Krupke, and W.K. Bischel, J. Opt. Soc. Am. B, 15, 996-1008 (1998).
11. P. N. Yocom, J. P. Wittke, and I. Ladany, Metal. Trans., 2, 763-767 (1971).
12. F.W. Ostermayer, Jr., Metal. Trans. 2, 747– 755 (1971).
13. S. Eufinger, W. J. van Ooij, and T. H. Ridgway, *Journal of Appl. Pol. Sci.*, **61**, 1503 (1996).

14. W. J. van Ooij, S. Eufinger, and T. H. Ridgway, *Plasma and Polymers*, **1**, 231 (1996)
15. R. W. Siegel, *Nanostructured Materials*. **3**, 1 (1993).
16. G. C. Hadjipanayis and R. W. Siegel, *Nanophase materials, Synthesis-properties-applications* (Kluwer Press, Dordrecht, 1994).
17. G. M. Whitesides, J. P. Mathias, and C. T. Seto, *Science*, **254**, 1312 (1991).
18. C. D. Stucky, and J. E. MacDougall, *Science*, **247**, 669 (1990).
19. H. Gleiter, *Nanostructured Materials*, **6**, 3 (1995).
20. Donglu Shi, S. X. Wang, Wim J. van Ooij, L. M. Wang, Jiangang Zhao, and Zhou Yu, "Interfacial Bonding via an Ultrathin Polymer Film on Al_2O_3 Nanoparticles For Low-Temperature Consolidation of Ceramics", *J. of Mat Res*. 2002, in presss.
21. Donglu Shi, S. X. Wang, Wim J. van Ooij, L. M. Wang, Jiangang Zhao, and Zhou Yu, "Multi-Layer Coating of Ultrathin Polymer Films on Nanoparticles on Alumina by a Plasma Treatment", Mat. *Res.Soc. Symp*. Vol. 635 (2001)
22. Peng He, Jie Lian[1], L. M. Wang[1], Wim J. van Ooij, and Donglu Shi, "Coating of Acrylic Acid Thin Films on ZnO Nanoparticles for Ion Exchange in Water", Mat. *Res.Soc. Symp*. (2002) in press
23. Donglu Shi, S. X. Wang, Wim J. van Ooij, L. M. Wang, Jiangang Zhao, and Zhou Yu, "Uniform Deposition of Ultrathin Polymer Films on the Surface of Alumin Nanoparticles by a Plasma Treatment", *Appl. Phys. Lett.*, 78, 1234 (2001)

SELECTIVE REMOVAL OF ATOMS AS A NEW METHOD FOR MANUFACTURING OF NANOSTRUCTURES FOR VARIOUS APPLICATIONS

B. Gurovich, E. Kuleshova, D. Dolgy, K. Prikhodko, A. Domantovsky, K. Maslakov, E. Meilikhov
Russian Research Center "Kurchatov Institute", Kurchatov sq. 1, Moscow 123182, Russia

Abstract: The paper demonstrates a possibility for effective modification of the thin-film material' chemical composition, structure and physical properties as result of selective removal of atoms by the certain energy ion beam. One of the most promising results of this effect consists in developing the new technology for 3D micro- and nano-structures production for various applications.

Key words: Selective removal, chemical composition

1. INTRODUCTION

The crossover at the high-technology market from the current micro-technologies to the future nanotechnologies is expected in the nearest 5-10 years. This process, revolutionary in its scope and consequences, will allow to derive all the benefits from both scaling and new circuit technology, based on the new physical principles, which can be materialized only in nanoscale devices. This commercial task depends directly on the research advances.

We have discovered and studied in detail the new fundamental effect of "selective removal of atoms by ion beams", which lays the foundation of the new technology for micro- and nano-devices production [1,2].

The physical basis of the method is as follows. Let us consider a situation that arises during interaction of a monochromatic ion beam of energy E and mass m with a two-atomic crystal consisting of atoms of different masses

B. Aktaş et al. (eds.), Nanostructured Magnetic Materials and their Applications, 13–22.
© 2004 *Kluwer Academic Publishers. Printed in the Netherlands.*

M_1 and M_2. The maximum energy transferred by the ions to atoms of a crystal is [3]:

$$E_{max}^{(1,2)} = \frac{4mM_{1,2}}{(M_{1,2} + m)^2} \cdot E,$$ (1)

where $E_{max}^{(1)}$ and $E_{max}^{(2)}$ are maximum energies which could be transferred by the accelerated ions to atoms with masses M_1 and M_2.

In deciding on a particular material for selective removal of atoms, metal compounds that are insulators in the initial state hold the greatest practical interest. Films behavior of different insulating di- and polyatomic materials under irradiation depends strongly on the ions energy, all other factors being the same. The observed radiation-induced modification of thin film properties was of clearly defined threshold character [4]. No modifications of structure, composition, electrical and magnetic properties of thin films were detected until the ion energy reached certain minimum value (which is individual for every studied material). It means that as long as the energy E_{max} transferred to material atoms by ions is low ($E_{max} < E_{d1}$ and $E_{max} < E_{d2}$, where E_{d1}, E_{d2} are the displacement threshold energies for atoms of the first and the second types, respectively), atom displacements off regular lattice positions do not take place.

With increasing the ions energy, conditions appear when $E_{d1} < E_{max} < E_{d2}$. Under this condition the selective removal of light atoms of the first kind (e.g. oxygen, nitrogen or hydrogen) is observed from the studied compounds (oxides, nitrides or hydrides, respectively). As a result the dramatic variations of the structure, composition, as well as electric, magnetic and optical properties appear as will be shown below.

With further increasing ion energy, the condition $E_{d1} \leq E_{d2} < E_{max}$ is reached and atoms of both types in diatomic compounds begin to displace off the regular lattice positions (but with various efficiency). The further increase in E_{max} has only small influence on the selectivity of atomic displacement in the studied energy range.

Thus, by varying the mass and the energy of ions, it is possible to achieve the situation in a two- or a multi-atomic crystal when the higher energy would be transferred to the atoms of low or high masses. If the maximum transferred energy exceeds the threshold value E_d for atoms of only one kind, then there exists a method of selective removal of only light (or only heavy) atoms from two- or a multi-atomic crystal. As a rule, this energy $E_d \approx 20-25$ eV, which exceeds considerably the sublimation energy [5].

The considered mechanism regarding the displacement of atoms of different kinds in a crystal refers equally to the same compounds in an amorphous state. Thus, it's clear that at the normal incident of the ion beam on a crystal surface, it's possible to achieve conditions when selective removal of only one kind of atoms is observed (for $E_{d1} \leq E_{max} < E_{d2}$). Under the condition $E_{d1} < E_{d2} < E_{max}$ (when the displacement selectivity is provided by the difference in the displacement rates of the various atom kinds) it's possible to remove the selected kind of atoms up to the needed properties level of the material composed of the residuary atoms of the second kind within a layer of a thickness comparable with the ion projective length in a two- or a multi-atomic crystal.

Let us formulate some obvious features of the considered physical mechanism of selective removal of atoms:

The rate of the process is proportional to the flux density of the incident ion beam.

The process is naturally a non-thermal one over a wide range of irradiation temperatures; this distinguishes it principally from a chemical reaction.

The process can be proceed in a layer below the surface even if covered by another material, if its thickness is less than the ion projective length in the layer. If, in addition, the threshold energy of atomic displacement in the additional layer is higher than the transferred energy from the ions, the directed displacement of atoms in that material will not occur. Otherwise, the atoms of material penetrate in the underlying layer and their transfer in the beam direction occurs over a distance comparable with the ion projective length in the "sandwich" considered.

The above-mentioned features of the proposed method determine its potential for efficient, purposeful, and spatially modulated modification of the composition, structure, physical, and chemical properties of materials. It will be shown below that such a modification of chemical composition can dramatically change the physical properties of a thin material layer, e.g. to produce an insulator-metal transition, to change magnetic or optical properties and so on. Thus a possibility exists to create a controlled volume "pattern" of areas with different physical properties.

2.   EXPERIMENTAL RESULTS AND DISCUSSION

Among the diatomic compounds, those are, for example, many metal oxides, as well as some hydrides and nitrides of metals. Though in the course of this work experiments were performed with compounds of all above-mentioned types, metal oxides were investigated most thoroughly. Qualitative features of the effects accompanying selective removal of atoms are identical in compounds of all above-mentioned types. The aim of experiments was to remove selectively oxygen (nitrogen or hydrogen) atoms by irradiation of the original insulator and to obtain finally a metal. Experiments have been performed with thin films of different thickness, which have been produced by reactive metals sputtering in the atmosphere of relevant gases (oxygen, nitrogen or hydrogen) [6].

In most cases, films have been irradiated by protons with energies ~1-5 keV. In addition, in some experiments films have been irradiated in electron microscope column by electrons with energy of 100-200 keV.

One could also expect that with selective removal of oxygen atoms out of initially nonmagnetic (or weak magnetic) oxides of ferromagnetic metals, films could transform into magnetic state. It is clear that in relevant experiments other di- or polyatomic compounds of ferromagnetic metals could be employed along with oxides.

Below represented are the results of our experiments performed with certain di- and polyatomic systems. Figure 1 illustrates typical changes of a chemical composition during the selective removal of atoms. Figure 2 shows the typical experimental results on both electric (a) and magnetic (b) property modifications of metal oxides during the proton irradiation. As a result, the insulator (MoO_3) transforms into metal (Mo), and non-magnetic material (Co_3O_4) becomes magnetic (Co).

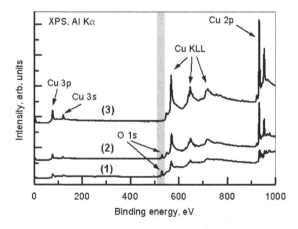

Figure 1. XPS-spectra of CuO-film in the initial state (1) and after proton irradiation of various doses (2,3), which demonstrate the reduction of oxygen concentration in the film during irradiation.

It's important to note that selective removal of atoms allows one to simultaneously change the physical properties of separate layers in a multi-layer structure. This is a principal advantage of the proposed technology compared to any other known technology or physical principle. As a result it enables the simultaneous (in parallel) production of structures with different shapes and properties in various layers by ion irradiation through the same mask. Such a procedure allows one to get an overlapping of the structure elements in various layers with an accuracy of about 1 nm. The latter feature is a crucial point in the production of multi-layer nanostructures.

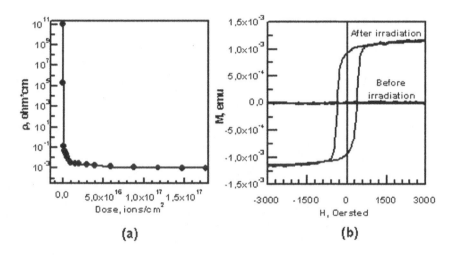

(a)   (b)

Figure 2. Typical behaviour of electric resistivity (a) and of magnetic hysteresis loop (b) of metal oxides during proton irradiation, which demonstrates the transition of insulators into metals (a), and non-magnetic materials into magnetic ones (b).

In Figure 3, the experimental results are shown which demonstrate the possibility of simultaneous changes of physical properties in various multi-layered structures with alternating functional and auxiliary layers (the scheme of the sandwich with different layers irradiated through the same mask - a, b). Also shown in Figure 3 is the dose dependence of Co_3O_4 and CuO layer resistances in the sandwich during their consecutive transformation under irradiation into Co and Cu, respectively (c).

It is very promising to use the proposed method for formation of magnetic patterned media with high areal density. To attain an extreme density of data storage with magnetic recording media, it is necessary to use the patterned media consisting of regularly positioned magnetic nanogranules ("bits") of identical form and orientation. The smallest possible size of the bit is defined by the so-called superparamagnetic limit. The minimum distance between the bits depends on dipole-dipole inter-bit interactionand cannot be less than its lesser size.

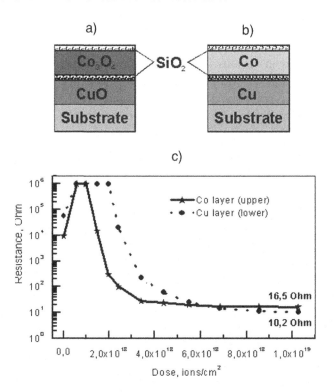

Figure 3. Schemes of multilayer structure in the initial state (a), after proton irradiation (b) and the dose dependence of Co_3O_4 and CuO layer resistances in the sandwich during their consecutive transformation under irradiation into Co and Cu, respectively (c).

On the other hand, one should not arrange the bits so closely that they nearly touch each other due to the above-mentioned magnetic inter-bit interaction. Thus, two physical reasons confine the accessible bit density: superparamagnetic limit and inter-bit-interaction. There is the optimal inter-bit distance (and, accordingly, the areal density of the data storage). According to calculations, the optimal gap between the nearest rectangular bits with the shape anisotropy factor of 5-6 amounts about the width of the bit. The calculation shows that with Fe-bits of 3x4x25 nm the storage density of about 500 $Gbit/cm^2$ (3000 $Gbit/inch^2$) could be attained.

Elongated (anisotropic) single-domain Co bits of small size (from 320x1600 nm down to 15x45 nm) in Co_3O_4 matrix have been produced through a mask prepared by electron lithography. Examples of the structures are shown in Figure 4. These images were obtained by atomic and magnetic force microscopy.

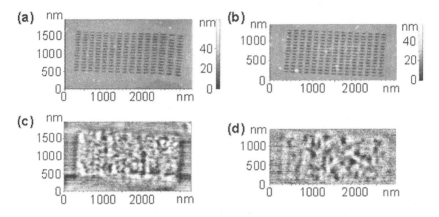

Figure 4. Examples of patterned magnetic media with areal densities: 45 Gb/inch^2 (bit size - 25x125 nm^2)– (a,c) and 57 Gb/inch^2 (bit size - 20x100 nm^2)– (b,d). (a,b)– AFM topography images, (c,d)– MFM images.

3.   PERSPECTIVES

Nowadays there is only submicron industrial technology for microelectronic device production (with the minimum design rules– 0.07-0.1 microns) based on optical lithography. There is the absolute necessity in industrial lithography for producing devices with smaller element size. Many experts consider nanoimprint lithography as the most realistic alternative to optical methods [7,8,9,10,11].

Nanoimprint lithography allows one to solve the first problem just now: the single-layered nanostructures with the element size of less than 10 nm [7,8], and with the areal density of elements ~700 Gb/inch^2 [12] have already been produced.

However, even nanoimprint lithography itself implies production of multilayer devices in a consecutive manner similar to optical lithography (i.e., consecutive creation of metal, dielectric and other layers and intermediate structures). Extending traditional consecutive principles of microdevice manufacturing to multilayer nanodevices demands the solution of two basic problems: a creation of structural elements with the sizes of 10 nm and less, and ensuring the overlapping of the structure elements in various layers with accuracy of 1-2 nm. Within the frameworks of optical lithography and of traditional consecutive layering principles, the solution of these problems is not obvious now, nor will it be in the foreseeable future.

It can be done by only the combination of nanoimprint lithography and the method of selective removal of atoms. The latter allows one to create various patterns simultaneously in several layers through the same mask. As a result, self-overlapping with 1 nm accuracy can be obtained. Besides this practical reason the application prospective of this method are conditioned by a number of other reasons.

Ion beams have a few important advantages:

Negligible back-scattering effects results in increasing spatial resolution of patterns on usual thin films deposited on massive substrates or thin layers on/in massive samples.

Short wavelengths of incident particles important for high resolution, could be obtained with low accelerating voltages.

It is important that during irradiation ions leave the material due to diffusion without any negative influence on material properties.

In conclusion, the method could be used to create directly the needed spatial modulations of atomic composition and physical properties of a material, such as metal or semiconductor patterns in insulators, magnetic drawings in nonmagnetic substances, light guides in opaque media, etc. and could be used for fabrication of micro- and nanoscale devices for various applications.

The main advantages of the proposed method are as follows:

It is a parallel processing technique with respective high throughput.

Possibility exists to form the various nanopatterns with needed physical properties in different layers through the same mask.

An intrinsic feature of the method is the self-overlapping (~1 nm) of elements in different layers of the structure.

The method can be easily combined with traditional CMOS technology to produce hybrid devices (the nanostructures being prepared by the proposed technology.)

We have already successfully tested this method for some elements production of the future nanodevices, the following results being obtained:

The resolution of 15 nm was achieved, which can be improved up to 3-4 nm in the nearest future.

The possibility of given relief production on the solid surface with 15 nm resolution has been demonstrated and the prototype of 3.5 inch stamp for imprint lithography was prepared with 1.5 micron resolution.

The possibility of parallel and simultaneous modification of material' properties in various layers of thin-film multilayer structure was experimentally shown.

It has been demonstrated that nanopatterns of areas with needed physical properties can be produced in various layers of thin-film multilayer structure through a single mask.

The possibility of patterned magnetic, conducting and optical medias production was demonstrated with areal density of elements up to 60 Gb/inch^2.

The examples of diode and transistor structures as well as nano-switchboards have been prepared.

4. CONCLUSION

In the present paper, the physical principles are described and the conditions are formulated which allow for selective removal of certain atoms from multiatomic solids by their thin films or thin multilayers irradiation with accelerated ions. The essential features of the process that are originated from the proposed physical mechanism were listed. We demonstrated experimentally the possibility of selective atom removal, confirmed the process mechanism, and investigated its most essential features. In the course of these experiments, we revealed that selective atom removal from multiatomic compounds is accompanied by radical alterations of the most important physical properties of materials, such as electrical, magnetic and optical. Investigations show that the modification of a material's properties is a result of changes in their atomic composition and structural transformations (phase transitions), which accompany the process. Besides, we have shown some prospective applications of the given method for production of micro- and nano 3D-structures for various purposes.

REFERENCES

[1] B.A. Gurovich et al.: Patent US 6,218,278 B1, priority May 1998.
[2] B.A. Gurovich, D.I. Dolgy, E.A. Kuleshova, E.P. Velikhov, E.D. Ol'shansky, A.G. Domantovsky, B.A. Aronzon, E.Z. Meilikhov, Physics Uspekhi **44** (1) (2001) 95.
[3] L.D. Landau, E.M.Lifshitz, Mechanics. 3rd edition, Vol.1, Butterworth-Heinemann, 1976.
[4] M.W. Thompson, Defects and Radiation Damage in Metals. Cambridge Press, 1969.
[5] Handbook of Physical Quantities. (Eds. I.S. Grigoriev, E.Z. Meilikhov), Published by CRC Pr, 1996.
[6] Handbook of Thin Film Technology (Eds. L.I. Maissel, R. Gland), New York: McGraw-Hill, 1970.
[7] S. Chou and P. Krauss, Microelectronic Engineering **35**, 237, (1997).

[8] S. Chou, P. Krauss, W. Zhang, L. Guo and L. Zhuang, J. Vac. Sci. Technol. B **15**(6) 2897, (1997).

[9] S. Zankovych, T. Hoffmann, J. Seekamp, J.-U. Brunch and C.M. Sotomayor Torres, Nanotechnology **12** 91, (2001).

[10] B. Heidari, I. Maximov and L. Montelius, J. Vac. Sci. Technol. B **18**(6), 3557, (2000).

[11] L.J. Heyderman, H. Schift, C. David, B. Ketterer, M. Auf der Maur, J. Gobrecht, Microelectronic Engineering **57-58**, 375,(2001).

[12] F. Carcenac, C. Vieu, A. Lebib, Y. Chen, L. Manin-Ferlazzo, H. Launois, Microelectronic Engineering **53**, 163, (2000).

NEW LASER ABLATION CLUSTER SOURCE FOR A SYNTHESIS OF MAGNETIC NANOPARTICLES, AND FMR STUDY OF Ni:MgF$_2$ COMPOSITES

A.L. Stepanov [a,b], R.I. Khaibullin [b,c], B.Z. Rameev [b,c], A. Reinholdt [a],
R. Pecenka [a], U. Kreibig [a]

[a] *Physikalisches Institut IA der RWTH, Sommerfeldstrasse 14, 52056 Aachen, Germany*
[b] *Kazan Physical-Technical Institute, Sibirsky Trakt 10/7, 420029 Kazan, Russia*
[c] *Gebze Institute of Technology, 41400 Gebze-Kocaeli, Turkey*

Abstract: A new technique for synthesis of Ni magnetic nanoparticles in diamagnetic matrix by using a novel laser universal cluster ablation source (LUCAS) is presented. The source has been developed specially for highly refractory materials with high melting temperatures. Ablation/evaporation laser pulses were applied to the Ni bulk target to form metal vapor in high pressure seeding Ar-gas, and then expansion of the material vapor through a nozzle with the carrier Ar gas was performed. The nickel nanoparticles deposited on a substrate surface were protected by a complementary electron sputtering of magnesium fluoride bulk material. Electron microscopy showed that a highly homogeneous dispersion of the crystalline Ni nanoparticles with a size of near 3 nm was formed in the deposited layer. The magnetic properties of the nanostructured Ni:MgF$_2$ composite were investigated by magnetic resonance technique in X-band (9.8 GHz) at room temperature. A highly anisotropic ferromagnetic resonance (FMR) responce of the Ni:MgF$_2$ composite was found. Broadening of the FMR signal, when the DC magnetic field is rotated towards the plane of the composite, was also observed. Effective media approach to the FMR in thin granular magnetic films was applied to analyze the resonance field and linewidth dependencies on orientation, and the value of the magnetization of the synthesized Ni:MgF$_2$ composite was extracted.

Key words: FMR; Laser ablation; nanocrystalline materials; 3d group ions

B. Aktaş et al. (eds.), Nanostructured Magnetic Materials and their Applications, 23–31.
© 2004 *Kluwer Academic Publishers. Printed in the Netherlands.*

1.   INTRODUCTION

Fine magnetic particles are the object of intensive researches because of their unique properties that considerably differ from those of the bulk. Several specific features related to nanometer size particles have been reported, for example, enhanced magnetization and anisotropy, high coercivity, single domain state and superparamagnetism phenomena, and others [1]. The properties of an ensemble of the magnetic nanoparticles are strongly dependent on their shape, crystallinity, spread in sizes, their space distribution and density in diamagnetic host, i.e. the parameters determined by conditions during the synthesis of nanoparticles.

Magnetic nanoparticles can be prepared by chemical methods (sol-gel, inverse micelle technique, liquid solution precipitation, *etc*) or physical ways (inert gas evaporation, sputtering, ion implantation, laser evaporation, *etc*). Among the physical ones, atomic vapor aggregation approaches are currently well advanced and frequently applied [2]. The traditional aggregation procedure leads usually to formation of particle gas or cluster smoke with random thermal distribution of particle motions at high collision rates in inert gas or reducing atmospheres. One can deposit the particle vapor on a substrate for synthesis of the desired nanoparticles and many-particle systems. Further development of the method is using of the cluster beam formed by adiabatic expansion of the material vapor (with or without a carrier gas) through a nozzle. Such beam is characterized by uniform cluster motion and low collision rates due to narrow distribution of cluster velocities. The laser-based Universal Cluster Evaporation/Ablation Source (LUCAS) was recently developed to form cluster beams of highly refractory materials or/and of materials with high melting temperatures [3]. Previous results showed clear advantages in preparation of metallic (Y, Cu, Fe) and semiconductor (Si, ITO) nanoparticles [3-5].

Nickel is one of the important 3d ferromagnetic materials which has widely been used to form magnetic nanoparticles. Recent studies of the magnetism in the nickel nanoparticles (NN) found them to be highly magnetic, with moment μ per atom that exceed the bulk value of $0.61\mu_B$ per atom due to surface-enhanced magnetism in the small particles [6]. The aim of the current work was to estimate the possibility for fabrication of the magnetic NN in diamagnetic MgF_2 matrix by using LUCAS, and to study the magnetic properties of the novel $Ni:MgF_2$ composite system.

2.   EXPERIMENTAL

An overview of the cluster beam deposition apparatus LUCAS is shown in Fig. 1. For the present purpose a pulsed Nd:YAG-laser LUMONICS JK 702H of 350 W working at 1.06 μm was used. Ablation/evaporation laser pulses were applied to a Ni bulk target of 99.99 % purity. The repetition rate and pulse duration of the laser were 80 Hz and 1 ms, respectively. Hot metal plasma was generated by laser ablation/vaporization of the target in presence of the Ar gas at pressure up to 0.6 bar. The nucleation and growth process can be modified by igniting a laser induced plasma discharge. The Ar gas cools the plasma and favours thus cluster condensation. The cluster–argon atom mixture expands via a 1.2 mm diameter nozzle of special conical shape out of the source chamber forming a well-defined cluster/particle beam. The cluster beam is introduced through differential pumping equipment of turbo-molecular pump with effective pumping into the experiment chamber with a background argon pressure of $5x10^{-6}$ mbar. The differential pumping equipment includes the pumping chamber. The pump chamber is pumped with speed of 1000 l/s and root-pump, with total pumping power of 780 m^3/h. The cooling of particles occurs during the expansion. The expanding cluster/particle beam passes through two 8 mm diameter skimmers into the experimental chamber with a second 1000 l/s turbo-molecular pump. The NN were deposited on some substrate, in our case, on a SiO_2 plate covered by a MgF_2 layer. The experiment chamber is equipped with a quartz balance to measure and control the absolute amount of deposited cluster material. The total mass of the layer and the deposition rate are determined. The depositions were performed at room temperature of the substrate. The NN on the substrate surface were protected by an electron sputtering of MgF_2 bulk material by thicknesses of 100 nm.

Transmission electron microcopy (TEM) analysis of the NN sizes was performed by a 100-keV Philips Tecnai-12 microscope. For this purpose, NN were separately deposited on carbon-coated copper grids.

The room temperature ferromagnetic resonance (FMR) spectra were recorded using a Bruker EMX electron spin resonance spectrometer with the frequency (ν_{res}) of 9.8 GHz for various orientations of the sample plane with respect to the applied magnetic field H. The first derivative dP/dH of the absorption P was measured as a function of the applied magnetic field. The resonance field H_{res} was considered as the value of the magnetic field at which the curve dP/dH intersects the baseline. The resonance linewidth ΔH_{pp} was determined as the difference between the

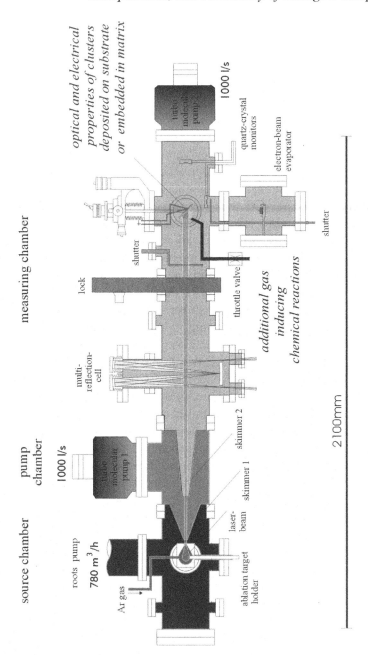

Figure 1. Overview of the LUCAS apparatus.

values of the magnetic field at minimum and at maximum of the d*P*/d*H* curve, correspondingly.

3.   RESULTS AND DISCUSSION

A bright-field TEM image of the NN's deposited by LUCAS is presented in Fig. 2. The picture reveals two important microstructural features of the Ni:MgF$_2$ composite material under study: (i) spatially homogeneous dispersion of NN (dark spots on the bright background) throughout the whole analysed sample area, and (ii) a narrow size distribution of the observed particles with the mean diameter of 3.2 nm, as can be seen from histogram in Fig.2. The particle size distribution was extracted from TEM examinations by considering more than 1000 particles. This distribution is characterised as normal distribution with a dispersion width of σ=1.6 nm that is narrower than ones for the NN prepared by chemical sol-gel method [7] or physical gas evaporation technique [8].

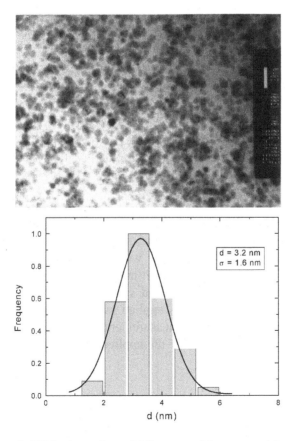

Figure 2. TEM plane view of Ni nanoparticles prepared by using LUCAS setup, histogram of the size distribution and normal fitting (solid line).

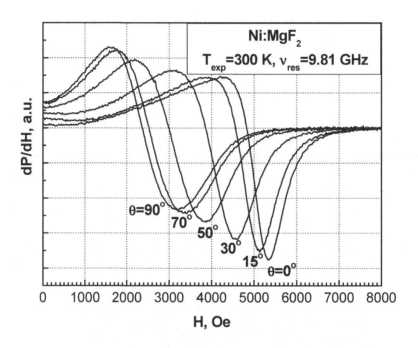

Figure 3. The angular dependencies of FMR spectra of Ni:MgF$_2$ composites. Here θ is the angle between the normal to the sample surface and the static magnetic field.

Fig. 3 shows magnetic resonance spectra registered in the Ni:MgF$_2$ composite for different orientations (θ_H) of the applied magnetic field with respect to the normal of the sample plane. A signal shift from the high-field region towards the low-field region, as well as the signal broadening and lineshape modification are observed with increasing the polar angle θ_H. The lowest resonance field H_{res}=2370 G and symmetric Gaussian-like line are observed when the magnetic field is applied in the sample plane (θ_H =90°). The subsequent azimuthal angular measurements show that the value of H_{res} and the lineshape are not changed if the applied magnetic field is rotated in the sample plane.

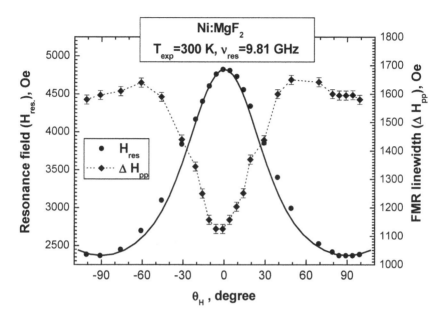

Figure 4. The angular dependence of resonance field and FMR linewidth in Ni:MgF2. Solid line presents the modeling results at the parameters of M_{eff}=136 G and g_{eff}=2.25.

The strong out-of-plane angular dependence of the magnetic resonance spectra, observed in the Ni:MgF$_2$ composite, is typical for FMR in the granular films of interacting magnetic nanoparticles [9-11]. The resonance field H_{res} in a granular film varies from below H_0 (for H in plane of the film) to substantially above H_0 (for H perpendicular to the film plane), where H_0 (defined as $h\nu_{res}/g\beta$) is the "mean resonance field", that is the average resonance field if the granular film consist of the ensemble of the *randomly-oriented*, *non-interacting* and *spherical-shaped* magnetic particles. Here, β is the Bohr magneton, and g is the g-factor value (a dimensionless number of about 2). FMR response in the granular films with strong magnetic coupling between nanoparticles (even without direct contact between them, as result of dipole-dipole interaction) can be well approximated by using Kittel's resonance conditions for the continuous ferromagnetic film. To fit the experimental values of the resonance field in the Ni:MgF$_2$ composite presented in Fig. 4, we used an equation of the Ref. [11]:

$$\left(\frac{h\nu_{res}}{g_{eff}\beta}\right)^2 = (H_{res} - 4\pi M_{eff}\cos 2\theta)\times(H_{res} - 4\pi M_{eff}\cos^2\theta)$$

(1)

Here the angle θ specifies the equilibrium orientation of the magnetization vector (M_{eff}) relative to the normal to the sample plane, and g_{eff} is the g-factor of the composite medium. The out-of-plane magnetization angle θ is related to the polar angle θ_H by the equilibrium condition:

$$H_r \sin(\theta - \theta_H) = 2\pi M_{eff} \sin 2\theta$$

(2)

The equations of (1) and (2) were numerically resolved with respect to H_{res} and θ for various θ_H value, using M_{eff} and g_{eff} values as the fitting parameters. The solid curve in Fig. 4 shows the best of the theoretical model to the experimental data, which gives the values of M_{eff} = 136 G and g_{eff} = 2.25. The determined value of g_{eff} is very close to the g-factor of bulk fcc Ni g_{Ni}= 2.22±0.02 [12] that indicates that the Ni nanoparticles are in the metallic state. Besides, it means that an effect of the single-particle anisotropies (both the magnetocrystalline and shape ones) is averaged to zero and may be neglected.

The typical size of the Ni nanoparticles about 3 nm, determined in the TEM measurements, is below the "superparamagnetic" limit [13]. Therefore, in the case of non-interacting particles the orientations of the total magnetic moments of individual grains are expected to fluctuate at room temperature with the frequency essentially higher than the FMR resonance frequency (ν_{res}), and observation of FMR signal is not possible. Therefore, the FMR response observed in our magnetic resonance experiments is necessarily imply that an essential part of the Ni nanoparticles forms "magnetic" agglomerates with interparticle distances smaller than the particle sizes [10]. This conjecture well corresponds to the TEM results (Fig.2), where such particle agglomeration is in fact observed. The magnetic moments of these agglomerated particles are strongly coupled to each other magnetically by dipolar forces. As a result, the "effective magnetic" size (or correlation size, in terms of Ref.[14]) of such agglomerates is far beyond the superparamagnetic limit. Thus, the agglomerated particles behave themselves as a ferromagnetic media, and FMR signal can be observed, as it is realized in our measurements.

4.   CONCLUSIONS

A new experimental approach for synthesis of magnetic nanoparticles using LUCAS is described shortly. As demonstration, Ni:MgF$_2$ composite material was prepared on silica substrates at room temperature. The nanostructure and magnetic properties of the composite were characterized by electron microscopy and ferromagnetic resonance. The homogenous 3.2 nm size Ni nanoparticles were observed by TEM. The $Ni:MgF_2$ composite material reveals an intensive FMR response caused by the magnetic nanophase of metallic nickel. FMR measurements show that the room temperature ferromagnetism results from the Ni particles, which are strongly coupled by dipolar forces in "magnetic" agglomerates. The observed out-of-plane magnetic anisotropy of the FMR spectra is analyzed on the basis of the effective media approach for the thin granular magnetic films.

ACKNOWLEDGEMENTS

We are grateful to the Alexander von Humboldt Foundation (Germany) and the Deutsche Forschungsgemeinschaft for support of A.L.S. 2. R.I.K. acknowledges the support of the TUBITAK under NATO PC-B Advanced Fellowship programme and the Russian Federation Government support of Leading scientific school grant № SS 1904.2003.

REFERENCES

1. Magnetic Properties of Fine Particles (Eds. J.L. Dormann and D.Fiorani), North-Holland, Amsterdam, 1992.
2. U. Kreibig, M. Vollmer: *Optical properties of metal clusters*. (Springer, Berlin 1995).
3. A.L. Stepanov, M. Gartz, G. Bour, A. Reinhold, U. Kreibig: Vacuum **67**, 223 (2002).
4. M. Gartz, C. Keutgen, S. Kuenneke, U. Kreibig: Eur. Phys. J. D **9**, 127-131 (1999).
5. A.L. Stepanov, G. Bour, M. Gartz, Yu.N. Osin, A. Reinhold, U. Kreibig: Vacuum **64**, 9-14 (2001)
6. I.M.L. Billas, A. Chatelain, W.A. de Heer: Science **265**, 1682 (1994)
7. F.C. Fonseca, G.F. Goya, R.F. Jardim, N.L.V. Carreno, E. Longo, E.R. Leite, R. Muccillo, Appl. Phys. A 76, 261-623 (2003).
8. S.-H. Wu, D.-H. Chen: J. Coll. Interf. Sci. **259**, 282-286 (2003).
9. U. Netzelmann, J. Appl. Phys., **68**, 1800-1807 (1990).
10. G.N. Kakazei, A.F. Kravets, N.A. Lesnik, M. Pereira de Azevedo, Yu.G. Pogorelov, J.B. Sousa, J. Appl. Phys. **85**, 5654-5656 (1999).
11. D.L. Griscom, J.J. Krebs, A. Perez, M. Treilleux, Nucl. Instr. Meth. B **32**, 272-278 (1988).
12. S.M. Bhagat, P. Lubitz, Phys. Rev. B **10**, 179 (1974).
13. C.P. Bean, J. Appl. Phys. **26**, 1381 (1955); L. Néel, Ann. Geophys. **5**, 99 (1949).
14. A. Butera, J. N. Zhou, J. A. Barnard, Phys. Res. B **60**, 12270-12278 (1999).

ION BEAM SYNTHESIS OF MAGNETIC NANOPARTICLES IN POLYMERS

R. I. Khaibullin[1,2], B. Z. Rameev[1,2], A. L. Stepanov[1], C. Okay[3], V. A. Zhikharev[1], I. B. Khaibullin[1], L. R. Tagirov[2,4], B. Aktaş[2]

[1]*Kazan Physical-Technical Institute of RAS, 420029 Kazan, Russia*
[2]*Gebze Institute of Technology, 41400 Gebze-Kocaeli, Turkey*
[3]*Marmara University, P.K. 81040, Göztepe/Istanbul, Turkey*
[4]*Kazan State University, 420008, Kazan, Russia*

Abstract: Different polymers (viscous-flow epoxies, viscoelastic silicone resins and solid state polyimides) were implanted with 40 keV Fe^+ or Co^+ ions to the doses of $0.1 \div 2.0 \times 10^{17}$ ions/cm^2. The influence of the dose and viscosity of polymer target on the process of nucleation and growth of metal nanoparticles in the implanted polymers as well as on the magnetic properties of ion-synthesized composites were investigated by electron microscopy and magnetic resonance. The implantation of the polymers with 40 keV ions causes a surface carbonization of polymer substrate and at the doses more than 0.25×10^{17} ions/cm^2 results in the formation of metal (iron or cobalt) nanophase in thin subsurface layer. Mean sizes, crystalline structure, shape and space packing of the ion-synthesized nanoparticles strongly depend on the dose, kind of implanted ions and the polymer viscosity during implantation. The ion synthesis of the isolated cubic or spherical particles with the mean sizes in the range of $2 \div 200$ nm, as well the formation of many-particles clusters, fractal-type agglomerates and single microscaled plates are observed in the implanted polymers under study. Ion-synthesized iron or cobalt nanoparticles reveal the magnetic resonance response, and at high doses their resonance signals demonstrate the typical features of ferromagnetic resonance in granular magnetic films. The values of magnetization and coercivity of the granular composite films were obtained from the analysis of FMR data. The non-linear dependencies of the composite magnetization on ion dose and on the viscosity of polymer target are presented and discussed in the frame of the magnetic percolation transition in the many-particles system.

Key words: ion implantation, magnetic nanoparticles, metal-polymer composites

B. Aktaş et al. (eds.), Nanostructured Magnetic Materials and their Applications, 33–54.

1.   INTRODUCTION

High-dose implantation of metal ions into dielectrics leads to the local excess of the dopant, which is unstable in the form of metal atoms dispersed homogeneously over lattice sites. The system relaxes into precipitates of metal which are variously termed colloids, granules or nanoparticles. The crystalline structure, sizes and morphology of the ion-synthesized metal nanoparticles as well the local particle concentration can be controlled to some extent by the regimes of ion implantation, by the degree of crystallinity and chemical composition of dielectric substrates and by post-implanted thermal treatments. In this way the different kinds of the metal nanoparticles may be formed in the near-surface layers of the irradiated dielectrics.

Ion synthesis of transition metal nanoparticles in traditional dielectric substrates, as sapphire, silica or silica-based glasses, has been studied extensively recently [1-3] because ion implantation is very suitable for fabrication of planar nanocomposite materials for magnetic date storage industry, magneto-sensor and magneto-optical electronics. However, only a few studies related to the ion synthesis of magnetic nanoparticles in polymers were performed [4-6]. One of the reasons is that ion implantation in polymers leads to the considerable change of their molecular structures (see, i.e. review [7] and reference therein). High-energy ions break of molecular bonds and lead to strong gas element depletion. Consequently, the implanted region of the polymer became amorphous carbon after implantation, and carbonisation process reveals the saturation level at doses of 10^{16} ions/cm^2 [7]. Carbonization of the implanted polymers is a typical process for all types of polymers. It was expected that magnetic nanoparticles subsequently forming in the irradiated carbonized layer will reveal similar structural and magnetic properties independent of the polymer type. In this paper, we summarize the basic results of our investigations related to ion synthesis of magnetic (iron or cobalt) nanophase in different polymers [8-16]. Beside the technique of ion synthesis of magnetic nanoparticles was also supplied to polymer substrates with same chemical composition but with variable viscosity. The results presented below show clearly that the dispersive parameters and the morphology of ion-synthesized iron (or cobalt) nanoparticles, as well their magnetic response are strongly dependent on the types of virgin polymer and the polymer viscosity during ion implantation.

2.   EXPERIMENTAL

The following polymer matrices were used in our experiments:

Epoxy adhesive (EA), $C_{18}H_{19}O_3Cl_2$: The epoxy substrates were prepared by hand deposition of viscous mixture of the commercial epoxy resin (ED-20, 93 weight %) and the hardener (polyethylenepolyamine, 7 %) on glass plates. The thickness of produced epoxy films was about 1 mm. During the curing (polymerization) process the epoxy mixture passes different relaxation states: from the viscous-flow through viscoelastic to the solid glassy state. The dynamic viscosity η, measured by a capillary viscometer, varied in the range 20-180 Pa·s during the first 250 min of curing process. (Fig. 1). The solid glassy-like EA substrates ($\eta \geq 10^5$ Pa·s) were received after 24 hours of the cure process;

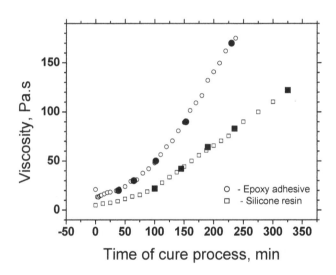

Figure 1. Viscosity of epoxy and silicone substrates versus time of their curing at room temperature. The second set of the samples was obtained when iron (or cobalt) implantation started at different values of substrate viscosity (black circles or squares).

Silicone resin (SR), $C_9H_{16}O_4Si_2$: The silicone substrates were prepared by centrifugation of viscous mixture of the commercial silicone resin (SKTN-F, 98.5 weight %) and the curing agent (diethyl-dicaprilat of tin, 1.5%) on the glass plates. The thickness of the produced silicone films was about 150 μm. Kinetics of cure (polymerization) process for the silicone composition was characterized by measuring their dynamical viscosity (η) using REOTEST-2 viscometer at room temperature. This composition exhibits a long time (~70 min) initial stage of cure process with a small change in viscosity from 7 Pa·s to 15 Pa·s that allows ones providing the necessary conditions for following ion irradiation. Viscosity of the composition is relatively rapidly varied in the range of 20-125 Pa·s with following the curing during the 350 min. (Fig. 1). The viscosity of the fully cured viscoelastic silicones was found to be in order of 10^3 to 10^4 Pa·s;

Polyimide (PI), $C_{22}H_{10}O_5N_2$: Polyimide is the low-friction and high-temperature resistant polymer. The commercial glassy-like polyimide foils with 40 μm thick were used as substrates for implantation.

Two specific sets of samples were prepared to investigate a process of ion-beam synthesis of magnetic nanoparticles in different polymer matrices in detail. *In the first set*, solid EA, SR and PI substrates were implanted with 40 keV Fe^+ and Co^+ ions in *a wide dose range* of 0.1-2.0×10^{17} ions/cm^2 at the current density 4 μA/cm^2. *In the second set*, the viscous SR or viscous EA substrates were implanted with the same ions at the fixed dose of 1.25 or 1.5×10^{17} ions/cm^2 in *a wide range of viscosity*. In other words, the silicone or epoxy targets had the different initial value of viscosity when ion implantation had started (see Fig.1). It is important to note that after implantation, the implanted viscous samples transformed in the solid state due to curing process. The implantation was carried out on ILU-3 ion-beam accelerator at room temperature with the residual vacuum of 10^{-5} Torr.

For microstructure studies the thin surface layers of irradiated polymers were obtained by chemical etching. The morphology and the crystalline phases of the synthesized metal granular films were studied in the plane of irradiation by electron diffraction and transmission electron microscopy (TEM) on the *TESLA-BS500* and *EM-125* microscopes. The specimens for cross-section TEM were prepared by using the microtome *LKB*. Magnetic resonance spectra of the metal-polymer composites were recorded at room temperatures with *Bruker EMX* spectrometers operating in X-band (≈9.5 GHz). As usual, the field derivative of microwave power absorption (dP/dH) was registered as a function of the applied magnetic field (H). The resonance field (H_0) was considered as a value of magnetic field, at which the first derivative curve intersects a baseline.

3.   RESULTS

3.1   The nucleation and growth of metal nanoparticles in the implanted polymers

The precipitation phenomenon (ion synthesis) of iron or cobalt crystalline nanoparticles in different polymers is clearly visualized by TEM observations in the bright field mode. The metallic nanoparticles exhibit dark contrast (black sports) in-plane TEM images of the surface layers of polymers implanted with metallic ions. The nucleation of the metal phase of iron or cobalt is observed at the dose of 0.25×10^{17} ions/cm^2 and this process does not depend on a type of virgin polymer substrates. However, with increasing dose, the morphology and growth stages of the metal nanoparticles depend on both the type of virgin polymer and the relaxation state (viscosity) of polymer substrate during ion irradiation.

Fig. 2 shows TEM in-plane images of near-surface region of polyimide foils implanted with Fe+ ions at two different doses. For the dose of 0.5×10^{17} ions/cm^2 (Fig. 2A), the formation of metal particles with a mean size of (112 ± 12) nm in diameter was observed. The size distribution is presented on the histogram of Fig. 2A. Two groups of particles with mean sizes of about 25 and 150 nm, respectively, may be distinguished. The areal density of nanoparticles was estimated to be about 3.5×10^8 cm^{-2}. For the high dose of 1.25×10^{17} ions/cm^2 (Fig. 2B), more homogeneously distributed metal particles with a mean size of 60 nm at areal density of $\sim 4.0 \times 10^9$ cm^{-2} are formed in the implanted layer. However, the size distribution is asymmetric with an extended tail toward larger nanoparticles, and it is close to a log-normal distribution as can be seen from histogram in Fig. 2B. The metal nanoparticles contain about 65% of all implanted iron atoms according to our estimations based on the mean particle sizes. Electron diffraction patterns (EDP) of iron-implanted polyimides exhibit 2-3 weak diffusive rings that may be related to metallic and ferromagnetic α-Fe phase. Cross-sectional TEM and AFM studies (Fig. 3) show that the iron nanoparticles are located under the polyimide surface at the depth from 30 up to 100 nm, and they induce the surface swelling as semi-spherical bumps on the surface.

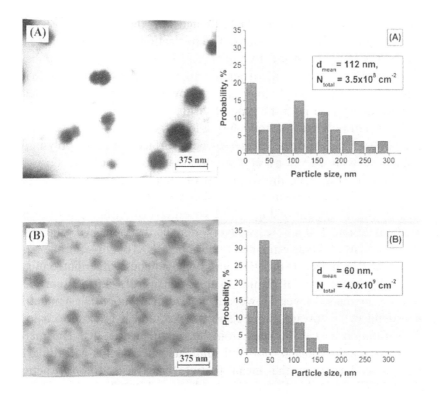

Figure 2. TEM in-plane images of iron nanoparticles (dark sports) formed by implantation with 40 keV Fe^+ ions in polyimide foils at ion current density of 4 μA/cm2 and doses: A) 0.5×10^{17}; B) 1.25×10^{17} ions/cm^2.

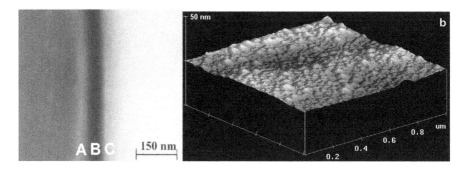

Figure 3. TEM cross-sectional (left) and AFM surface (right) views of polyimide foil implanted with 40 keV Fe^+ ions to a dose of 1.25×10^{17} ion/cm^2. Region (A) corresponds to about 30 nm thick near-surface modified layer. Black strip (B) (about 70 nm thick) is a layer consisting of carbonised polymer filled by iron nanoparticles. Grey strip (C) corresponds to the region with the iron atoms diffused inward polymer.

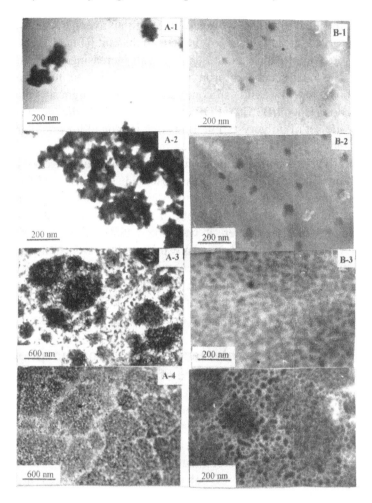

Figure 4. TEM in-plan images of iron nanoparticles formed by implantation with 40 keV Fe⁺ ions in viscous (A1-A4) and solid (B1-B4) silicone substrates at different dose: 1) 0.3×10^{17}, 2) 0.6×10^{17}, 3) 1.0×10^{17}, 4) 1.5×10^{17} ions/cm^2.

Figure 4 shows the nucleation and the subsequent growth of iron nanoparticles in the viscous (A) and solid (B) silicone substrates with increasing implantation dose. The iron nanostructures (granular films) formed in viscous silicone substrate with initial η=20 Pa·s consist of nanoparticle agglomerates. These agglomerates present a set of connected iron nanoparticles and they are already formed at low doses (Fig. 4:A1,A2). By increasing the dose, the mean size of the agglomerates increases and they begin to touch each other and to form a connecting network of needle-like iron nanoparticles (Fig. 4:A3,A4). In the irradiated solid-state silicone substrate, the isolated drop-like iron particles with an average size of 20-25 nm were observed at low dose of 0.3×10^{17} ions/cm^2 (Fig. 4:B1). These nanoparticles are presented in great quantity with increasing the dose value (Fig. 4:B2). The coalescence of iron nanoparticles due to high particles density is observed in solid silicone substrates at highest values of implantation dose (Fig. 4:B3,B4). EDP on iron implanted silicone substrates exhibited a wide set of polycrystalline rings that is typical for granular metal films. The analysis of the EDP indicates that most of the observed rings correspond to the b.c.c lattice of metallic α-Fe. Additionally, weak rings of iron oxides and silicides have also been observed. Data from TEM cross-section studies showed that synthesized iron granular films are about 30-50 nm thick and they are buried at the depth of 15-20 nm from the surface of the irradiated polymer. However, it was noted that metal nanostructures formed in a viscous silicone substrate are thinner than iron granular films synthesized in a solid polymer for a given implantation dose. Thus structural investigations of iron-implanted silicone substrates show that the kinetics of metal phase growth and morphology of the synthesized iron nanoparticles may be strong depended on polymer viscosity during ion irradiation.

3.2 The influence of polymer viscosity upon morphology, sizes and crystal structure of metal nanoparticles

The epoxy and silicone substrates being under different stages of curing process were implanted with Co$^+$ ions at the fixed dose (1.2-1.4×10^{17} ions/cm^2) to investigate, in detail, the influence of the polymer viscosity on the ion synthesis of metal nanoparticles.

Fig. 5 shows TEM images of cobalt nanoparticles formed in epoxy substrates with different initial viscosity, as well as in fully cured solid epoxy. As it is seen, the needle-like (Fig. 5a) or uniform spherical (Fig. 5b) nanoparticles are formed in viscous substrates at small epoxy viscosity of 20 and 30 Pa·s, respectively. The band of ultra-small cobalt particles (Fig. 5c, $\eta=50$ Pa·s) or metal nanoparticles of cubic–like shape (Fig. 5d, $\eta=90$ Pa·s) are observed in epoxy substrates implanted at subsequent stages of curing process. In solid-state epoxy drop-like grains (Fig. 5e) were grown as a result of cobalt implantation. The particular interesting result was obtained in the experiment when cobalt ions were implanted into the viscous matrix ($\eta=170$ Pa·s) at the enhanced dose of 1.8×10^{17} ions/cm^2 and the current density of 10 μA/cm^2. The formation of the cobalt micron-size cobalt plates with a regular hexagonal shape (Fig. 5f) is observed in contract to the nanoscaled particles presented above. Analysis of electron diffraction measurements (see insets in Fig. 5a-53f) indicates that the polycrystalline structure of the pure α-Co phase (a-c) or the mixed α-Co and β-Co phase (d), as well as monocrystalline structure α-Co microsize plates (f) are typical for the epoxy samples implanted in viscous state. The cobalt nanoparticles formed in the solid substrate reveal amorphous structure (e). Data on TEM cross-section showed that the ion-synthesized composite Co/epoxy layers (films) are about 40-50 nm of thickness and occur at the depth of 20 nm from the principal surface of the irradiated polymer.

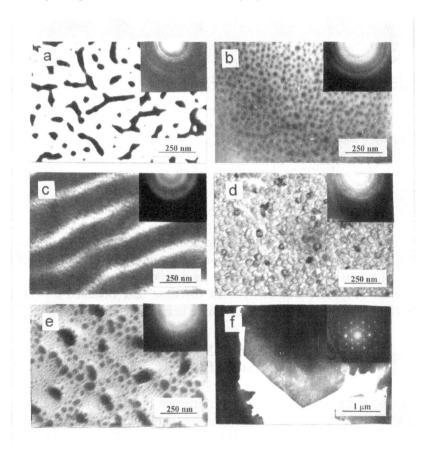

Figure 5. TEM in-plan images of cobalt nanoparticles formed by implantation with 40 keV Co$^+$ ions in epoxy substrates at different viscosity: a) 20 Pa·s, b) 30 Pa·s, c) 50 Pa·s, d) 90 Pa·s, f) 170 Pa·s, e) solid-state epoxy polymer. The implantation parameters: (a), (d), (e) - D = 1.5×10^{17} ion/cm^2 and (b), (c) D = 1.2×10^{17} ion/cm^2 at j = 4 μA/cm^2; for (f) - D = 1.8×10^{17} ion/cm^2, j = 10 μA/cm^2.

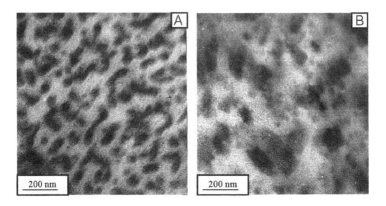

Figure 6. TEM images of cobalt nanoparticles formed by implantation with 40 keV Co$^+$ ions in viscous (A) and solid (B) silicone substrates at a dose of 1.25×10^{17} ions/cm^2.

Figs. 6 show the TEM images of cobalt granular films produced in viscous ($\eta = 20$ *Pa·s*) and viscoelastic (fully cured $\eta \sim 10^3$ *Pa·s*) silicone substrates implanted with Co+ ions to dose of 1.25×10^{17} ion/cm^2. It is seen that the formation of worm-like metal particle takes place in viscous silicone substrates (Fig. 6A) as it was similar in epoxy matrix with the same viscosity (Fig. 5a). In the cured solid-state silicone matrix the ion-synthesized cobalt granular film consists of widely separated groups of fine particles with sizes of 40-50 nm (Fig. 6B). The analysis of the EDP date obtained on cobalt particles indicates the hexagonal α-Co phase in implanted polymers only.

3.3   Magneto-resonance study of ion-synthesized metal-polymer nanocomposite films

The magnetic resonance signals were detected in all polymers implanted with 40 keV Co$^+$ or Fe$^+$ ions. However the implantation dose threshold, above which the magnetic resonance response appeared, is depended on both the kinds of implanted ions and the types of polymer substrate. For example, the cobalt-epoxy nanocomposites exhibit a signal at low-temperature only for implantation doses higher than 1.4×10^{17} ion/cm2, while the cobalt implanted silicone substrates or the cobalt implanted and subsequent annealed polyimides show a room-temperature resonance signals at much smaller dose of about 0.6×10^{17} ion/cm^2. The magnetic resonance spectra of all cobalt implanted polymers and their more detailed analysis may be found in [10, 12-13,16]. Here we present only the most important results related to iron implanted polymers where magnetic response is more intensively and the implantation dose threshold is about of 0.25×10^{17} ion/cm^2.

Figure 7 show the magnetic resonance spectra of Fe-implanted polyimide foils implanted with 40 keV Fe$^+$ to different doses. As seen in figure, the magnetic response increases with the dose. The resonance signal is almost independent on the orientation of applied magnetic field with respective to sample plane at low dose, but it becomes strongly anisotropic at higher doses. The shift of resonance in low-field region of magnetic spectra for parallel orientation, and vice versa in high-field region for perpendicular ones is clear observed with increasing dose.

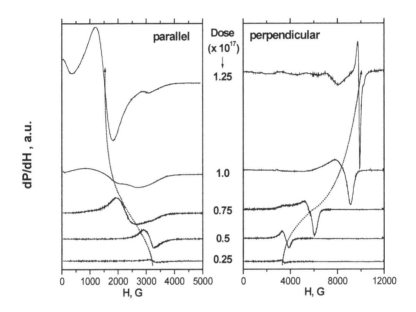

Figure 7. FMR spectra of polyimide foils implanted with 40 keV Fe+ ions at different doses. The left panel corresponds to parallel orientation and the right ones – to perpendicular orientation of the applied magnetic filed with respect to the sample plane. The value of implantation dose is shown between panels.

The effective anisotropy (ΔH_a) of resonance field is an anisotropy parameter often used in FMR spectroscopy, and ones is defined as $\Delta H_a = H_0(\theta=0°) - H_0(\theta=90°)$. Here, the angle θ is the polar angle between the normal of the substrate surface and the direction of the applied magnetic field. The polar angular dependencies of resonance field in Fe-implanted polyimide are presented in Fig. 8. It is clearly seen that out-of-plane anisotropy ΔH_a increases with implantation dose and at higher doses the observed dependencies resemble the behavior of resonance field in continuous ferromagnetic film. The subsequent azimuthal measurements support the film geometry for Fe-implanted samples of polyimide. The value of resonance field is not changed when the applied magnetic field is rotated in the sample plane. From these measurements we conclude that iron-implanted polyimides exhibit uniaxial out-of-plane anisotropy, composite magnetization lies parallel to the substrate surface, i.e. "in-plane", and the observed resonance response is ferromagnetic resonance (FMR) in thin granular film.

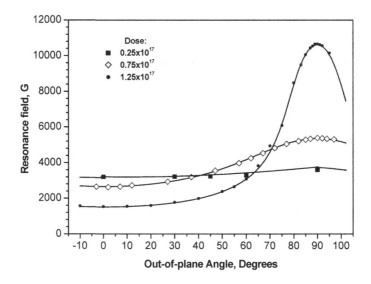

Figure 8. Angular dependencies of resonance field in polyimides implanted with 40 keV Fe^+ ions at different doses.

Figure 9 shows FMR spectra of silicone substrates implanted with 40 keV Fe^+ ions to the given dose of 1.25×10^{17} ions/cm^2 and with different values of the viscosity at initial moment of iron implantation. It is seen from the figure, that FMR signal position strongly and non-monotonic depends on polymer viscosity. At first, the shifts of the resonance signal to the low- field side of spectrum for parallel orientation and to the high-field side of spectrum for perpendicular orientation, respectively, are observed with an increasing of polymer viscosity. This means, that out-of-plane magnetic anisotropy of the iron/silicone composites increases with viscosity. The shifts of FMR signals for both orientations reach their maximal values at the viscosity of about 70 Pa·s, and then the absorption peaks begin displacing in opposite direction. Thus, magnetic measurements show that by varying of both the implantation dose and the polymer viscosity one could control the

FMR response and magnetic out-of plane anisotropy of the ion-synthesized metal-polymer composite films.

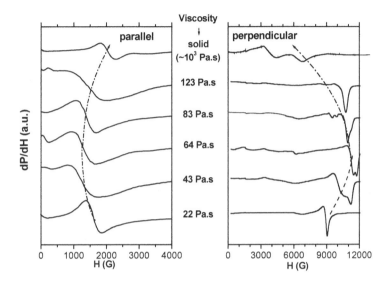

Figure 9. FMR spectra of silicone substrates implanted with Fe^+ ions to the dose of 1.25×10^{17} ion/cm^2 at different polymer viscosity. The left panel corresponds to parallel orientation and the right ones – to perpendicular orientation of the applied magnetic filed with respective to sample plane. The viscosity of silicone substrate at initial moment of implantation is given between panels.

For both polyimide and silicone resin substrates implanted with 40 keV Fe^+ to doses of more than 1.0×10^{17} ions/cm^2 the hysteresis of the FMR signal at low magnetic field was observed. As example, hysteretic behaviour of the FMR absorption for iron-implanted polyimide is shown in Fig.10. After first recoding of FMR spectrum in the sample plane ($\theta = 90°$) we turned then the sample in the plane on 180° and recorded the spectrum again. The changing in FMR line absorption is revealed clearly in zero-field region of magnetic spectra. This changing is caused by the reversal of the remanent magnetization with increasing the magnetic field applied in reverse direction. According to the presented spectra the value of the magnetic field (coercive field) at which the remanent magnetization completely reverses is of 270 G.

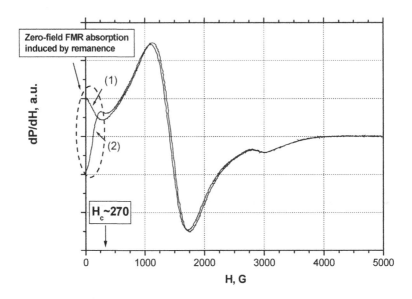

Figure 10. FMR spectra of polyimide implanted with Fe$^+$ ions to the dose of 1.25×10^{17} ions/cm2 for parallel orientation of the applied magnetic filed with respective to sample plane: (1) - initial one; (2) - spectrum after the in-plane turn of the sample on 180°.

4.   DISCUSSION

4.1   Mechanisms of nucleation and growth of metal particles in implanted polymers

Energetic ions are stopped in implanted materials by interactions with electrons and by ion-target nucleus collisions. For 40 keV Co$^+$ or Fe$^+$ ions incident on a polymer, the ion stopping is mainly due to nucleus collisions, and the ion depth range is of the order of 50 nm and the straggling of 15 nm [17]. These data are in accord with our TEM cross-sections studies of the implanted polymers (see, e.g. Fig. 3). The ion-target nucleus collisions lead to a breaking of molecular bonds, strong gas element depletion, cross-linking and surface carbonization of polymers. On the other hand, the interaction of the ions with the electrons of the target causes changes in the ionic charge. The cross-section for electron capture becomes very large at low ion velocities. There is a high probability for the implanted metal ions to be in a neutral charge state when they stopped in the carbonized layers of polymer.

It is known that nucleation of new phase during ion implantation start at concentrations exceeding the solubility limit ($N_{lim.}$) of the implanted impurity in given matrix. For metal impurity implanted in polymers, e.g. in polyimide [18], this value is about $N_{lim.} \approx 10^{20}$ atom/cm^3. According to simple estimation the number of Co or Fe atoms in the implanted layer with thickness of 100 nm is of the order of 10^{21} cm^{-3} for ion doses on the order of 10^{16} ion/cm^2. Thus the metal nucleation has to occur in the implanted polymers at the early stage of implantation. The formed nucleations are effective drains for subsequently implanted impurity atoms. This implies that the subsequent growth of the metal nanoparticles or their aggregates, particle shape and crystal structure are controlled by diffusion of implant or nucleation, particle coalescence processes and the supermolecular structure of polymer substrate at initial moment of ion irradiation.

In order to understand mechanisms of ion synthesis of different types of metal nanostructures in viscous and solid polymer substrates, one should take into account the microstructure and phase composition of the implanted carbonized layer, as well as the features of epoxy (or silicone) polymerization during cure process. However the detailed analysis of the carbonized phase formed in the viscous or solid polymer has not been done still, and the naive concepts of radiation-induced carbonization and polymerization process may be considered only.

It was experimentally found that the mean size of metal nanoparticles formed in the solid polymers have a tendency to decrease with an increasing of the specific density of virgin substrates: silicone resin (ρ=1.0 g/cm^3), epoxide (1.2 g/cm^3) and polyimide (1.43 g/cm^3). Possible explanations of this result are that the density of carbonized layer with the implanted nanoparticles seems to be proportional depended on polymer density. It is clearly that a small density of the carbonized layer is favorable factor for the growth of large particles.

Not only the density, as well as the microstructure of the carbonized layer may depend on the type and the relaxation state of virgin polymer substrate. During cure process the epoxy (or silicone) substrate passes directly from the liquid-like viscous-flow state of the initial composition to the solid (viscoelastic or glassy-like) state of polymer with changing its viscosity. As shown in Fig. 1, the viscosity of the organic substrate changes a little (on order of 10-20 Pa·s) within first 100 minutes of the early stage of the cure process. At this moment the epoxy (or silicone) macromolecules are only opening and the polymer system may be considered as a low-viscosity uniform molecular liquid. Then, on the second stage, the macromolecules of initial composition start to bind with each other and to form the microblocks and supermolecular network structure of polymer. The viscosity of epoxy (or silicone) substrate rapidly increases by 10 Pa·s (or 5 Pa·s, for silicone) every

10 minutes of cure process. At last after 24 hours of the cure process we have a final product – fully cured solid polymer.

All said above enables one to suppose that implantation at early stage of cure process, when the polymer substrate (or the carbonized layer) may be considered as molecular liquids, leads to formation of metal particles in their supersaturated solution. At given conditions the homogeneous mechanism of nucleation and the growth of nanoparticles by diffusion of impurity atom to nucleation determines the formation of the spherical-shaped cobalt nanoparticles with the α-Co lattice structure given in Fig.5b. This assumption is supported by our analysis [9] of dose dependence of disperse parameters of cobalt films which have been synthesized in viscous epoxy $(\eta \cong 30 \; Pa \cdot s)$. In polymer matrices with minimal viscosity $(\eta \cong 20 \; Pa \cdot s)$ the nucleation or metal clusters may have high mobility. Under this condition the fast-growing metal clusters may join in the many-particles agglomerates or coalesce into more long formations due to magnetostatic interactions. This mechanism describes the agglomeration of iron nanoparticles in viscous silicone substrates (Fig 4:A1-A4), as well as the growth of need-like or worm-like cobalt nanoparticles in the viscous epoxy (Fig.5a) and silicone (Fig.6a) matrices. On the subsequent stages of cure process, the supermolecular network structure is formed in polymer substrate. The formation of bond structures (Fig.5c) and growth of cubic particles with unstable fcc-lattice structure (Fig.5d) is expected to occur due to the influence of the forming cubic network of epoxy matrix. Detailed mechanism of the formation of monocrystal microscaled plates of α-Co in viscous epoxy substrate (Fig.5f, $\eta \cong 170 \; Pa \cdot s$) is not settled yet. However it is known that the highest current density of ions and high-dose implantation lead to large longitudinal stresses in the implanted layers. We believe that anisotropic crystallization of cobalt implants can be induced by these internal stresses.

In fully cured solid substrates, the rigidity of the polymer matrix drastically decreases the diffusion of metal atoms and nucleation. We assume that the dominant nucleation mechanism in these samples is heterogeneous. Structure and radiation polymer defects serve as effective centers of nucleation. As a result, the size, shape and crystal structure of metal particles is mainly determined by a system of defects and low value of diffusion coefficient of cobalt atoms. The presence of radiation defects in solid polymer matrices is often the reason of the formation of two-maximum size distribution functions of the ion-synthesized particles observed, e.g., in Fe-implanted polyimide (Fig. 1A) or in solid epoxy substrate implanted with cobalt ions (Fig. 5e, [9]). According to [19], low mobility of impurity atoms usually leads to the growth of irregular shaped metal aggregates with amorphous structure, as it is shown in Fig.5e.

4.2 Magnetization of ion-synthesized metal-polymer nanocomposite films

As was already noted the observed out-of-plane anisotropy of resonance field in ion-synthesized nanocomposites is typical for FMR in thin magnetic film. An ensemble of iron nanoparticles formed in the implanted layer of polymer may behave as a thin ferromagnetic continuum due to strong magnetic dipolar coupling between the nanoparticles [20]. The magnetic percolation transition in thin granular films may be observed in FMR measurement even without direct contact between nanoparticles. The transition occurs when the concentration of magnetic nanoparticles is high enough for interparticle coupling strength to be comparable in magnitude with the Zeeman energy of nanoparticles in the external magnetic field:

$$\frac{m_i \cdot m_{i+1}}{r_{i,i+1}^3} \approx m_i \cdot H_{mean} \tag{1},$$

where m_i is the magnetic moment of an individual nanoparticle, $H_{mean} = h\nu_{res}/g\beta$. – "Mean resonance field" of magnetic nanoparticles (for X-band $H_{mean} \approx 3300$ G), ν_{res} - resonance frequency, β - Bohr magneton and g is the g -factor value (a dimensionless number ~ 2).

In the ferromagnetic continuum approximation the resonance field for two limiting orientations of magnetic field with respective to sample plane may be determined by Kittel set of equations [21]:

$$h\nu_{res.} = g_{eff}\,\beta \cdot (H_o - 4\pi M_{com.})\,(\text{for H} \perp \text{sample plane, } \theta = 0^\circ) \tag{2}$$

$$h\nu_{res.} = g_{eff}\,\beta\sqrt{H_o \cdot (H_o + 4\pi M_{com.})}\,(\text{for H} \perp \text{sample plane, } \theta = 90^\circ) \tag{3}$$

Here ν_{res} and H_0 are experimental values of the resonance frequency and the magnetic resonance field, respectively. These equations allow one to calculation the values of g_{eff} and $M_{com.}$ which are the g-factor and the magnetization of metal-polymer composite film. For all iron-implanted samples the calculated values of g_{eff}-factor (2.1 ± 0.1) are close to the g-factor of bulk iron film. This fact clearly indicates that most of implanted iron atoms are in a metallic state that is in an agreement with our structural investigations.

The calculated values of the composite magnetization in polyimide, viscous and solid silicone samples implanted with Fe^+ ions are presented in

Figures 11 and 12. There is the set of features in the dependencies of magnetization on the implantation dose and on the viscosity of polymer. The first feature is the inflections of dose dependencies shown in Fig. 11 and observed at dose values of about 0.6×10^{17} ions/cm^2 in solid silicone substrate (curve A) and polyimide (curve C). We suppose that these inflections reflect the points of the magnetic percolation transition in the many-particle system, as it was above discussed. This supposition is in the agreement with the dose dependency of magnetization in silicone substrates implanted with iron ions in viscous state with viscosity of 20 Pa·s (curve B in Fig.11). There is not inflection in the dose dependency of magnetization due to the agglomeration of the iron nanoparticles observed in viscous silicone substrates at all doses (see, Fig. 4, A1-A4).

Another feature is the different slopes of the dose dependencies of composite magnetization for the various polymer substrates implanted with iron ions. The magnetizations of the polyimide or the viscous silicone substrate increase more rapidly with increasing implantation dose in the comparison with the magnetization of Fe-implanted solid silicone. Here it is important to note that the composite magnetization is proportional to the iron filling factor of a metal-polymer film [20], i.e. the relative volume that is occupied by synthesized iron particles in the irradiated carbonized layer of polymer. Hence the iron filling factor depends on both the type of virgin polymers and the relaxation state of substrate during ion implantation. It is observed that at given implantation dose the metal filling factor is larger in the composite films formed by iron implantation in viscous silicone matrix compared to the silicone polymer implanted in solid state. This fact suggests that quantity of the iron ions, which entered into the metallic phase and form the magnetic particles, depends on the viscosity of polymer.

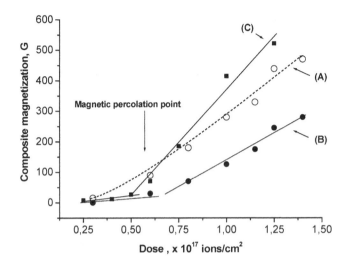

Figure 11. Dose dependence of the magnetization for metal/polymer composite films formed by implantation with 40 keV Fe$^+$ ions in viscous (A) and in solid (B) silicone substrates, as well as in solid polyimide foil (C). Viscous silicone substrates had an initial viscosity of 20 Pa·s when iron implantation started.

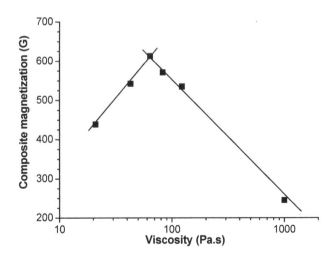

Figure 12. Dependence of composite magnetization on initial viscosity of the silicone substrate implanted with Fe$^+$ ions to the fixed dose of 1.25×10^{17} ion/cm^2/

Figure 12 shows the dependence of composite magnetization on the viscosity of silicone substrate implanted with Fe+ ions to the fixed dose of 1.25×10^{17} ions/cm^2. The observed dependence has peak-like shape, and a maximal magnetization of 600 G corresponds to silicone sample implanted with the viscosity of 65 Pa·s. We suppose that it is optimal value of silicone viscosity at which iron filling factor in the implanted layer of polymer have a maximal value. Note that this value is is two or three times higher than magnetization of sample implanted in solid state (the last point in the figure 12). The specificity of the ion synthesis of metal particles in viscous media determines the non-monotonic behavior of magnetization with the increase of the silicone viscosity. As it was above noted, the magnetization of the metal/polymer composite film is proportional to the metal filling factor. At low values of the viscosity a part of the implanted iron ions may leave the irradiated region due to fast diffusion and not participate in the formation of magnetic nanoparticles. On the other hand, the rigidity of the fully cured silicone substrate at high viscosity drastically decreases the diffusion mobility of iron ions and also hampers the coagulation of implanted ions in the metal granules. Both processes result in the reduction of the magnetic phase fraction in the implanted region and, consequently, in the decrease of magnetization. It is clear that there is an optimal value of viscosity (in our case η is about 70 Pa·s.) at which the iron filling factor and the effective magnetization of the composite film should have maximal values.

5.   CONCLUSION

We implanted epoxy adxesive, silicone resin and polyimide films with 40 keV Co$^+$ or Fe$^+$ ions in a wide dose range of 0.1-2.0×10^{17} ions/cm^2 to investigate an influence of the type and the relaxation state of virgin polymers on ion synthesis of magnetic nanoparticles. The implantation of the polymers with metal ions to doses more than 0.25×10^{17} ions/cm^2 results in the formation of iron or cobalt precipitates (metal nanoparticles) in thin subsurface carbonized layer of polymer, and this process does not depend on a type of virgin polymer substrate. The subsequent growth of metal nanophase in a polymer with increasing the implantation dose was investigated by transmission electron microscopy. It was found that various metal nanostructures (isolated particles with different morphology and sizes, multi-particles clusters, fractal-type agglomerates) and single-crystalline microsize plates are formed at different initial viscosity of the targets, kinds of implanted ions and doses of implantation. The iron or cobalt nanostructures, embedded in the carbonized subsurface layer, reveal the magnetic resonance response. The magnetic nanoparticles synthesized at high doses are strongly coupled by interparticle dipole-dipole interactions, and their magnetic resonance signal demonstrates typical features of ferromagnetic resonance in granular magnetic films. The out-of-plane anisotropy and magnetization of the ion-synthesized metal-polymer composite films increase with the implantation dose, and the nanostructured films formed at highest dose values reveal the characteristics of a continuous magnetic film: the remanence and the coercivity. The magnetic properties of the metal-polymer composite films formed by metal implantation in viscous polymer substrate non-monotonically depend on the substrate viscosity during ion irradiation. The viscosity value corresponding to the maximal magnetization of iron/silicone composite film was experimentally found.

Our investigations reveal possibility of controlling the dispersive parameters and morphology of ion-synthesized magnetic nanoparticles by varying of the type of polymer substrate, polymer viscosity and the ion dose. The iron and cobalt nanoparticles synthesized in the carbonized polymers exhibit diverse magnetic properties (magnetic anisotropy, magnetization, coercivity). The new planar composite materials formed by implantation of transition metal ions in polymer look promising for practical applications.

ACKNOWLEDGEMENTS

This work was supported by "Russian Federal Programme for support of Leading scientific schools", grant No. SS 1904.2003.2, and by Gebze Institute of Technology, grant No. 02-A-01-02-03. R.I.K. gratefully acknowledges the NATO-TUBITAK Advanced Fellowship Programme for support of his work at Gebze Institute of Technology. A.L.S. is grateful to the Alexander von Humboldt Foundation in Germany and the Lisa Meitner program of the Austrian Science Foundation. L.R.T. thanks the BRHE grant REC-007 and the Programme "Universities of Russia" (URFI) for the partial support. The authors wish to thank Dr. V. Popok from Göteborg University for the discussion of paper.

REFERENCES

[1] A. Meldrum, R.F. Haglund, L.A. Boatner, C.W. White, Adv. Mater. 13 (2001) 1431
[2] A. Meldrum, L.A. Boatner, C.W. White, Nucl. Instr. and Meth. B 178 (2001) 7
[3] F. Gonella, Nucl. Instr. and Meth. B 166-167 (2000) 831

[4] N. C. Koon, D. Weber, P. Penrsson, and A. I. Shindler, Mat. Res. Soc. Symp. Proc. 27 (1984) 445

[5] K. Ogava, United State Patent, No 4.751.100 (1988)

[6] V. Petukhov, V. Zhikharev, M. Ibragimova, E. Zheglov, V. Bazarov, I. Khaibullin, Sol. St. Commun. 97 (1996) 361

[7] D.V. Sviridov, Russian Chemical Review, 71 (2002) 315

[8] R.I. Khaibullin, S.N. Abdullin, A.L. Stepanov, Yu.N. Osin, I.B. Khaibullin, Tech. Phys. Lett. (USA) 22 (1996) 112

[9] S.N. Abdullin, A.L. Stepanov, Yu.N. Osin, R.I. Khaibullin, I.B. Khaibullin, Surf.&Coat. Technol. 106 (1998) 214.

[10] I.B. Khaibullin, R.I. Khaibullin, S.N. Abdullin, A.L. Stepanov, Yu.N. Osin, V.V. Bazarov, S.P. Kurzin, Nucl. Instr. and Meth. B 127-128 (1997) 685

[11] R.I. Khaibullin, Y.N. Osin, A.L. Stepanov, I.B. Khaibullin, Nucl. Instr. and Meth. B 148 (1999) 1023

[12] R.I. Khaibullin, V.A. Zhikharev, Y.N. Osin, E.P. Zheglov, I.B. Khaibullin, B.Z. Rameev, B. Aktas, Nucl. Instr. and Meth. B 166-167 (2000) 897

[13] B.Z. Rameev, B. Aktas, R.I. Khaibullin, V.A. Zhikharev, Y.N. Osin, I.B. Khaibullin, Vacuum 58 (2000) 551

[14] R.I. Khaibullin, V.N. Popok, V.V. Bazarov, E.P. Zheglov, B.Z. Rameev, C. Okay, L.R. Tagirov, B. Aktas, Nucl. Instr. and Meth. B 191 (2002) 810

[15] R.I. Khaibullin, B.Z. Rameev, V.N. Popok, E.P. Zheglov, A.V. Kondyurin, V.A. Zhikharev, B. Aktas, Nucl. Instr. and Meth. B 206, 4 (2003)1115

[16] B.Z. Rameev, F. Yildiz, B. Aktas, C. Okay, R.I. Khaibullin, E.P. Zheglov, J.C. Pivin, L.R. Tagirov, Microelect. Engin. 69 (2003) 330

[17] J.F. Ziegler, J.P. Biersack, U. Littmark, The Stopping and Range of Ions in Solids, (Pergamon Press, NY, 1985).

[18] J.H. Das, J.E. Morris, J. Appl. Phys. 66 (1989) 5816.

[19] L.N. Aleksandrov, in Kinetika kristallizatsii i perekrikstallizatsii poluprovodnikovyh plenok, ed. L.N. Spiridonova (Nauka, Novosibirsk, 1985) p.225 (in Russian).

[20] G.N. Kakazei, A.F. Kravets, N.A. Lesnik, M. Pereira de Azevedo, Yu.G. Pogorelov, J.B. Sousa, J. Appl. Phys. 85 (1999) 5654.

[21] J. Kittel, in: Introduction to Solid State Physics, John and Sons Inc., 4th edition, 1978, p. 618.

PART 2: MAGNETIC TUNNEL JUNCTIONS

IMPACT OF GEOMETRY AND MATERIAL STACKING ON THE PROPERTIES OF MAGNETIC TUNNELLING JUNCTIONS

G.Reiss*[1], H. Brückl[1], A. Hütten[1], H. Koop[1], D. Meyners[1], A. Thomas[1,2], S. Kämmerer[1], J. Schmalhorst[1], M. Brzeska[1]

[1]*University of Bielefeld, Department of Physics, Nanodevice group, P.O. Box 100 131, 33501 Bielefeld, Germany*
[2]*MIT, Francis Bitter Magnet Lab., NW 14-2128, 170 Albany St., 02139 Cambridge, MA, USA*

Abstract: The discoveries of antiferromagnetic coupling in Fe/Cr multilayers by Grünberg, the Giant MagnetoResistance by Fert and Grünberg and a large tunnelling magnetoresistance at room temperature by Moodera have triggered enormous research on magnetic thin films and magnetoelectronic devices. Large opportunities are especially opened by the spin dependent tunnelling resistance, where a strong dependence of the tunnelling current on an external magnetic field can be found. Within a short time, the quality of these junctions increased dramatically. We will briefly address important basic properties of these junctions depending on the material stacking sequence of the underlying thin film system with special regard to the ferromagnetic electrodes. Next, we discuss scaling issues, i.e. the influence of the geometry of small tunnelling junctions especially on the magnetic switching behaviour down to junction sizes below $0.01\mu m^2$. The last part will give a short overview on applications beyond the use of the tunnelling elements as storage cells in MRAMs. This concerns mainly field programmable logic circuits, where we demonstrate the clocked operation of a programmed AND gate. The second 'unconventional' feature is the use as sensing elements in DNA or protein biochips, where molecules marked magnetically with commercial beads can be detected via the dipole stray field in a highly sensitive and relatively simple way.

Key words: Magnetism, Thin Films, Tunnelling, MRAM

B. Aktaş et al. (eds.), Nanostructured Magnetic Materials and their Applications, 57–70.

1.   INTRODUCTION

In recent years the interest in magnetic tunnel junctions (MTJs) has increased due to their high potential as memory cells in magnetic random access memories (MRAMs) or read heads in hard disk drives [1-4]. Nevertheless, the magnetic switching behavior of MTJs with lateral extensions below one micron is not yet understood in detail. Distorted switching curves (astroids) obtained from magnetoresistance curves were reported by, e.g., Klostermann et. al. [5]. Moreover, identically prepared tunnel junctions often show different junction specific switching behaviour [6, 7]. On the one hand, the physical origin of these variations is unknown up to now, on the other hand, they limit the technical applicability of the MTJs.

In this work we first present investigations of sub-μ magnetic tunnel junctions. First we will discuss the influence of the stacking sequence on the magnetic and the related Tuinnelling MagnetoResistance (TMR) properties. A detailed study by atomic and magnetic force microscopy (MFM) [8], in combination with micromagnetic numerical simulations tries to give a deeper insight in the properties shown by individual MTJs. The lithographic steps in the fabrication process inevitably lead to imperfect boundaries of the MTJs with a roughness on the nanometer scale. The impact of these structural imperfections on the magnetic switching behaviour will be discussed.

On MTJs smaller than 200 nm, the resolution of the MFM of around 30nm and the small thickness of the ferromagnetic electrodes hinders a reliable imaging of the magnetic states during switching the soft electrode. On such small patterns, we therefore employed conducting force microscopy [6], (c-AFM), where we form a contact between the tip and the top electrode. With this technique, minor loops and complete astroids can be obtained on MTJs as small as around 50nm.

The next section will address new materials integrated into MTJs. As an example, we chose Co_2SiMn. As full Heusler alloy, this is one of the materials with possibly large spin polarization [9]. First examples show, that this material can be used as ferromagnetic electrode; the values of the room temperature TMR, however, is still below the numbers obtained with conventional ferromagnets. Reasons for that will be discussed and possible improvements suggested.

The paper will close with the discussion of possible applications beyond MRAM's. This concerns mainly field programmable logic circuits, where we demonstrate the clocked operation of a programmed AND gate [10]. The second 'unconventional' feature is the use as sensing elements in DNA or protein biochips, where molecules marked magnetically with commercial beads can be detected via the dipole stray field in a highly sensitive and relatively simple way [11, 12].

2. STACKING SEQUENCE AND GEOMETRY

2.1 Stacking sequence

2.1.1 Conventional tunnelling elements

First attempts to form reliable MTJ's tried to use one relatively hard and one relatively soft ferromagnetic electrode. This, however, turned out to be not particularly stable, because the domain splitting of the soft electrode causes large stray fields which induce a deterioration of the hard magnetic material [13]. The same is true, if the hard electrode is additionally stabilized by an antiferromagnetically coupled trilayer as, e.g., Co / Cu / Co. This leads typically to minor loops as shown in Fig. 1 [14].

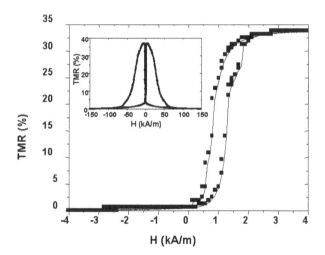

Figure 1. Resistance as a function of an external magnetic field (minor loop) for a tunnelling junction Co/Cu/Co/Al2O3/Ni80Fe20

This minor loop, however, is not suitable for storing one bit due to the obvious instabilities produced by the interaction of the soft with the hard magnetic electrode. Within a short time, however, major breakthroughs concerning the demands on magnetic stability could be obtained in the last years. This can be best illustrated by comparing results from 1999 from fig. 1 and from 2002 (Fig. 2).

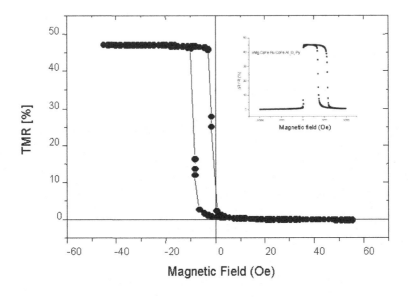

Figure 2. Resistance as a function of an external magnetic field (minor loops) for a tunnelling junction IrMn/CoFe/Ru/CoFe/Al2O3/Ni80Fe20. The TMR increased from about 15% (Fig. 1) to 45% in this example (the inset shows the corresponding major loop).

Fig. 2 gives the example for a junction biased by the combination of antiferromagnetic IrMn with a CoFe/Ru/CoFe trilayers. Using such layer systems, TMR values between 45% and larger than 50% combined with reproducible switching behaviour can now been routinely obtained [15].

2.1.2   Alternative electrode materials

One possibility to further increase the TMR values, which is frequently discussed, is the use of highly spin polarized materials. Heusler alloys with a predicted gap at the Fermi level for only one spin direction are thus very promising candidates for this effort.

We prepared thin films of the (so called full) Heusler alloy Co_2SiMn which is theoretically predicted to have a magnetic moment of around $5\mu_B$

per unit cell at room temperature and a gap for one spin direction of around 0.4eV [9]. These values critically depend on the degree of ordering of especially Mn and Co and should decrease with increasing disorder. On thin films on a Ta buffer, we obtained [16] a magnetic moment of 4.7 μ_B per unit cell after annealing at around 400°C. This relatively low annealing temperature is related with the presence of the Ta buffer layer, which already induces a texture in the Heusler film during the growth at room temperature.

These films are then integrated in tunnelling elements with a stacking sequence of Ta / Cu / Ta / Co_2SiMn / Al2O3 / CoFe / MnIr / Ta / Cu / Au. The problem in producing these elements is the need of a relatively high temperature for ordering the Heusler alloy and only about 275°C for inducing the exchange bias by the IrMn and improving the barrier properties. Thus the first annealing step at high temperature was done after the oxidation of the barrier, the second one after the complete stack was prepared. The AFM imaging of the topography showed a rms roughness of about 0.2nm, i.e. the moderate annealing did not lead to an increase of the surface roughness. Figures 3a, b and c show first results of a magnetization measurement of this stack (a) and the tunnelling characteristic (I / curve, b).

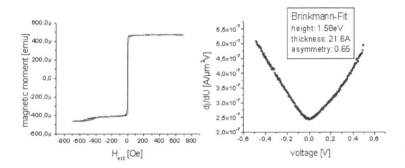

Figure 3. Magnetic moment as a function of an external magnetic field (minor loop) for a tunnelling junction including the Heusler alloy Co2SiMn as softmagnetic electrode (a), the corresponding *I/V* curve of the tunnelling element (b).

Although the magnetization shows a distinct and separated switching of the pinned and the Heusler-electrode (Fig. 3a), the current voltage characteristic is not yet at optimum for obtaining large TMR values. This is shown by evaluating the *I/V* curve of fig. 3b with a Brinkmann-fit [17], giving the barrier parameters of a barrier height of only 1.6eV, a large barrier asymmetry of 0.6eV and an incorrect barrier thickness. This points to an underoxidation of the Al which usually strongly suppresses the TMR signal.

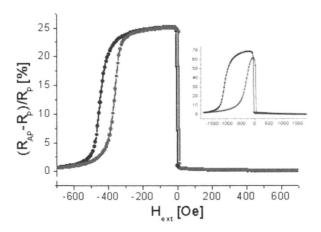

Figure 4. The TMR for a tunnelling junction including the Heusler alloy Co_2SiMn as a soft magnetic electrode at room temperature and at 11K (inset).

Although the I/V curve does not point to an optimum for obtaining large TMR values, the tunnelling elements with Heusler alloy electrode do produce a reasonably large TMR value of around 25% at room temperature and of 70% at 10 K. Especially the low temperature TMR is among the largest as compared to conventional 3d ferromagnets.

Thus there should be considerable further improvements possible by further work concentrating on the preparation of better tunnelling barriers on top of the Heusler alloy. This and an intense investigation of the interface are thus on the way.

2.2   Influence of the geometry

Junctions with different shapes and sizes were investigated: rectangular junctions ranging from 700 nm × 700 nm to 700 nm × 1400 nm and elliptical patterns with 500 nm short axes and 850 nm long axes. The patterns were covered by a 15 nm thick Ta layer which minimizes stray field effects of the homemade, CoCr covered MFM tip and, hence, tip induced perturbations of the soft layer magnetization. Sufficient signal-to-noise ratio and small perturbations were obtained for a CoCr thickness of 30 nm.

For the MFM investigations, a modified Nanoscope III from Digital Instruments was operated in the Lift-Mode™. The magnetic field was generated by two pairs of coils surrounding the microscope. MFM images of the magnetization of the patterned NiFe soft electrodes were recorded at different external fields.

As an example, we discuss here the results obtained both with MFM measurements and with corresponding micromagnetic numerical simulations [18] for elliptical junctions [8]. Typical examples are shown in Fig. 5:

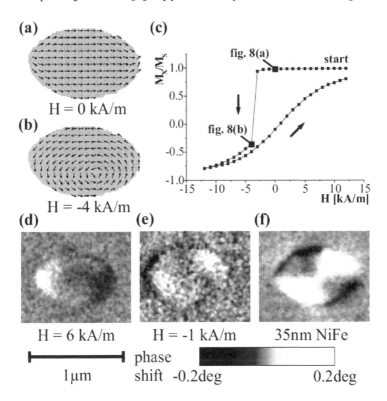

Figure 5. Vortex nucleation in 500nm x 850nm elliptical junctions. (a)-(b) Typical magnetization configurations. (c) Related magnetization curve. (d)-(e) MFM images of elliptically patterned magnetic tunnel junctions recorded at different stages of the minor loop. Additionally an experimental MFM image of a vortex state in a 35nm thick NiFe ellipse is shown (f).

In the calculations for elliptical patterns a common feature is found (Fig. 5). Elliptically patterned electrodes often show a high remanent magnetization [MX=MS ¼ 0:98, Fig. 5(a),(c)]. The shape, however, favors vortex formation due to minimization of the stray field energy.

Consequently, the magnetization reversal of elliptical junctions is often dominated by vortex nucleation and vortex motion with high saturation fields [Fig. 5(b)-(c)]. The results of the simulations are experimentally proven by MFM investigations of the elliptically shaped MTJs, where the complete layer stack including the antiferromagnet was patterned. In saturation or near saturation the NiFe electrodes show a high magnetic contrast at their end points [Fig. 5(d)]. At H = -1kA/m the magnetization shows four opposite regions with bright or dark contrast, which is typical for a vortex state [Fig. 5(e)]. Thicker films with higher contrast show similar patterns more pronounced due to the larger stray fields [Fig. 5(f)].

Here, it should be noted, that considerably different magnetic behaviour was found for nominally identical shapes. This could be traced back to the individually shaped edges of the patterns which result from the grainy structure of the films. Frequently, domain wall pinning was found at kinks or bumps at the edge as small as around 10 nm.

MTJs smaller than around 300 nm did not produce enough signal for a reliable MFM measurement. We therefore customized a commercial AFM for electrical measurements with a diamond coated conducting tip. With this instrument, we were able to take TMR minor loops of junctions with sizes down to around 50nm.

Whereas on rectangular and elliptical MTJs results similar those of the MFM investigations were obtained for sizes down to abut 100 nm, truncated elliptically shaped patterns turned out to show a reproducible switching behavior as illustrated in fig. 6a by an astroid for a nominally 200 nm wide MTJ.

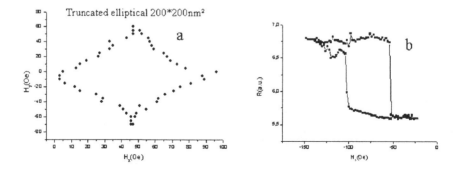

Figure 6. Astroid of a truncated elliptically shaped MTJ with a reproducible switching behavior (a) and a single minor loop for a circular MTJ with a diameter of 50nm (b).

On junctions with sizes well below 100nm we were up to now only able to take minor loops. In fig. 6b, a typical result for a 50nm circular MTJ is shown. For these ultrasmall elements, we never observed steps in the minor loops or unusually large saturation fields which would point to domain switching or vortex formation, respectively. This and the shape of the minor loops (Fig. 6b) therefore suggest a single domain behavior at these small sizes, which could be a considerable advantage regarding downscaling issues in MRAM applications.

3. APPLICATIONS BEYOND MRAM

In this part, we will show, that TMR-effects can be of considerably broader interest than only as storage cells for MRAMs. First, we concentrate on a closely related use in field programmable logic gate arrays.

3.1 Field programmable logic gate arrays

In logic devices, field programmable means that the function performed by a gate array can be changed during the operation of the processing unit. Up to now, this is done by CMOS technique with, e.g., floating gates, were programming is time consuming and requires high voltages. Using a bridge configuration of TMR cells as shown in Fig. 7 could overcome these drawbacks.

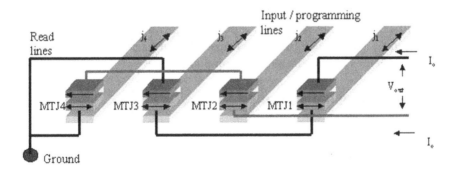

Figure 7. Bridge configuration of 4 MTJs. The input/programming lines produce a magnetic field able to rotate the soft magnetic electrode's magnetization and thereby changing the output voltage V_{out} which then is the logic function of the inputs.

Here, the input is represented by currents on two input lines, which can change the magnetization state of the MTJs soft electrodes. Two neighbouring lines are used to set the resistance state of the other two MTJs which 'programmes', i.e. defines the value V_{out} obtained as logic function of the two inputs.

In Fig. 8, we show a clocked operation of such an arrangement with a programmed "AND" function [10]:

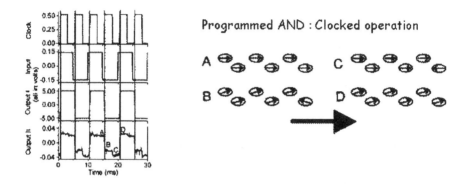

Figure 8. Clocked operation of six MTJs in a bridge arrangement. Output 1 is a rectified signal of the original V_{out} (output II). The right side shows the magnetization corresponding to the different output states.

This feasibility study thus demonstrates, that MTJs can be used in logic gate arrays and that these devices can be programmed during operation. The large advantages of this type of "Magnetic Logic" are the scalability, which should be similar to the MRAM, and the speed. Programming these gate arrays will be as fast as performing the logic operation with typical time scales down to the nsec regime. This opens in turn new perspectives for innovative schemes like reconfigurable computing [19].

3.2   Detecting biomolecules

As last example, we now turn to a completely different field: In biotechnology and medical applications, molecules like DNA or proteins are frequently marked by magnetic spheres called "beads". These beads are commercially available with sizes down to around 100nm and an already functionalized shell. Functionalized means in this context, that, e.g., streptavidin molecules are attached to the beads which are able to bind very specifically and tightly to, e.g., biotin molecules which in turn can be specifically attached to DNA or proteins.

This opens the possibility to measure the presence or absence of these biomolecules by detecting the magnetic beads with an MTJ. Baselt et.al. [11] already described this technique using Giant MagnetoResistance sensors. Figure 9 shows the principle of this method.

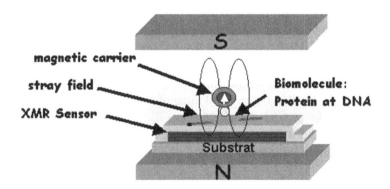

Figure 9. Principle of the magnetic biochip: Biotin marked DNA-molecules hybridize with complementary strands attached to the surface. After that, streptavidin coated beads bind to the biotin. By applying a magnetic field perpendicular to the sensor surface, only the in plane components of the dipole stray field of the beads are detected.

In fig. 10, we show the result of the measurement of different bead concentrations with a 100µm wide TMR cell [12].

TMR Biochip Sensor:

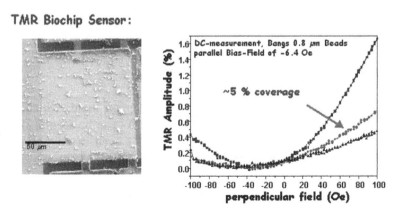

Figure 10. A TMR sensor surface covered by magnetic beads (left) and the TMR signal measured during applying a magnetic field perpendicular to the sensor surface and an in plane field which is close to the switching field of the soft electrode.

As can be seen in Fig. 10, reasonable signals as in dependence of the perpendicular field are obtained at a surface coverage of only a few percent, if an in plane field is additionally applied which brings the soft electrode close to switching. Comparisons with the established optical method of marking with fluorescent molecules showed, that the magnetic biosensor can be more sensitive at low concentrations of the analyte molecule, which is the most interesting area of application.

In order to test the ultimate sensitivity and to check out, if even single molecule detection can be possible with the TMR sensor, we again customized an AFM. Here, we use the homemade MFM-tips for imaging a TMR cell and simultaneously detect the resistance change of the MTJ during scanning. This on one hand mimics the presence of one bead at the sensor surface and gives on the other hand quantitative information about the resistance change in dependence of the specific position of the magnetic particle. Note, that the MFM-tips are magnetized perpendicular to the MTJ surface prior to the measurement. Figure 11 shows a typical result of this experiment [20]:

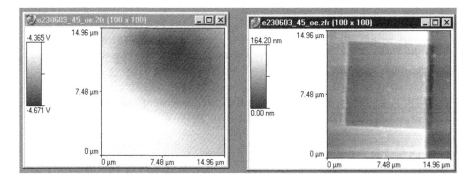

Figure 11. AFM topography of a 10m rectangular TMR cell (right) and the resistance change of the MTJ (left) during imaging with an MFM tip.

In this example, we applied an additional in-plane field, which again drives the soft (top-) electrode close to switching. As can be seen in fig. 11, reasonably large signals are produced by the MTJ even by this relatively large cell size due to the magnetic stray field of the MFM tip, which was covered by 20 nm CoCr. Thus the detection of single beads and thereby of single molecules should be possible with this type of biosensor.

This method, however, provides also new information about the TMR cell itself, because the response of the MTJ to a locally concentrated field can be measured in different states of the magnetization process, i.e. for different values of the external in plane field. Both experimental work as well as simulations are thus on the way in order to evaluate the potential of this technique for a detailed characterization of MTJs especially considering different shapes and edge roughnesses.

4.   SUMMARY

In this contribution, we have discussed aspects of stacking sequence and geometry on the properties of magnetic tunnelling junctions. New materials as, e.g., Heusler alloys can be integrated into the standard stacking sequence of the most advanced MTJs, thereby allowing to reliably test the amount of TMR obtainable. Although the very large spin-polarization of Heusler alloys predicted theoretically has not yet been seen in 'real' MTJs, promising results have been already obtained and further improvements seem to be straightforward.

Concerning the influence of the geometry, subtle effects of edge roughness on the magnetization switching seem to be a critical point, although certain shapes have been successfully designed for a single domain like switching. At sizes below 100nm, no signs of domain splitting or vortex formation have been seen up to now, which is very promising for the further downscaling of MTJ storage devices.

Beyond this application in MRAMs (or read heads for disk drives), MTJs can be used for realizing a field programmable magnetic logic, where programming is as fast as the logic operation itself, opening thereby the fascinating field of reconfigurable computing. Moreover, MTJs are able to not only detect bits on hard disks but also magnetic micro- and nanoparticles which are already in use for biotechnological and medical applications. Proof of principle experiments even demonstrate, that a detection of single molecules should be possible by using MTJs in magnetic biochips. Thus a possible production of MRAM chips could boost much more possible applications still ahead.

ACKNOWLEDGEMENTS

The authors gratefully acknowledge transfer of samples and fruitful discussions with J. Wecker, G. Gieres (Siemens AG), P. Jutzi (Dept. of Chemistry, U Bielefeld), A. Pühler and A. Becker (Dept. of Biology, U Bielefeld). This work was supported by the BMBF (grants 13N7382 and 13N7859) and the DFG (SFB 613).

REFERENCES

1   J. S. Moodera, L. R. Kinder, T. M. Wong, and R. Meservey, Phys. Rev. Lett. 74, 3273 (1995)

2   T. Miyazaki and N. Tezuka, J. Magn. Magn. Mater. 139, 231 (1995)

3   S. S. P. Parkin, K. P. Roche, M. G. Samant, P. M. Rice, R. B. Beyers, R. E. Scheuerlein, E. J. O'Sullivan, S. L. Brown, J. Bucchigano, D. W. Abraham, Y. Lu, M. Rooks, P. L. Trouilloud, R.A. Wanner and W. J. Gallagher, J. Appl. Phys. 85, 5828 (1999)

4   G. Reiss, H. Brückl, A. Hütten, J. Schmalhorst, M. Justus, A. Thomas, S. Heitmann, phys. stat. sol. (b) 236, 289 (2003)

5   U.K. Klostermann, R. Kinder, G. Bayreuther, M. Rührig, G. Rupp, J. Wecker, J. Magn. Magn. Mat. 240, 305 (2002)

6   H. Kubota, G. Reiss, H. Brückl, W. Schepper, J. Wecker, Jpn. J. Appl. Phys. 41, L180 (2002)

7   Yu Lu, R.A. Altman, A. Marley, S.A. Rishton, P.L. Trouilloud, Appl. Phys. Lett. 70, 2610 (1997)

8   D. Meyners, H. Brückl, G. Reiss, J. Appl. Phys. 93, 2676 (2003)

9   S. Ishida, T. Masaki, S. Fujii, S. Asano, Physica B 245, 1 (1998)

10 R. Richter, L. Br, J. Wecker, G. Reiss, Appl. Phys. Lett., 80, 1291, (2002)

11 D. R. Baselt, G. U. Lee, M. Natesan, S.W. Metzger, P. E. Sheehan, and R. J. Colton, Biosens. Bioelectron., 13, 731 (1998)

12 J. Schotter, P.B. Kamp, A. Becker, A. Phler, D. Brinkmann, W. Schepper, H. Brückl, G. Reiss, IEEE Trans. Magn. 38, 3365 (2002)

13 S. Gider, B.-U. Runge, A. C. Marley and S. S. P. Parkin, Science 281, 797 (1998)

14 J. Schmalhorst, H. Brückl, G. Reiss, G. Gieres, M. Vieth and J. Wecker, J. Appl. Phys., 87, 5191 (2000)

15 A. Thomas, H. Brückl, M. D. Sacher, J. Schmalhorst, G. Reiss, J. Vac. Sci. Technol. B 21, 2120 (2003)

16 S. Kämmerer, S. Heitmann, D. Meyners, D. Sudfeld, A. Thomas, A. Hütten, G. Reiss, J. Appl. Phys. 93, 7945 (2003)

17 H.Brückl, J.Schmalhorst, G.Reiss, G.Gieres, J.Wecker, Appl. Phys. Lett. 78, 1113 (2001)

18 OOMMF program (release1.1), NIST (Gaithersburg, USA), available at http://math.nist.gov/oommf for public use

19 W. C. Black, Jr., B. Das, J. Appl. Phys. 87, 6674 (2000)

20 M. Brzeska, M. Justus, J. Schotter, K. Rott, G. Reiss, H. Brückl, submitted to Acta Physica Polonica

INTERLAYER EXCHANGE COUPLING OF FERROMAGNETIC FILMS ACROSS SEMICONDUCTING INTERLAYERS

D.E. Bürgler, R.R. Gareev, M. Buchmeier, L.L. Pohlmann, H. Braak, R. Schreiber, and P. Grünberg
Institut für Festkörperforschung, Forschungszentrum Jülich GmbH, D-52425 Jülich, Germany d.buergler@fz-juelich.de

Abstract: We review our observations of surprisingly strong antiferromagnetic interlayer exchange coupling across Si-rich $Fe_{1-x}Si_x$ spacers, which becomes stronger with increasing Si content, x, in the spacer. We show that the nominally pure ($x = 1$) spacers that mediate the strongest coupling act at the same time as a tunneling barrier for electric transport in current-perpendicular-to-plane geometry by verifying the validity of the necessary and sufficient Rowell criteria for inelastic tunneling. Moreover, we present first data on the coupling across spacers that contain Ge as an alternative semiconductor material.

Key words: Magnetic interlayer coupling, magnetic multilayer, Si spacer, tunneling barrier, Rowell criteria

1.   INTRODUCTION

Interlayer exchange coupling (IEC) across metallic interlayers has been extensively investigated, and it is now well established that IEC displays a damped oscillation between the ferro- and antiferromagnetic (AF) state as a function of the interlayer thickness [1]. Typical coupling strengths are of the order of 1 mJ/m^2. Theoretically, it was shown that the coupling is due to the formation of standing electron waves in the interlayer, which result from spin-dependent electron interface reflectivity. When applying the same theoretical framework to insulating or semiconducting interlayers, however, there are only evanescent waves in the spacer, which exponentially decay with distance from the interfaces with the metallic, magnetic layers

B. Aktaş et al. (eds.), Nanostructured Magnetic Materials and their Applications, 71–77.

[2]. Accordingly, IEC is also expected to decay exponentially when the thickness of a non-conducting interlayer increases. For amorphous insulators like a-SiO$_2$ and a-Al$_2$O$_3$ IEC has never been observed experimentally, but there is a recent report of AF coupling with a strength of about 0.25 mJ/m^2 in epitaxial Fe/MgO/Fe structures for very thin (< 7 Å) MgO thicknesses [3]. Thus, for semiconducting interlayers, one would expect IEC to be weaker than in metals, but stronger than in insulators and therefore to be observable to larger spacer thicknesses than for insulators.

The results presented below are in contrast to this simple picture and indicate that IEC across semiconducting spacers cannot be described by the present theoretical models.

2.   SAMPLE GROWTH AND CHARACTERIZATION

Fe/Fe$_{1-x}$Si$_x$/Fe(001) trilayers ($x = 0.6...1$) are prepared by molecular beam epitaxy (MBE) onto single-crystalline, 150 nm-thick Ag(001) buffer layers grown at 375 K on UHV-annealed GaAs(001)-wafers using a 1 nm-thick Fe seed layer (see Fig. 1).

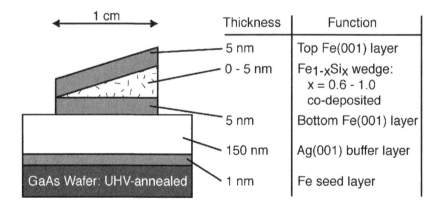

Figure 1. Structure of the MBE-grown samples. A thick Ag(001) layer serves as a buffer and enables epitaxial growth of all subsequent layers.

The Fe$_{1-x}$Si$_x$ interlayers are prepared by co-deposition of Fe and Si from two separate crucibles with well controlled rates. Trilayers with Ge-wedge spacers have the same structure as shown in Fig. 1, except that we studied samples with and without 4 Å-thick Si boundary layers at the interfaces between Fe and Ge (not shown in Fig. 1). The buffer layers and the bottom Fe layers are grown according to optimized procedures involving elevated temperatures [4], whereas the $Fe_{1-x}Si_x$ or Ge wedges, the Si boundary layers, and the top Fe layer are grown at room temperature.

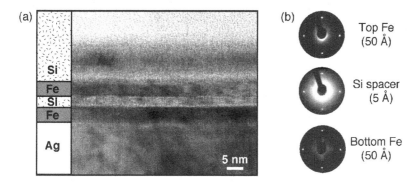

Figure 2. (a) Cross-sectional TEM image of an Fe/Si/Fe trilayer covered with a thick Si protection layer. The Si interlayer appears as light grey band with sharp boundaries towards the adjacent Fe layers [7]. (b) LEED pattern of bottom and top Fe layers and a 5 Å-thick Si spacer layer in between.

The wedge-shaped spacers allow studying the spacer thickness dependence of IEC on a single sample with high accuracy by using the magnetooptical Kerr effect (MOKE) and Brillouin light scattering (BLS), as both techniques are based on a focused laser beam with a diameter of about 100 μm. The bilinear (J_1) and biquadratic (J_2) coupling constants are determined from fitting MOKE hysteresis loops and the field dependence of BLS data using the procedures described in [5]. Further details concerning sample preparation, characterization, and data analysis can be found in [4,5,6]

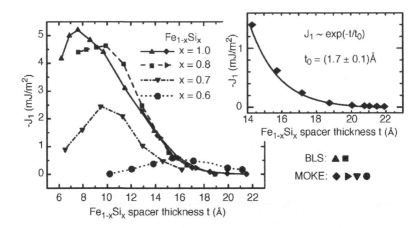

Figure 3. J_1 versus $Fe_{1-x}Si_x$ spacer thickness t for different nominal Si contents $x = 0.6$, 0.7, 0.8, and 1.0 [6]. The data are derived from MOKE and BLS measurements as indicated by the different symbols. The inset shows a fit to an exponential decay, that yields a decay length $t_0 = 1.7$ Å.

Figure 2(a) displays a cross-sectional transmission electron microscopy (TEM) picture of a Fe/Si/Fe structure with a nominally pure, 5 nm-thick Si interlayer. The Fe/Si/Fe trilayer is covered by additional Si for protection. Both the Fe/Si and the Si/Fe interfaces of the Fe/Si/Fe structure are sharp excluding appreciable interdiffusion. The epitaxy throughout the system is further confirmed by low-energy electron diffraction (LEED) images of all layers [Fig. 2(b)].

3. $Fe_{1-x}Si_x$ ($x = 0.6-1$) INTERLAYERS

Figure 3(a) shows the bilinear coupling strength J_1 as a function of the $Fe_{1-x}Si_x$ interlayer thickness for different nominal Si contents x. The most obvious feature is the drastic increase of $|J_1|$ by almost one order of magnitude when x is increased from 0.6 to 1.0. The coupling strength of more than 5 mJ/m^2 for a nominally pure Si spacer is very surprising, as one would expect it to be weaker than across metallic spacers, where values around 1 mJ/m^2 are more typical.

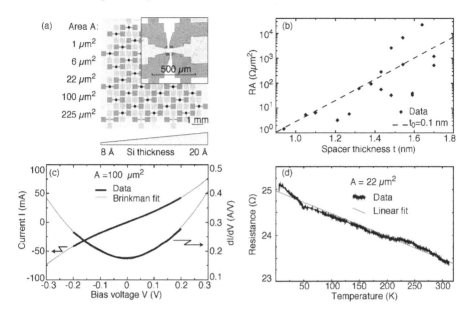

Figure 4. (a) Wafer layout for the CPP-measurements. The Si spacer thickness and the junction area A vary in the horizontal and vertical direction, respectively. The Rowell criteria for tunneling are fulfilled [8]: (b) exponential thickness dependence of RA, (c) parabolic voltage dependence of dI/dV, and (d) small and negative temperature coefficient of R. The inset in (a) shows a photo of a single contact.

For this reason, we have further investigated the electronic nature of the interlayer by transport measurements in current-perpendicular-to-plane (CPP) geometry across lithographically defined, micron-sized contacts [see inset of Fig. 4(a)]. In order to prove that the transport occurs via tunneling, one has to show that the three Rowell criteria are fulfilled: (i) exponential increase of the resistance times area product RA with barrier thickness t, (ii) parabolic voltage dependence of the $dI/dV - V$ characteristics, i.e. the $I - V$ curve can be fitted by the Brinkman formula, and (iii) the junctions resistance slightly decreases with increasing temperature. In Figs. 4(b)-(c) we show the validity of all the three criteria. We can conclude, that strong coupling across Si spacers coexists with transport via tunneling. Therefore, the Si spacer layers that mediate the AF coupling are non-conductive and act as a tunneling barrier.

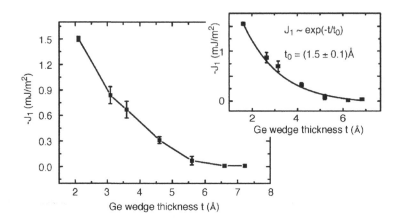

Figure 5. J_1 versus Ge wedge thickness t for a sample with two 4 Å-thick Si boundary layers [9]. The inset shows a fit to an exponential decay, that yields a decay length $t_0 = 1.5$ Å.

4.   Ge-CONTAINING INTERLAYERS

We show in Fig.5 first results from samples that contain Ge in the spacer. It turns out that AF IEC with $|J_1| > 1$ mJ/m^2 is observable when we prevent direct contact between Fe and Ge by introducing 4 Å-thick Si boundary layers at both interfaces to Fe. With one of these boundary layers missing, the trilayers behave like a single ferromagnetic film. Most likely, the boundary layers suppress Fe-Ge interdiffusion and the formation of ferromagnetic Fe-Ge compounds throughout the whole spacer [10]. The coupling decays exponentially with a decay length $t_0 = 1.5$ Å, which is very similar to t_0 for nominally pure Si spacers indicating a common origin of IEC across Si and Ge.

5.   SUMMARY

Epitaxial Si spacer layers mediate strong AF IEC and at the same time act as a tunneling barrier for electron transport in CPP-geometry. First results with Ge spacers also reveal sizable AF coupling and, therefore, indicate that relatively strong AF IEC might be a common feature of well-ordered, epitaxial semiconductor spacer layers. A quantitative theoretical description of strong AF coupling across semiconductor spacers –

representing the intermediate case between metallic and insulating spacers – is highly desired.

ACKNOWLEDGMENTS

The authors wish to thank C.L. Jia, Y.L. Qin, M. Siegel, and S. Stein for their collaboration on transmission electron microscopy and transport measurements. This work was supported by the HGF-Strategiefondsproject "Magnetoelectronics".

REFERENCES

[1] D.E. Bürgler and P. Grünberg and S.O. Demokritov and M.T. Johnson, Interlayer Exchange Coupling in Layered Magnetic Structures, in: Handbook of Magnetic Materials, K.H.J. Buschow, 13 Elsevier,1-85,Amsterdam,2001
[2] P. Bruno, Phys. Rev. B (1995) 411-439
[3] J. Faure-Vincent and C. Tiusan and C. Bellouard and E. Popova and M. Hehn and F. Montaigne and A. Schuhl, Phys. Rev. Lett. 89 (2002) 107206
[4] R.R. Gareev and D.E. Bürgler and M. Buchmeier and D. Olligs and R. Schreiber and P. Grünberg, Phys. Rev. Lett. 87 (2001) 157202
[5] M. Buchmeier and B.K. Kuanr and R.R. Gareev and D.E. Bürgler and P. Grünberg, Phys. Rev. B (2003) 184404
[6] R.R. Gareev and D.E. Bürgler and M. Buchmeier and R. Schreiber and P. Grünberg, J. Magn. Magn. Mater. (2002) 237-239.
[7] D.E. Bürgler and M. Buchmeier and S. Cramm and S. Eisebitt, R.R. Gareev and P. Grünberg and C.L. Jia and L.L. Pohlmann and R. Schreiber and M. Siegel and Y.L. Qin and A. Zimina, Journal of Physics: Cond. Mat. v.15 (2003) S443-S450
[8] R.R. Gareev and L.L. Pohlmann and S. Stein and D.E. Bürgler and P.A. Grünberg, Journ. Appl. Phys., v.93 (2003) 8038-8040
[9] R.R. Gareev and D.E. Bürgler and R. Schreiber and H. Braak and P.A. Grünberg, Appl. Phys. Let., v.83, 1806-1808, 2003.
[10] J.J. de Vries and J. Kohlhepp and F.J.A. den Broeder and P.A. Verhaegh and R. Jungblut and A. Reinders and W.J.M. de Jonge, J. Magn. Magn. Mater. v.165, 435-438, 1997

SPIN-FILTER SPECULAR SPIN VALVES WITH HIGHER MR RATIO AND THINNER FREE LAYERS

G. Pan, Z.Lu and A. Al-Jibouri [*]

CRIST, School of Computing, Communication and Electronics, University of Plymouth, Plymouth, Devon, PL4 8AA, UK
[*]*Nordiko Ltd., Havant, Hampshire, PO9 2NL, UK.*

Abstract: Spin-valves with higher magnetoresistance (MR) ratio and thinner free layers are required for applications of nanometer-sized reply heads. In traditional spin valve structures, reducing the free layer thickness below 5 nm normally results in a reduction in MR ratio. In this work we have developed a spin-filter specular spin-valve with structure "Ta 3.5 nm/NiFe 2 nm/IrMn 6 nm/CoFe 1.5 nm/NOL1/CoFe 2 nm/Cu 2.2 nm/CoFe t_F/Cu t_{SF}/Nol2/Ta 3 nm, which is demonstrated to maintain MR ratio higher than 12% even when the CoFe free layer is reduced to 1 nm. A semi-classical Boltzmann transport equation was used to simulate MR ratio. Results will be presented in comparison with the experimental measurement. An optimized MR ratio of 15% was obtained when t_F was about 1.5 nm and t_{SF} about 1.0 nm as a result of the balance between the increase in electron mean free path difference and current shunting through conducting layer. MR ratio of up to 20% is obtainable for such a structure if Cu conductor spacer is reduced to 1.8 nm. It is found that the Cu enhancing layer not only enhances the MR ratio but also improves soft magnetic properties of CoFe free layer due to the low atomic intermixing observed between Co and Cu. The CoFe free layer of 1–4 nm exhibits a low coercivity of ~3 Oe even after annealing at 270 °C for 7 h in a field of 1 kOe. HRTEM cross sectional images showed that the NOL1 introduced from oxidation of the original bottom-pinned CoFe layer is actually a mixture of oxides and ferromagnetic metals. Un-oxidized CoFe grains epitaxially grown right across the nano-oxide layer are present. This led us to believe that the specular reflection of spin-polarized electrons is achieved by these oxidised regions and that the remained un-oxidised ferromagnetic CoFe in NOL1 shall decrease the specular scattering on the NOL1 surface, but it provides the direct exchange paths between the IrMn and pinned CoFe layer.

Key words: Nano-oxide layer, specular valve, spin filter, spin valve.

B. Aktaş et al. (eds.), Nanostructured Magnetic Materials and their Applications, 79–89.

1.   INTRODUCTION

As magnetic recording density in hard drives progress towards 100 Gbit/in^2, it is necessary to enhance spin valve magnetoresistance (MR) ratio and reduce its free layer thickness [1]. The higher MR ratio is important for compensating the signal reduction due to the increasingly smaller recorded bits, and the thinner free layer is essential to maintaining the sensitivity and linearity of the transfer curves of the increasingly smaller sensor elements. However, in a traditional spin valve structure, reducing the free layer thickness below ~5 nm results in a rapid reduction in MR ratio due to the reduced difference in the mean free path between spin-down and spin-up electrons. Recently, a spin-filter structure, having free layer composed of a very thin magnetic layer and an adjacent Cu enhancing conduction layer [2,3], has been reported. This enhancing layer has a good conductivity and an electronic structure that matches that of the free layer, allowing a large mean free path difference for the spin-down and spin-up electrons. A MR ratio of up to 10% could be maintained in this structure even if the free layer is very thin (1.0-3.0 nm). Another development is to insert nano-oxide layers (NOL) in the pinned and free ferromagnetic layers [4,5] to enhance the MR ratio of up to 13%. The enhanced MR originates from specular reflection of electron at the oxide interfaces while conserving their spin direction. However, the free layer thickness for optimum MR in the specular spin-valve case was about 3 nm. And the coercivity of the free layer is higher than that of traditional spin valves, due to the oxidation of the free layer by the top specular layer.

In this paper, a new type of spin-filter specular spin valves (SFSSV) with structure Ta 3.5 nm/NiFe 2 nm/IrMn 6 nm/CoFe1.5 nm/NOL1/CoFe 2 nm / Cu 2.2 nm/ CoFe t_F / Cu t_{SF} /NOL2/Ta 3 nm was investigated. It was found in our work that the Cu enhancing layer not only enhances MR ratio at thinner free layers but also improves the soft magnetic properties of free layer. We will discuss the experimental results based on semiclassical microscopic transport simulation and high resolution cross sectional TEM examination of the spin valve structures.

2.   EXPERIMENTAL METHOD

Spin-filter specular spin valves with structure Ta 3.5nm/NiFe 2nm/IrMn 6nm/ CoFe 1.5nm / NOL1 / CoFe 2nm/Cu 2.2 nm/ CoFe t_F / Cu t_{SF} /NOL2/Ta 3nm were deposited in an argon pressure of 0.2Pa on 6 in. [100] Si wafer. The depositions were done at Nordiko on a 9606 physical vapor deposition (PVD) system with a base pressure below 4.5×10^{-5} Pa. The process conditions were optimized separately for each of the target materials. During the deposition, a uniform magnetic field of 50 Oe was applied to the substrates to induce the uniaxial magnetic anisotropy of ferromagnetic layer. In order to obtain the exchange biasing of the overlaying CoFe by IrMn, a twofold layer consisting of Ta and a very thin layer of NiFe was used. The oxidation was done at atmosphere pressure and room temperature. After completing deposition, the samples were annealed for 10 min ~10 h at 270°C under an argon atmosphere in an applied magnetic field of 1 kOe. This procedure leads to an increase of the IrMn (111) texture, resulting in well-defined exchange field.

The magnetic properties were measured using vibrating sample magnetometer (VSM). MR properties are measured with a standard four-point geometry.

Cross-section specimens for HREM observation were prepared by using a standard method and studied in a CM200-FEG TEM with a point resolution of 0.25 nm and line resolution of 0.1 nm.

3. RESULTS AND DISCUSSIONS

The typical MR curves of spin-filter specular spin valves and conventional spin valves are shown in Fig. 1, where Fig. 1(a) shows the high field MR(h) loops and Fig. 1(b), the low field MR(H) loops. These results show that the MR ratio has a large enhancement (~ 14.5%) in this spin-filter

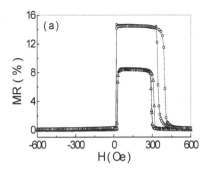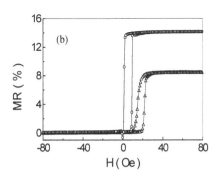

Figure 1. MR(H) loops for spin-filter specular spin valve (circle), Ta 3.5nm/NiFe 2nm/IrMn 6nm/ CoFe 1.5nm / No11 / CoFe 2nm/Cu 2.2nm/ CoFe 1.5nm / Cu 1.0nm / Nol2/Ta 3nm, and traditional spin valve (triangle), Ta 3.5nm/NiFe 2.0nm/IrMn 6.0nm/CoFe 3.5nm/Cu 2.2nm/CoFe 4.0nm/Ta3.0nm, at high field (a) and at low field (b).

specular spin valve even when the thickness of CoFe free layer goes down to 1.5 nm. The interlayer coupling field (H_{int}) between free layer and pinned layer is about 3.3 Oe, the coercivity of the free layer (Hc_l) is about 3.6 Oe and the exchange field H_{ex} is about 380 Oe compared to the traditional spin valves (MR~ 8%, $H_{int} \sim 15$ Oe, $H_{cl} \sim 5$ Oe and $H_{ex}\sim 300$ Oe). In spin-filter specular spin valves, a considerable large exchange field between the antiferromagnetic IrMn and the pinned layer is still obtained, and a decrease in the ferromagnetic coupling between the free layer and pinned layer is also observed.

MR ratio as a function of CoFe free layer thickness (1.0~4.0 nm) for spin-filter specular spin valves Ta 3.5 nm/NiFe 2 nm/IrMn 6 nm/ CoFe 1.5 nm / Nol1 / CoFe 2 nm/Cu 2.2 nm/ CoFe t_F / Cu 1.0 nm /Nol2/Ta 3 nm are shown in Figure 2. For t_F of 1.0 ~ 5.0 nm, spin-filter specular spin valves showed MR ratio above 12.5%, peaked at $t_F = 1.5$ nm ($MR\sim14.5\%$).

In order to understand the experimental MR data, we extended a semiclassical model based on the Boltzmann transport equation initiated by Camley and Barnas to compute the MR ratio in spin-filter specular spin valves [6.7]. In our model, the conductivity per spin $\sigma^{\uparrow(\downarrow)}$ is given by

$$\sigma^{\uparrow(\downarrow)} = \frac{ne^2}{2mv_f} \frac{3}{4\pi} \int d^3 \widehat{v} \widehat{v}_x^2 \lambda_{eff}^{\uparrow(\downarrow)}(\vec{v},z) \qquad (1)$$

where n is the total conduction electron density, e and m are the electronic charge and mass, respectively, and v_f is the Fermi velocity. $\lambda_{eff}^{\uparrow(\downarrow)}$ is the effective electron mean free path. Here we assume a dependence of the mean free path on the angle θ between the electron velocity v and the magnetization M [8]:

$$\lambda_{eff}^{\uparrow(\downarrow)} = \lambda_0^{\uparrow(\downarrow)}(1 - a^{\uparrow(\downarrow)} \cos^2 \theta - b^{\uparrow(\downarrow)} \cos^4 \theta)$$

The parameters $a^{\uparrow(\downarrow)}$ and $b^{\uparrow(\downarrow)}$ are a measure for the anisotropy of the scattering, which are determined by the amplitude of MR voltages. Similar to Rijks et al [8], we set $b^{\uparrow(\downarrow)} = 0$ and adjust $a^{\uparrow(\downarrow)}$ to fit the MR ratio. In the numerical calculation, we use the values $\lambda_0^{\uparrow} = 10$ nm, $\lambda_0^{\downarrow} = 0.55$ nm, $a^{\uparrow} =0.0327$, $a^{\downarrow} = -0.00556$, $b^{\uparrow}= b^{\downarrow}= 0$, and $\rho_{CoFe}=15$ μΩ–cm for CoFe film. $\lambda_0^{\uparrow} = 20$ nm, $\lambda_0^{\downarrow} = 20$ nm, and $\rho_{Cu}=2.8$ μΩ-cm for the Cu film. By selecting coefficients of reflection at the interfaces of NOL insulator/ferromagnetic layers $R=0.81$, the calculated MR ratio can fit the experimental results shown

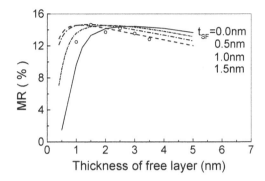

Figure 2. The experimental (circle) and calculated (line) MR ratios as a function of t_F for spin-filter specular spin valves Ta 3.5nm/NiFe 2nm/IrMn 6nm/ CoFe 1.5nm / No11 /CoFe 2nm/Cu 2.2nm/ CoFe t_F / Cu t_{SF} /No12/Ta 3nm.

in Fig. 2. As shown in Figure 2, the MR ratio decreases rapidly when the thickness of free layer goes down 1.5 nm for the spin valve without an enhancing layer. However, the MR ratio remains high in spin-filter specular spin valve with an enhancing layer larger than 1.0 nm even when the thickness of free layer goes down to 0.5 nm. It clearly shows that the presence of the enhancing layer results in an improvement in MR ratio for the thin free-layer spin valves.

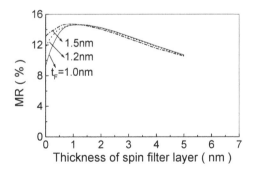

Figure 3. The calculated MR ratios as a function of t_{SF} for spin-filter specular spin valve Ta 3.5nm/NiFe 2nm/IrMn 6nm/ CoFe 1.5nm / No11 / CoFe 2nm/Cu 2.2nm/ CoFe t_F / Cu t_{SF} /No12/Ta 3nm.

The calculated MR ratio as a function of the enhancing layer thickness for spin-filter specular spin valves is shown in figure 3. The optimal Cu enhancing layer thickness is about 1nm to 1.5nm depending on the thickness of the free layer. This thickness dependence is a result of the balance between the increase in electron mean free path difference and current shunting through the enhancement layer. Therefore, the simulated results can be used as a guideline in the optimization of spin valve structures.

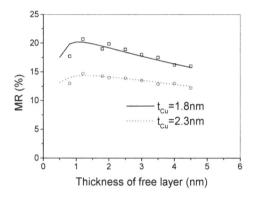

Figure 4. Experimental and simulated results on the ddependence of MR on free layer thickness for two different Cu conducting spacer thicknesses.

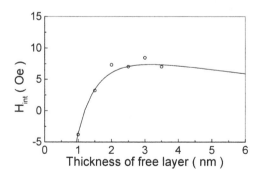

Figure 5. The calculated (line) and experimental (circle) interlayer coupling field H_{int} as a function of t_F for spin-filter specular spin valve Ta 3.5nm/NiFe 2nm/IrMn 6nm/ CoFe 1.5nm / Nol1 / CoFe 2nm/Cu 2.2nm/ CoFe t_F / Cu 1nm /Nol2/Ta 3nm.

Fig. 4 shows the dependence of MR ratio on the free layer thickness for two different thicknesses of Cu conducting spacer. As can been seen, the MR ratio of the spin valves increases with the decrease of Cu conducting spacer thickness. The optimal MR ratio for Cu conducting spacer of 1.8 nm is around 20%.

The interlayer coupling field H_{int} as a function of the free layer thickness t_F is shown in Figure 5. H_{int} has a negative value (H_{int} = -3.8 Oe), when t_F=1nm, then becomes positive with increasing t_F. The behavior can be understood by considering the negative RKKY coupling and magnetostatic coupling effect [2,9]. For the case of the bottom spin valves assuming a columnar structure with conformal waviness, the interlayer coupling field with thickness t_F can be written as, [10]

$$H_{int} = \frac{J_0}{\mu_0 M_F t_F t_{Cu}^2} \sin(\frac{2\pi t_{Cu}}{\Lambda}+\Psi)\frac{t_{Cu}/L}{\sinh(t_{Cu}/L)}$$
$$+\frac{\pi^2 h^2 M_P}{\sqrt{2}\lambda t_F}[1-\exp(-2\pi\sqrt{2}t_F/\lambda)][1-\exp(-2\pi\sqrt{2}t_P/\lambda)]\exp(-2\pi\sqrt{2}t_{Cu}/\lambda)$$

(3)

where Λ and Ψ are the period and phase of the oscillation, L is an attenuation length inversely proportional to the temperature. M_F and M_P are the saturation magnetizations of free layer and pinned layer respectively. t_F, t_P, and t_{Cu} are the thickness of free layer, pinned layer and spacer, respectively. h and λ are the amplitude and wavelength of the correlated interface waviness. Here, we use the parameter values $M_F = M_P = 1543$ Oe, h=0.6 nm, λ=18 nm to fit the experimental data. As shown in figure 4, the calculation values can fit the experimental results. It means that the interlayer coupling field contributed from RKKY coupling is compared to $1/t_F$ dependence, while the contribution from magnetostatic coupling decays slower than this $1/t_F$ dependence. Therefore, for thin free layer less than 1.5 nm, the negative RKKY antiferromagnetic coupling becomes stronger than the positive magnetostatic coupling. With increasing the thickness of free layer, magnetostatic coupling becomes dominant as the RKKY coupling becomes weak. The interlayer coupling field has a peak at about 3 nm. With further increasing the thickness of the free layer, the magnetostatic coupling becomes weaker, the interlayer coupling field is decreased.

The coercivity of the free layer H_{c1} as a function of the free layer thickness t_F is shown in Figure 5. Unlike the specular spin valves without an enhancing layer, the coercivity of free layer has an increase due to the oxidation of the free layer by the top specular NOL layer. Here, by inserting a Cu enhancing layer, the soft magnetic properties of CoFe free layer can be improved due to the low atomic intermixing observed between CoFe and Cu.

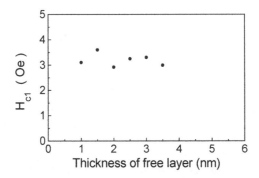

Figure 6. The coercivity of free layer H_{c1} as a function of t_F for spin-filter specular spin valve Ta 3.5nm/NiFe 2nm/IrMn 6nm/ CoFe 1.5nm / Nol1 / CoFe 2nm/Cu 2.2nm/ CoFe t_F / Cu 1nm /Nol2/Ta 3nm.

The CoFe free layer of 1~4 nm exhibited coercivity as low as ~3 Oe even after annealing at 270°C for 7 hours in a static field of 1 kOe. Such a thin soft CoFe free layer is particularly attractive for high density read sensor application.

Cross-sectional TEM lattice imaging examination (Fig. 7) of the specular spin-filter spinvalve multilayer structure has shown that the NOL1 introduced in the bottom-pinned CoFe layer is inhomogeneous and consists of a regions of oxides and regions of ferromagnetic metals, as marked in I, II and III of Fig. 7, respectively. It appears that most of the oxidation occurs in the grain boundaries of the CoFe layer and the epitaxial growth process of the CoFe grains was not interrupted by the oxidation process because large un-oxidized grains are present right across the NOL layer, region II for example. Owing to the fact that each of the sub-layer of the SV structure has a texture characteristic as (111) planes perpendicular to the thin film normal, we can study the structure inhomogeneous according to the difference in interspacing of (111) planes from each part of NOL1. As 6 nm thick IrMn layer has a good (111) texture, the (111) interspacing of IrMn from the HREM images was used as a reference. The spacing measurement was carried out by using the Mac. Digital Micrograph 2.5 (PPC). The error is about ± 0.05 Å for this method. A series of measurements showed that the interspacing of (111) plane corresponding to the area denoted by I, II, and III in Fig. 7(a) was: 2.15 Å, 1.99 Å, 2.46 Å, respectively. In contrast from a homogeneous oxide phase, the interspacing of (111) planes in different part of NOL1 was not equal, which indicates that the oxidation of the original

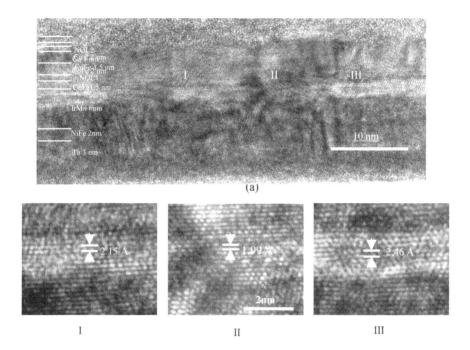

Figure 7. Cross-sectional HREM images of a specular spin filter spin valve in low magnification (a), and high resolution lattice images (b) of the marked region I, II and III in (a), respectively.

ferromagnetic $Co_{90}Fe_{10}$ was not uniform. According to the data from the international tables of crystallography for all of the three elements Co, Fe, and O, the only possible two phases in region I is CoO and/or FeO. In region II the interspacing of (111) plane with 1.99 Å can only be indexed as that of ferromagnetic CoFe metal or pure Co and Fe. For region III, the lattice spacing of 2.46 Å can be attributed from various possibilities of $CoFe_2O_4$, Fe_2O_3, Fe_3O_4 and Co_3O_4. Though it is hard to identify these four oxide phases in region III by using HREM lattice images, we can conclude from above analysis that the NOL1 is not a homogeneous oxide layer but with some regions, as the one denoted by number II, remained as ferromagnetic metals. This led us to believe that the specular reflection of spin-polarized electrons is achieved by these oxidised regions and that the remained un-oxidised ferromagnetic CoFe in NOL1 shall decrease the specular scattering on the NOL1 surface, but it provided the direct exchange paths between the IrMn and pinned CoFe layer. If the NOL1 were oxidized uniformly and no ferromagnetic metal remained, the exchange coupling between IrMn and pinned CoFe would attenuate greatly. Our results on the direct exchange coupling between pinning and pinned layers in SVs through the remained ferromagnetic metals across NOL1 agree well with the Rutherford backscattering spectrometer results [4] Compared with non-NOL SVs, the discontinuous NOL1 not only enhanced the specular scattering, but also maintain the strong direct exchange coupling between IrMn and pinned CoFe, thus enhancing the GMR effect. However, it is not yet clear whether the GMR of these specular spin-valves can be further enhanced by making the NOL more continuous.

4.   CONCLUSION

Spin-filter specular spin valves with structure Ta 3.5 nm/NiFe 2 nm/IrMn 6 nm/ CoFe1.5 nm/Nol1/CoFe 2 nm/Cu 2.2 nm/ CoFe t_F/ Cu t_{SF} /Nol2/Ta 3nm were deposited. It is shown that MR ratio in this kind of spin valves has been significantly improved. MR ratio remains higher than 12 % even when the free layer CoFe goes down to 1 nm. An optimised MR ratio of ~ 14.5% has been obtained when t_F was about 1.5 nm and t_{SF} about 1.0 nm. It is found that the Cu enhancing layer not only enhances the MR ratio but also improves soft magnetic properties of CoFe free layer. The free layer CoFe of 1~ 4 nm exhibited coercivity of ~3 Oe after annealing at 270 °C for 7 hours in field of 1 kOe. Furthermore, the interlayer coupling field H_{int} can be controlled by balancing the RKKY and magnetostatic coupling. Cross-sectional HREM studies showed that NOL1 in the pinned CoFe layer was not oxidized completely, but with a mixture of oxides and ferromagnetic metals. The ferromagnetic metals in NOL1 formed ferromagnetic paths over the NOL1, allowing direct exchange coupling to exist.

ACKNOWLEDGMENT

This work was partially supported by the Royal Society China Joint Project Q758 and the UK EPSRC grant GR/M46808.

REFERENCES

1 A. Veloso and P. P. Freitas, J. Appl. Phys. **87**, 5744 (2000).
2 Y. Huai, G. Anderson, and M.Pakala, J. Appl. Phys. **87**, 5741 (2000).
3 H. J. M. Swagten, G. J Strijkers, P. J. H. Bloemen, M. M. H. Willekens, and W. J. M. de Jonge, Phys. Rev. B **53**, 9108 (1996).

4 M. F. Gillies and A. E. T. Kuiper, J. Appl. Phys. **88**, 5894 (2000), and references therein.

5 A. Veloso, P. P. Freitas, P. Wei, N. P. Barradas, J. C. Soares, B. Almeida, and J. B. Sousa, Appl. Phys. Lett. **77**, 1020 (2000).

6 M. F. Gillies, A. E. T. Kuiper, and G. W. R. Leibbrandt, J. Appl. Phys. **89**, 6922 (2001).

7 W. Y. Lee, M. Carey, M. F. Toney, P. Rice, B. Gurney, H. C. Chang, E. Allen, and D. Mauri, J. Appl. Phys. **89**, 6925 (2001).

8 Y. Sugita, Y. Kawawake, M. Satomi, and H. Sakakima, J. Appl. Phys. **89**, 6919 (2001).

9 C. Lai, C. J. Chen, and T. S. Chin, J. Appl. Phys. **89**, 6928 (2001).

10 S. Sant, M. Mao, J. Kools, K. Koi, H. Iwasaki, and M. Sahashi, J. Appl. Phys. **89**, 6931 (2001).

SPIN-POLARISED TUNNELING EFFECTS OBSERVED ON THE MAGNETITE (001) AND (111) SURFACES

S.F. Ceballos, N. Berdunov*, G. Mariotto, I.V. Shvets
SFI Laboratories, Trinity College, Dublin 2, Ireland

Abstract: The (001) and (111) surface structure of magnetite have been studied using Auger electron spectroscopy, low energy electron diffraction (LEED) and scanning tunneling microscopy (STM). A clean surface was obtained by a combination of in-situ annealing in oxygen atmosphere, Argon sputtering and subsequent annealing in UHV for Fe_3O_4 (001) and (111) surfaces. A ($\sqrt{2} \times \sqrt{2}$)R45° superlattice was found by LEED and STM on the Fe_3O_4 (001) surface. Evidence of the formation of a superstructure on the surface of Fe_3O_4 (111) single crystal is given. The superstructure, which has a periodicity of 42 Å and the three-fold symmetry, has been observed by means of STM and LEED. Atomic resolution STM images have been achieved on the Fe_3O_4 (001) and (111) surface using MnNi tips, which are interpreted in terms of spin-polarized effect.

Key words: Spin Polarised Scanning tunneling microscopy, magnetite.

1.   INTRODUCTION

Magnetite has been the subject of intensive studies by the scientific community during the last decades. Efforts to understand its magnetic and electronic properties have intensified in the past few years due to its half-metallic nature, which makes magnetite a promising material for applications in spin electronics.

Magnetite is an inverse spinel material. The crystal structure of magnetite is based on a face-centered cubic (f.c.c.) unit cell, containing 32 O^{2-} anions and 24 mixed valence Fe cations, with a lattice parameter of a=8.39 Å. The formula can be written as $YA[XY]_BO_4$, where $X = Fe^{2+}$, $Y = Fe^{3+}$ and A and

B. Aktaş et al. (eds.), Nanostructured Magnetic Materials and their Applications, 91–98.

B denote tetrahedral and octahedral sites, respectively. This formula indicates that one half of the ferric Fe^{3+} ions occupies 8 of the 64 available tetrahedral interstices, and the other half of the ferric ions, together with an equal amount of ferrous Fe^{2+} ions, occupy 16 of the 32 available octahedral interstices. Stoichiometric magnetite undergoes a metal-insulator transition at ~ 120 K, known as the Verwey transition temperature T_V. At room temperature, the electrical conductivity of Fe_3O_4 is $\approx 200 \ \Omega^{-1} \ cm^{-1}$ and it gradually decreases with decreasing temperature. When cooled down below T_V, the conductivity abruptly decreases by about two orders of magnitude [1,2]. The change of conductivity is accompanied by a change in the crystallographic structure, whose symmetry is lowered from cubic to monoclinic. Verwey *et al.* [3] first proposed that the transition is due to the ordering of the Fe^{+3} and Fe^{+2} ions on the octahedral sites, and that alternating planes contain either Fe^{+3} or Fe^{+2} only. The charge order on the magnetite surface even less understood due to an additional complexity. According to the classification of the surface ionic or partially ionic crystals, given by Tasker [7,8], the (001) and (111) surfaces of magnetite are polar and must therefore reconstruct. The $(\sqrt{2} \times \sqrt{2})R45°$ surface reconstruction is the most commonly observed reconstruction on the magnetite (001) surface. It has been observed on natural and artificial crystals [9,10] and also on thin films grown by molecular beam epitaxy (MBE) [11,13]. Different terminations have been reported for (111) surface depending on the preparation procedure [14,16]. In the present investigation we employ spin-polarization STM to discriminate the magnetic and electronic properties of the magnetite surface on atomic scale and associate it with a possible charge order occurred on the surface at room temperature. To achieve a spin-polarised contrast in tunnelling current we used tips of MnNi antiferromagnetic alloy prepared as described by Ceballos *et al.* [17] In a joint experiment an in-plane magnetic field of 60mT has been applied. Spin contrast can be seen in the STM images for both surfaces.

2.   EXPERIMENTAL

A set of synthetic single crystals has been used in these experiments. The crystals were aligned with a precision of 1° with respect to the (001) and (111) respectively. The crystals were first characterized by powder x-ray diffraction. Four-point resistance vs. temperature measurements were also performed; Verwey transition temperatures of 108 K for the (001) and 120 K for the (111) crystals were measured respectively. The crystals were mechanically polished using diamond paste with grain size down to 0.25 μm before being introduced into the UHV system. This system is equipped with facilities for LEED, AES and STM analysis. STM measurements were carried out at room temperature in constant current mode using a home-built instrument. A bias voltage from +0.6 V to + 1 V was applied to the (001) sample and from −1.0 V to 1.0 V to the (111) sample. Not stable tunneling was achieved with negative bias voltage on the (001) surface. A tunnelling current of 0.1 nA and 0.3 nA was typically used. MnNi tips were used for the STM experiments. MnNi is antiferromagnet within a composition range of about ±2 % of the equiatomic composition. MnNi are a good candidate for SP-STM studies since magnetostatic interactions between tip and sample are eliminated producing a very stable tunneling [17].

3. RESULTS AND DISCUSSION

3.1 Fe₃O₄ (001) surface

The crystal was prepared *in situ* by a subsequent Ar^+ ion sputtering, annealing in UHV and annealing in an oxygen partial pressure. The crystal was annealed in a resistive heater equipped with a K-type thermocouple to monitor the anneal temperature. This preparation method, described in detail elsewhere [18], was found to produce a contaminant-free surface, as confirmed by Auger analysis. LEED analysis showed a sharp pattern, where a $(\sqrt{2} \times \sqrt{2})R45°$ mesh was visible. STM images showed a surface characterized by multiple terraces with straight edges aligned along the [110] and [1-10] directions. The step height between the terraces was a multiple of 2.1 Å corresponding to the separation between A-A planes or B-B planes. Atomically resolved images of the surface reveal atomic rows aligned parallel to the step edges. These rows are rotated by 90° on terraces separated by an odd multiple of 2.1 Å, indicating that the surface consists of octahedrally terminated terraces. The corrugation along the rows alternates between bright points and dark points, where a 0.2 Å corrugation was measured at the bright points, while a 0.1 Å corrugation was measured at the darker points. The periodicity of the bright points along the [110] rows is 12 Å and the separation between bright and dark points is 6 Å (see Figure 1).

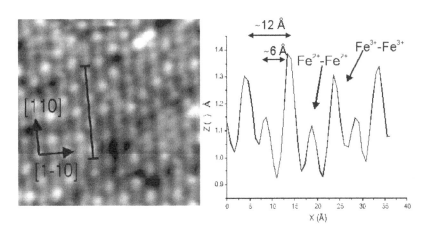

Figure 1. 65×65 A^2 STM image of an synthetic Fe_3O_4 (001) single crystal. Taken with a MnNi tip, I_t = 0.1 nA, V_b = 1.0 V. Line profile taken along the [110] direction.

The contrast in the STM images is attributed to the Fe^{2+} and Fe^{3+} ions at the B-sites. The variation of the tunneling current depends on m_s, which differs locally for Fe^{2+} and Fe^{3+} in magnetite being $4\mu_B$ and $5\mu_B$ respectively, making these ions distinguishable in STM images using a magnetic tip. The full width at half maximum (FWHM) of the points is ≈ 3 Å. The observed periodicity and the FWHM of the imaged points suggest that the MnNi probe is imaging Fe dimers as opposed to individual Fe cations. The structure observed could not be explained in terms of a tetrahedral terminated surface, since only one species of Fe cations is present in the A sublattice. A half-filled A-layer model proposed by Kim *et al.* [19] would not be suitable in this case due to mere topographical considerations [18]. As the sample is positively biased the empty states of the sample are being probed. We therefore believe that the bright points correspond to Fe^{3+}-Fe^{3+} dimers and that the dark points correspond to Fe^{2+}- Fe^{2+} dimers. The nature of the Verwey transition has been long investigated and, although no definite evidence has been given, it is believed that the transition is due to an ordering of Fe cations in the B sites, leading to a reduction in the electrical conductivity. Anderson [20] showed that the repulsion energy due to the cations in octrahedral sites is minimized provided every tetrahedron formed by the nearest-neighbor octahedral sites is occupied by two Fe^{2+} and two Fe^{3+} ions, imposing the so-called short-range order. Our results provide evidence of charge ordering at the surface of magnetite and of an intimate link between charge ordering and the $(\sqrt{2} \times \sqrt{2})R45°$ mesh observed by

LEED and STM. A ($\sqrt{2} \times \sqrt{2}$)R45° reconstruction has been observed by other groups and different models have been proposed to explain this reconstruction. Whether the models proposed support a tetrahedral or octahedral termination of the surface, all explain the ($\sqrt{2} \times \sqrt{2}$)R45° reconstruction in terms of structural changes of the surface to stabilize the surface energy and provide the observed surface symmetry. We note that the model proposed to explain our results led to a non autocompensated surface [21,22], but it has to be taken into account that a number of polar non-autocompensated surfaces has been observed before [23-27]. Yet, our results do not suggest that the autocompensation model should be discarded. It is indeed possible that the surface was not in its ground state.

3.2   Fe₃O₄ (111) surface

The magnetite sample was annealed in UHV at 950±20 K followed by a short anneal in an oxygen atmosphere of 10^{-6} mbar at 950K for 15 min, which was in turn followed by cooling in the oxygen atmosphere. This sample preparation procedure leads to the formation of a well-defined hexagonal superlattice with a periodicity of 42±3 Å. This superstructure is highly regular and covers almost the entire sample surface. The high-resolution STM image in Fig. 2(a) shows the atomic arrangement within the superstructure. One can see that the superstructure consists of three distinct areas, marked as areas I, II and III. Detailed analysis shows that area I has a periodicity of 3.1±0.1 Å, while areas II and III have a periodicity of 2.8 ± 0.1 Å along the [011] direction, which is consistent with the LEED pattern. As we have shown in the earlier publication [28], the superstructure depicts an oxygen-terminated magnetite bulk, which reconstructs due to the electron-lattice instability, a polaron- or a charge density wave-like. Thus, the STM image in Fig. 2(a) represents a lattice of oxygen sites on the top of iron layer. A number of the defects, seen in Fig. 2(a), represent the missing oxygen atoms (oxygen vacancies).

In our SP-STM experiments a magnetic field of 60 mT was applied parallel to the surface during the STM scan. We have verified *ex-situ* by vibrating sample magnetometer (VSM) that this magnetic field is strong enough to fully demagnetize the sample. The contrast achieved in STM images when the magnetic field switched off/on is demonstrated in Fig. 2 (left, right) respectively. As it can be seen, the appearance of the superstructure and corrugation between different areas is almost unaffected by the magnetic field. However, the major changes occur in proximity of the oxygen vacancies. Three bright spots appear in the vicinity of the defects as can be seen in Fig. 2(b). A 6 Å separation between the spots and their positions correspond to those of Fe ions in the layer underneath the topmost oxygen lattice.

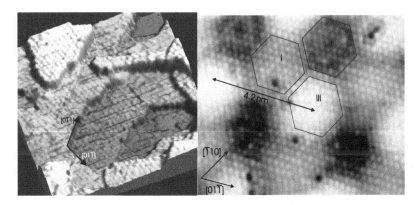

Figure 2. [(a) 120×120nm² STM image of overoxidized Fe_3O_4 (111) surface representing a large terraces covered by superstructures of about 42 Å period; (b) 10×8 nm² STM image of superperiodic pattern seen in Fig. 3b. Areas II, III have 3.1±0.1 Å interatomic periodicity, and area I has 2.8 ± 0.1 Å periodicity (V_{bias} = -1.0V).

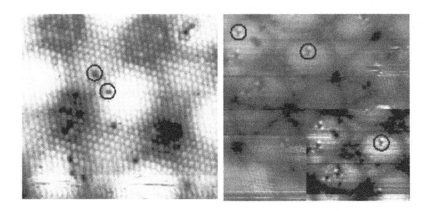

Figure 3. STM images of the superstructure formed on the over-oxidized Fe_3O_4 (111) surface obtained without (left) and with magnetic field (right). Blue circles mark the missing oxygen defects in the topmost surface layer. When an external in-plane magnetic field of 60 mT was applied, three bright points appeared around the defects.

We can conclude that by applying an in-plane magnetic field we were able to alter the magnetic moments of individual atoms around these defects, thus showing a spin contrast in STM images.

4.   CONCLUSIONS

The Fe_3O_4 (001) and (111) surfaces have been studied with AES, LEED and STM. A sharp ($\sqrt{2} \times \sqrt{2}$)R45° surface reconstruction was observed by LEED and STM on the clean Fe_3O_4 (001). The symmetry observed has been explained in terms of charge ordering at the octahedral sites where a formation of dimers has happened. A 42 Å periodicity superstructure has been observed on the oxidized Fe_3O_4 (111) surface. The superlattice is interpreted in terms of an oxygen terminated surface representing an electronic effect rather than a mosaic of several iron oxide phases. Antiferromagnetic tips made of MnNi have been used to obtain atomic resolution images for both Fe_3O_4 surfaces. There is strong evidence for the interpretation of these results in terms of spin-polarized tunneling effect.

ACKNOWLEDGEMENTS

Financial assistance from the Science Foundation Ireland (SFI) agency, contract No. 00/PI.1/C042 is gratefully acknowledged.

REFERENCES

[1] P.A. Cox, Transition Metal Oxides, An Introduction to their Electronic Structure and Properties (Clarendon Press, Oxford, 1995).

[2] R. Aragon, R.J. Rasmussen, J.P. Shepherd, J.W. Koenitzer, J. M. Honig, J. Magn. Magn. Mater. 47-54, 1335 (1986).

[3] E.J. W. Verwey, P. W. Haayman and R. Romeyn. J. Chem. Phys. 15, 181 (1947).

[4] S. Iida, K. Mizushima, M. Mizoguchi, K. Kose, K. Kato, K. Yanai, N. Goto and S. Yumoto. J. Appl. Phys. 53, 2164 (1982).

[5] M. Mizoguchi. J.Appl. Soc. Jpn. 44 1501 (1978).

[6] J.M. Zuo, J.C.H. Spence and W. Petuskey, Phys. Rev. B 42, 8451 (1990).

[7] P.W. Tasker, J. Phys. C 12, 4977 (1979).

[8] P.W. Tasker, Philos. Mag. A 12, 4977 (1979).

[9] S. Iida and K. Mizushima and M. Mizoguchi and K. Kose and K. Kato and K. Yanai and N. Goto and S. Yumoto. J. Appl. Phys. 53, 2164 (1982).

[10] M. Mizoguchi. J. Phys. Soc. Jpn. 44, 1501 (1978).

[11] J M Gaines and P J H Bloemen and J T Kohlhepp and C W T Bulle-Lieuwma and R M Wolf and R M Reinders and R M Jungblut and P A A van der Heijden and J T W M van Eemeren and J. aan de Stegge and W J M de Jonge. Surf. Sci. 373 85 (1997).

[12] F.C. Voogt, T. Fujii, P.J.M. Smulders, L. Niesen, M.A. James, T. Hibma, Phys. Rev. B 60, 11 193 (1999).

[13] B. Stanka, W. Hebenstreit, U. Diebold, S.A. Chambers, Surf. Sci. 448, 49 (2000).

[14] Condon, N. G., Leibsle, F. M., Parker, T., Lennie, A. R., Vaughan, D. J., and Thornton, G., Phys. Rev. B **55**, 15885 (1997).

[15] G. Ketteler, G. Weiss, W. Ranke, R. Schologl. Phys. Chem. Chem. Phys. 3, 1114, (2001)

[16] W. Weiss, M. Ritter. Phys. Rev. B. 59, 5201 (1999).

[17] S.F. Ceballos, G. Mariotto, S. Murphy, I.V. Shvets, Surf. Sci. 523 131 (2003).

[18] G. Mariotto, S. Murphy, I.V. Shvets, Phys. Rev. B. 66 245426 (2002).

[19] Y. J. Kim, Y. Gao, and S. A. Chambers, Surf. Sci. 371, 358 (1997).

[20] P. W. Anderson. Phy. Rev. 102, 1008 (1956).

[21] M.D. Pashley. Phy. Rev. B. 40 10 481 (1989).

[22] J. LaFemina. Crit. Rev. Surf. Chem. 3, 297, (1994).

[23] O. Warren and P. Thiel. J. Chem. Phys. 100, 659 (1994).

[24] A. Barbieri, W. Weiss, M. V. Hove, and G. Somorjai, Surf. Sci. 302, 259 (1994).

[25] Q. Guo and P. Moeller, Surf. Sci. 340, L999 (1995).

[26] N. Condon, F. Leibsle, P. Murray, T. Parker, A. Lennie, D. Vaughan, And G. Thornton, Surf. Sci. 397, 278, (1998).

[27] S.F. Ceballos, G. Mariotto, S. Murphy, K. Jordan and I.V. Shvets. Surf. Sci. Accepted.

[28] N. Berdunov, G. Mariotto, S. Murphy and I.V. Shvets. Europhys. Lett. 63, (6), 867 (2003).

3

PART 3: NANOSTRUCTURED FERROMAGNET/SUPERCONDUCTOR STRUCTURES

TRIPLET SUPERCONDUCTIVITY IN SUPERCONDUCTOR/FERROMAGNET MULTILAYERED STRUCTURES

F. S. Bergeret [a], F. Volkov [a,b], K.B. Efetov [a,c]

[a] *Ruhr-University Bochum, D-44780 Bochum, Germany*
[b] *Institute of Radioengineering and Electronics of the Russian Academy of Sciences, Mokhovaya str.11, Moscow 103907, Russia*
[c] *L.D. Landau Institute for Theoretical Physics, 117940 Moscow, Russia*

Abstract: We demonstrate that the multilayered superconductor-ferromagnet structures can be used in order to detect the triplet component of the superconducting condensate. The latter one is odd in frequency and is generated if the magnetizations of different ferromagnetic layers are non-collinear. This triplet condensate coexists in superconductors with the conventional singlet pairing but decays very slowly in the ferromagnet. If the thickness of the ferromagnetic layers is large enough the Josephson coupling is only due to the triplet component. Depending on the mutual direction of the ferromagnetic moments the coupling can be both of 0 and π type.

Key words: Joserhson   effect,   superconductor/ferromagnet   contacts,   triplet superconductivity

In spite of many theoretical and experimental works on the proximity effect in conventional-superconductor-ferromagnet (S/F) structures, only recently it was pointed out the existence of a triplet component (TC) of the superconducting condensate function in such structures [1]. This type of condensate is odd in frequency and may penetrate into the ferromagnetic region over the length $\xi_T = \sqrt{D/T}$ (we assume the system is diffusive, D is the diffusion coefficient). This length is in general much larger than the penetration length of the usual singlet component which is given by $\xi_J = \sqrt{D/J}$, where J is the exchange field of the ferromagnet. The triplet component is generated in S/F structures with non-homogeneous magnetization. For example, in structures with a magnetic domain wall close

B. Aktaş et al. (eds.), Nanostructured Magnetic Materials and their Applications, 101–108.

to the S/F interface [1,2], or in a multilayered structure in which the magnetizations of the F layers are non-collinear [3].

Triplet superconductivity is a rather exotic phenomenon which has been observed until now only in superfluid He^3 and in the superconducting material Sr_2RuO_4 [4]. It is expected to be very sensitive to disorder [4,5], which makes its observation even more difficult. Another possibility was suggested by Berezinskii long ago [6]. He conjectured that the triplet superconductivity might be possible if the condensate function is even in the momentum, but odd in the frequency. Attempts to find conditions for the existence of such an odd superconductivity were done in several works later [7], but the results were not encouraging (in Ref. [7] a singlet pairing odd in frequency and even in the momentum was considered).

The type of superconductivity analyzed in our paper complements the three known types of superconductivity: s-wave and d-wave singlet superconductivity that occur in ordinary superconductors and in high T_c superconductors respectively, and the p-wave superconductivity with triplet pairing observed in Sr_2RuO_4.

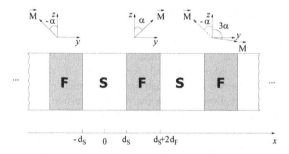

Figure 1. Schematic diagram of a S/F multilayered structure. On top: the magnetization $\vec{M}$ in different F layers. The $\vec{M}$ vector drawn by solid lines correspond to chiralities of the same sign. Opposite chirality is obtained if, for example, the magnetization of the right F layer is rotated in the opposite direction (dashed line).

We consider a S/F multilayered structure (Fig. 1) in which the new superconducting state might be observed. In this state, the superconductivity in the S regions is caused by the singlet component (SC) and takes place in the plane of the layers, while the transverse superconductivity through the F layers is mainly due to the TC (we are assuming that the thickness d_F of the F layer is larger than ξ_J). Moreover, the relation between the condensate current I_S and a phase difference φ depends in a crucial way on chirality of the magnetic moment $\vec{M}$ in space. The (in-plane) magnetizations $\vec{M}$ of the neighboring F layers are not parallel to each other and the angle between them is 2α. It will be shown that in such a structure the TC may arise if the thickness of the superconducting layers $2d_S$ is not too large as compared to the coherence length $\xi_S = \sqrt{D_S/2\pi T}$. The TC penetrates into the F layer over a long distance $\xi_T = \sqrt{D_F/2\pi T}$ and ensures a Josephson coupling between the nearest S layers. In the case under consideration the relation between the condensate current I_S and the phase difference φ has the conventional form $I_S = I_c \sin \varphi$. However, the sign of the critical current I_c depends on the chirality, namely, it is positive if the rotation angle $2\alpha_i$ of the magnetization at the S_i layer has the same sign for neighboring S layers (S_i and S_{i+1}) and it is negative if the signs of the rotation are opposite for neighboring S layers ($\alpha_i \alpha_{i+1} < 0$).

We consider the simplest case of a dirty system for which the Usadel equations can be applied to calculate the condensate function f, i.e., we assume that the condition $\Delta < J < \tau^{-1}$ is fulfilled, where Δ is the order parameter in S (in the ferromagnet $\Delta = 0$), J is the exchange energy and τ is the momentum relaxation time. In order to illustrate our results in a clear way we make further assumptions. First, we assume that the thickness of the F layers $2d_F$ exceeds the length ξ_T. In this case, one can find the condensate function for a FSF system assuming that it decays in the F regions to zero and neglecting the influence of the adjacent S layers (a weak Josephson coupling). The Josephson coupling is determined by an overlap of exponentially decaying condensate functions induced by the nearest S layers. We also assume that the temperature T is close to the critical temperature of the superconducting transition T_c^* of the FSF structure. In this case the Usadel equation can be linearized and what remains to do is to find the condensate function $f(x)$ by solving the linearized Usadel equation in the S and F regions. Then the Usadel equation for the matrix condensate function f takes the form [1,2]

$$D_S \partial_{xx}^2 f - 2|\omega| f = 2i\Delta, \quad \text{S layer} \tag{1}$$

$$D_F \partial_{xx}^2 f - 2|\omega| f + iJ\text{sgn}\omega\{[\hat{\sigma}_3, f]_+ \cos\alpha + \\ + \hat{\tau}_3[\hat{\sigma}_2, f]_- \sin\alpha\} = 0, \quad \text{F layer} \tag{2}$$

Here $D_{S,F}$ are the diffusion coefficients in the S and F layers respectively, $\Delta = i\hat{\tau}_2 \otimes \hat{\sigma}_3 \Delta$, $\omega = \pi T(2n+1)$ is the Matsubara frequency, J

is the exchange energy (strictly speaking, $\vec{J}$ is a vector directed along the magnetization vector $\vec{M}$), $\hat{\tau}_i$ and $\hat{\sigma}_i$ are the Pauli matrices in the Nambu and spin space, respectively. The brackets $[...]_{\pm}$ denote an anti-commutator and a commutator. We assume that the $\vec{M}$ vector lies in the (y,z) plane and α is the angle between the z–axis and $\vec{M}$. These equations were used previously by the authors to analyze other problems [1,8]. Eqs. (1)-(2) have to be supplemented by boundary conditions. In the case of a perfect S/F interface (the reflection coefficient is very small) they have the form [9]

$$f(d_S + 0) - f(d_S - 0) = 0, \tag{3}$$

$$\gamma \partial_x f(d_S + 0) - \partial_x f(d_S - 0) = 0,$$

where $\gamma = \sigma_F/\sigma_S$, $\sigma_{F,S}$ are the conductivities in the F and S regions (for simplicity we do not take into account a dependence of the conductivity σ_F on the spin directions).

The condensate matrix function f has following form

$$f(x) = i\left(\hat{f}_1(x)\hat{\tau}_1 + \hat{f}_2(x)\hat{\tau}_2\right), \tag{4}$$

where $\hat{f}_{1,2}$ are matrices in the spin space. In the F layers they can be written as

$$\hat{f}_1 = B_1(x)\hat{\sigma}_1 + B_2(x)\hat{\sigma}_2, \tag{5}$$

$$\hat{f}_2 = B_0(x) + B_3(x)\hat{\sigma}_3.$$

In the S layers the functions $\hat{f}_{1,2}$ have the same form, but the coefficients B should be replaced by A.

Let us discuss briefly some properties of the condensate function f (Eq. (4)). In the case of a normal metal ($J = 0$) only the function B_3 in Eq. (5) is nonzero. Thus, B_3 corresponds to the usual singlet component of the condensate. If $J \neq 0$ another the nonzero components of the condensate function arise. We distinguish two cases: a) if the magnetizations of the F layers are collinear beside B_3 also B_0 is non zero. The latter corresponds to the projection $S = 0$ of the triplet component. Obviously this component decays over the same short length ξ_J as the singlet one; b) more interesting is the case when the magnetizations of the different F layers are non-collinear, i.e $\alpha \neq 0$ in Eq.(2). In this case the function $B_1(x)$ in Eq.(5) is also nonzero. This corresponds to the projections $S = \pm 1$ of the triplet component and is responsible for the long-range proximity effect (cf. ref.[1]).

Since we are interested in the coupling between the S layers induced by the overlap of the long-range components of f we determine the function $B_1(x)$ which is given by [3]

$$B_1(x) = b_{1\omega} \exp(-\kappa_\omega(x - d_S)) + b_{1+} \exp(-\kappa_+(x - d_S))$$

$$+ b_{1-} \exp(-\kappa_-(x - d_S)), \tag{6}$$

where $\kappa_\pm = \xi_J^{-1}(1 \pm i)$, $\kappa_\omega = \sqrt{2\omega/D_F}$ and the coefficients b_1 must be determined from the boundary conditions. Only the first term corresponds to the long-range component and it is given by the expression

$$b_{1\omega} = -\frac{\Delta}{|\omega|} \sin \alpha \cdot \kappa_J \cdot \mathrm{sgn}\omega \cdot \{\cosh \theta_S [\kappa_\omega \tanh \theta_S + \kappa_S/\gamma]$$

$$\times [1 + \gamma\kappa_+/(\kappa_S \tanh \theta_S)][1 + \gamma\kappa_-/(\kappa_S \tanh \theta_S)]\}^{-1} \tag{7}$$

Contrary to SC determined by the solution $B_3(x)$, the TC is an odd function of x. This is seen from Eqs. (6), (7) and the fact the α has different sign at $x > d_S$ and $x < d_S$. In the S layer the corresponding solution has the form: $A_1(x) = a_1 \sinh(\kappa_S x)$. In the limit $T \ll J$, the TC penetrates into the F layer over the length of the order $1/\kappa_T = \sqrt{D_F/2\pi T}$ which is much greater than the SC penetration length $1/\kappa_+$. One can see from Eq. (7) that the amplitude $b_{1\omega}$ of the long-range TC is an odd function of the Matsubara frequency ω and is symmetric in the momentum space as in the case of the TC which arises in a ferromagnet with a non-homogeneous M near a S/F interface [1]. This new type of the condensate, odd in ω and even in the momentum p, has been proposed by Berezinksii [6] in order to explain the pairing mechanism in He^3 (it was proven later that the condensate in He^3 is in fact even in ω and odd in p). The solution we present here corresponds to this hypothetical pairing, which means that we have found a concrete realization of the idea. It follows from the geometry of the considered structure, that the odd-triplet superconductivity we found exists in the transverse direction.

Let us now calculate the Josephson current between the neighboring S (S_1 and S_2) layers assuming that the condition $\xi_J \ll \xi_T < 2d_F$ is fulfilled. Then, the Josephson current is due to the overlap of the TC induced near each S/F interface. In this case the TC in F is described by the expression

$$f_{trip}(x) = i\hat{\tau}_1 \otimes \hat{\sigma}_1 b_{1\omega} \exp(-\kappa_\omega(x-d_S)) +$$

$$+ \check{S} \cdot i\hat{\tau}_1 \otimes \hat{\sigma}_1 \cdot \check{S}^+ \tilde{b}_{1\omega} \exp(\kappa_\omega(x-d_S-2d_F)). \tag{8}$$

Here, the matrix $\check{S} = i\cos(\varphi/2) + i\hat{\tau}_3\sin(\varphi/2)$ allows us to take into account the phase difference φ between the neighboring S layers (we assume that the phase of the S_1 layer is zero). The coefficients have different signs ($b_{1\omega} = -\tilde{b}_{1\omega}$) if the magnetization M at both S layers rotates in the same direction, and $b_{1\omega} = \tilde{b}_{1\omega}$ in the opposite case (different signs of chiralities). The condensate current I_S between S_1 and S_2 is given by the formula

$$I_S = (L_y L_z)\sigma_F (\pi i T/4e) Tr(\hat{\tau}_3 \otimes \hat{\sigma}_o) \sum_\omega f \partial_x f. \tag{9}$$

Substituting Eq.(8) into Eq.(9), we get $I_S = I_c \sin\varphi$, where

$$I_c = -(2\pi T/eR_F) \sum_\omega \kappa_\omega b_{1\omega} \tilde{b}_1 \omega \exp(-2d_F \kappa_\omega), \tag{10}$$

and $R_F = ((L_y L_z)\sigma_F)^{-1}$ is the resistance of the F layer in the normal state. Substituting Eq.(7) into Eq.(10), we find for the critical current

$$eR_F I_c = (4T/\pi)(\Delta/T)^2 \frac{(\kappa_{So}d_S)^4}{[(\kappa_{So}d_S + \gamma\kappa_J)^2 + (\gamma\kappa_J)^2]^2} \times$$
$$\times \frac{\sin\alpha_1 \cdot \sin\alpha_2}{[\kappa_T d_S + 1/\gamma]^2} \exp(-2d_F \kappa_T) \tag{11}$$

where $\kappa_{So} = \sqrt{2\pi T/D_S}$.

If the magnetization vector at each S layer rotates in the same direction ($\alpha_1 = \alpha_2$), then the critical Josephson current I_c is positive (0-contact). If $\vec{M}$ rotates at S_1 and S_2 in opposite directions ($\alpha_1 = -\alpha_2$), then the critical current I_c is negative (π – contact). In latter case a phase difference π is established between neighboring S layers in a multilayered S/F structure. We would like to note that the mechanism of the π – contact considered here is completely different from that suggested in Ref. [10] and observed in Ref. [11]. In our case, the negative critical current is caused by the TC, but not by the SC as in Ref. [10], and, in addition, it is realized only if the chiralities at the S_1 and S_2 are different. The possibility of switching between the 0 and π -states by changing the angle α may find applications in superconducting devices.

In conclusion, we have demonstrated that in superconductor-ferromagnet structures a non-collinear alignment of the exchange fields in the ferromagnetic layers generates a triplet component of the superconducting condensate, and this component is odd in the frequency. The odd-triplet condensate penetrates into the ferromagnet over the long distances and is not sensitive to impurities. In the structure we suggest, the Josephson effect between superconductors separated by the ferromagnet is possible, and the critical current can be measured. The Josephson contact can be both of 0 and π -type depending on the arrangement of the magnetic moments.

ACKNOWLEDGMENT

We would like to thank SFB 491 for financial support.

REFERENCES

[1] F. S. Bergeret, A. F. Volkov and K. B. Efetov, Phys. Rev. Lett. **86** (2001) 4096.
[2] F.S. Bergeret, K.B. Efetov, and A.I. Larkin, Phys. Rev. B **62** (2000) 11872.
[3] A. F. Volkov, F.S. Bergeret and K. B. Efetov, Phys. Rev. Lett. **90** (2003) 117006.
[4] A. P. Mackenzie, R. K. W. Haselwimmer, A.W.Tayler, G.G.Lonzarich, Y.Mori, S.Nishizaki, and Y.Maeno, Phys. Rev. Lett. , 161 (1998)
[3] A.I.Larkin, Zh.Eksp.Teor.Fiz. Pis'ma Red. , 205 (1965) [Sov.Phys.JETP Lett. , 130 (1965)]
[6] V. L. Berezinskii, JETP Lett. , 287 (1975).
[7] A.V. Balatsky and E. Abrahams, Phys. Rev. B , 13125 (1992); A.V. Balatsky, E. Abrahams, D.J. Scalapino, and J.R. Schrieffer, Phys. Rev. B , 1271 (1995) .
[8] F. S. Bergeret, A. F. Volkov and K. B. Efetov, Phys. Rev. B **64** (2001) 134506.
[9] A. V. Zaitsev, Sov. Phys. JETP **59** (1984) 863; M. Y. Kuprianov and V. F. Lukichev, Sov. Phys. JETP **64** (1988) 139.

[10] L.N.Bulaevskii, V.V.Kuzii, and A.A.Sobyanin, Sov. Phys. JETP Lett. **25** (1977) 299;
     A.I.Buzdin, L.N.Bulaevskii, and S.V.Panyukov, Sov. Phys. JETP Lett. **35** (1982) 178.
[11] V.V.Ryazanov *et al.*, Phys. Rev. Lett. **86** (2001) 2427.

INDUCED SPIN-SPLITTING AT SUPERCONDUCTOR-FERROMAGNET INTERFACE

V. Krivoruchko

Donetsk Physics & Technology Institute, Donetsk-114, 83114 Ukraine
e-mail: krivoruc@krivoruc.fti.as.donetsk.ua

Abstract: We focus on some aspects of proximity effect in hybrid structures of a massive superconductor (S) ($d_S \gg \xi_S$) and thin ferromagnet (F) ($d_F \ll \xi_F$); here $\xi_{S,F}$ are the superconducting coherence lengths, and $d_{S,F}$ are the thicknesses of the S and F layers, respectively. The analysis is based on a microscopic theory of proximity effect for metals in the dirty limit conditions. It is shown that for SF hybrid structures the exchange field extends a distance of order ξ_S into the superconductor. Magnetic properties of the S metal are manifested in: (i) the superconducting phase variation at the SF interface; (ii) the dependence of the critical current of SFIFS tunnel structure on mutual orientation of the layers magnetization; (iii) the additional suppression of the S layer order parameter near the SF interface; (iv) the spin-splitting of quasi-particle density of states (DOS) of the S layer; (v) the appearance of local bands inside the energy gap; and directly in (vi) the induced equilibrium magnetization of the S layer.

1. INTRODUCTION

Proximity effects are phenomena stipulated by a 'penetration' of an order parameter (of some state) from one material into another, which does not possess such type of the order itself, due to the materials being in contact. The leakage of superconducting correlations into a non-superconducting material is an example of the superconducting proximity effect. For a normal nonmagnetic metal (N) in contact with a superconductor (S) the proximity effect has been intensively studied and well understood many years ago [1-3]. When a normal nonmagnetic metal is replaced by a ferromagnet (F), the physics of the proximity effect is much more

B. Aktaş et al. (eds.), Nanostructured Magnetic Materials and their Applications, 109–119.

interesting. In this case, proximity induced superconductivity is spatially inhomogeneous and the order parameter contains nodes where the phase changes by π [3-5]. The π-phase state of an SFS weak link due to Cooper pair spatial oscillation was first predicted by Buzdin *et al.*, [4,5]. Experiments that by now have been performed on SFS weak links [6-8] and SIFS tunnel junctions [9] directly prove the π-phase superconductivity. Planar tunneling spectroscopy also reveals a π-phase shift in the order parameter, when superconducting correlations coexist with ferromagnetic order [10]. However, experiments on superconducting transition temperature SF multilayer behavior, $T_C(d_F)$, are not so conclusive (see, e.g. [11,12]).

There is another interesting case of a thin F layer, $d_F \ll \xi_F$, being in contact with an S layer. As far as the thickness of the F layer is much less then corresponding superconducting coherence length, there is spin spiltting but there is no order parameter oscillation in the F layer. Meanwhile, it was recently predicted [13-16] that for the SFIFS tunnel structures with very thin F layers one can, if there is a parallel orientation of the F layers magnetization, turn the junction into the π-phase state with the critical current inversion. If there is an antiparallel orientation of the F layers internal fields, the direction of the current is preserved but an enhancement of supercurrent can be observed at low enough temperature.

While the induced superconductivity of the F layer is intensively discussed, much less attention has been paid to a modification of the electron spectrum of the superconductor due to a leakage of magnetic correlation into a region near the SF interface. (An exception is a nonequilbrium case, that we do not consider here). Meanwhile, as a rule, the Curie temperature of a ferromagnet is by order of magnitude larger than the superconducting critical temperature T_C, and the magnetic correlations may be more robust then the superconducting ones, i.e., for SF hybrid structures a 'leakage of magnetism' into the S metal, as one can expect, should be quite strong [17,18].

2.   THE MODEL AND FORMALISM

Let us consider the bilayer of a massive superconducting ($d_S \gg \xi_S$) and a thin ferromagnetic ($d_F \ll \xi_F$) metals, with arbitrary transparency of the SF interface. Here $\xi_S = (D_S/2\pi T_C)^{1/2}$ and $\xi_F = (D_F/2H_e)^{1/2}$ stand for superconducting coherence lengths, $D_{S,F}$ are the diffusion coefficients, $d_{S,F}$ are the thicknesses of the S and F layers, respectively; H_e is the exchange energy of the ferromagnet. (Henceforth, we have taken the system of units with $\hbar = k_B = 1$.) We assume the 'dirty' limit for both metals, i.e., $\xi_{S,F} \gg l_{S,F}$ where $l_{S,F}$ are the electron mean free paths. All quantities are assumed to depend only on a single coordinate x normal to the interface surface of the materials. As is well know, the superconductivity of 'dirty' metals is usually described by the quasiclassical Usadel equations. It is convenient to take into account the normalization of the Green functions explicitly and to introduce modified Usadel functions Φ_S, Φ_F, defined by the relations $\Phi_S = \omega F_S / G_S$, $\Phi_F = \tilde{\omega} F_F / G_F$, etc. (see, e.g., Refs. [1-3]). Here $G_{\sigma\sigma'}(x,\omega)$ and $F_{\sigma\sigma'}(x,\omega)$, are the normal and anomalous Green functions, integrated over energy and averaged over the Fermi surface; they are defined in a standard way. Other notations are: $\tilde{\omega} = \omega + iH_e$, and $\omega \equiv \omega_n = \pi T(2n+1)$, $n = \pm 1, \pm 2, \pm 3, \ldots$ is Matsubara frequency. Then we can recast the equations in terms of these functions. For the S metal ($0 \le x \le d_S$) we have:

$$\Phi_S = \Delta_S + \xi_S^2 \frac{\pi T_C}{\omega G_S}[G_S^2 \Phi_S']', \qquad G_S = \frac{\omega}{(\omega^2 + \Phi_S \tilde{\Phi}_S)^{1/2}}, \qquad (1)$$

with the superconducting order parameter $\Delta_S(x)$ determined by the usual self-consistency equation. Here the prime denotes differentiation with respect to a coordinate x. For the F metal ($-d_F \le x \le 0$) we have

$$\Phi_F = \xi^2 \frac{\pi T_C}{\tilde{\omega} G_F}[G_F^2 \Phi_F']', \qquad G_F = \frac{\tilde{\omega}}{(\tilde{\omega}^2 + \Phi_F \tilde{\Phi}_F)^{1/2}}. \qquad (2)$$

We put for the F metal a vanishing value of the bare superconducting order parameter, while the pair amplitude $F_F \ne 0$ due to a proximity with the superconductor. The equations for the functions $\tilde{\Phi}_S$ and $\tilde{\Phi}_F$ have a form analogous to (1), (2). We write our formulas for the F metal, Eq. (2), using the effective coherence length of normal nonmagnetic metal with the diffusion coefficient D_F, $\xi = (D_F/2\pi T_C)^{1/2}$, to have a possibility to analyze both limits $H_e \to 0$ (SN bilayer) and $H_e \gg \pi T_C$.

The Eqs. (1), (2) should be supplemented with the boundary conditions in the bulk of the S metal and at the external surface of the F layer. Far from the SF interface for the S layer we have the usual boundary conditions in bulk of the S metal: $\Phi_S(\infty) = \Delta_S(\infty) = \Delta_0(T)$, where $\Delta_0(T)$ is the BCS value of the order parameter. At the external surface of the F metal $\Phi_F'(-d_F) = 0$. The relations at the SF interface we obtain [19] by generalizing the results of Kupriyanov and Lukichev [20] for the interface between two superconductors. There are two parameters which enter the model: $\gamma = \rho_S \xi_S / \rho_F \xi$ is the proximity effect parameter, which characterizes an intensity of superconducting correlations induced in the F layer on account of its proximity to the S layer, and vice versa, an intensity of magnetic correlation induced into the S layer on account of its proximity to the F layer ($\rho_{S,F}$ are the resistivities of the metals in the normal state); and $\gamma_{BF} = R_B / \rho_F \xi$ describes the electrical quality of the SF boundary, here R_B is the product of the SF boundary resistance and its area.

We draw attention to the feature important for further conclusions: *there is only one space scale - a respective superconducting coherence length, - that encounters into the differential equations (1) and (2). I.e., the coordinate dependencies of superconducting and magnetic properties of the system will have one and the same characteristic scale.*

The problem of the proximity effect for the SF structure with thin F metal, $d_F \ll (\xi_F, \xi)$, can be reduced to consideration of a boundary value problem for the S layer. Indeed, the differential equation (2) can be solved by iteration procedure with respect to the parameter d_F / ξ_F [1-3,19,21]. Then, introducing the effective boundary parameters, $\gamma_M = \gamma d_F / \xi$ and $\gamma_B = \gamma_{BF} d_F / \xi$, instead of γ and γ_B , the problem of the proximity effect for a massive superconductor with a thin F layer reduces to solving the equations for a semi-infinite S layer with specified boundary conditions. We analyze here the limits: (a) $\gamma_M \ll 1$, small strength of the proximity effect - low suppression of the order parameter in the S layer near the SF boundary, and (b) $\gamma_M \gg 1$, strong suppression of the order parameter near the SF boundary; we made no specific assumption about γ_B in the formulae below. If $\gamma_M \ll 1$, we obtain for $\Phi_S(\omega, x)$ the form

$$\Phi_S(\omega, x) = \Delta_0 \{ 1 - \gamma_M \beta \tilde{\omega} \frac{\exp(-\beta x / \xi_S)}{\gamma_M \beta \tilde{\omega} + \omega A(\omega)} \} \qquad (3)$$

$A(\omega) = \left[1 + \gamma_B \tilde{\omega} \left(\gamma_B \tilde{\omega} + 2\omega / \beta^2 \right) / (\pi T_C)^2 \right]^{1/2}$; $\beta = [(\omega^2 + \Delta_0^2) / \pi T_C]^{1/2}$. In the opposite limit, $\gamma_M \gg 1$, the behavior of $\Phi_S(\omega, x)$ near the SF boundary, $0 < x \ll \xi_S$, is given by

$$\Phi_S(\omega, 0) = B(T) \{ (\pi T_C + \gamma_B \tilde{\omega}) / \gamma_M \tilde{\omega} \} \qquad (4)$$

Here $B(T) = 2T_C [1 - (T/T_C)^2][7\zeta(3)]^{-1/2}$ and $\zeta(3) \cong 1.2$. An important feature of the results obtained for the SF structure is that the $\Phi_S(\omega, x)$ function for the S layer, Eqs. (3) and (4), directly depends on the exchange field; i. e. *the exchange correlation has been induced into the S layer.*

3.   SPIN-SPLITTING AT SF BOUNDARY

Comparing the results for an SF structure with those for an SN bilayer, one can find a fundamental aspect, that leads to new physical consequences; namely, the $\Phi_S(\omega, x)$ is *a complex function near the SF interface* [5,15,16,21]. As a result, a 'superconducting phase rotation' (a phase jump for our approximation of a thin F layer $d_F << \xi_F$) occurs at the SF interface. The most prominent consequence of this is that such phase variation provides the possibility to change the phase of superconducting order parameter from a 0 to a $\pi/2$ value even if there is no order parameter oscillation in the F layer.

To illustrate this, consider the S layer phase dependence on the exchange field at the SF interface ($x = 0$). Let us take, for simplicity, a structure with favorable for magnetic effects interface parameters: $\gamma_M >> 1$ and $\gamma_B = 0$. Then, as follows from Eq. (4), in the limit $H_e >> \pi T_C$ the superconducting correlation function, $F_S(\omega, 0) \sim \Phi_S(\omega, 0)$, acquires an additional $\pm \pi/2$ phase shift in comparison with the similar function for the SN bilayer. If, for example, we deal with a SFIFS junction with strong enough ferromagnetism of both magnetic layers (both electrodes have $\pi/2$ phase shift), that makes preferable the π phase superconductivity of the system for parallel directions of the exchange fields; for antiparallel magnetizations orientation and low temperature the critical current can be enhanced [13-16,21].

Parallel orientation of the layers magnetization. Let us present here an analytical consideration for the case of a vanishing interface resistance, $\gamma_B = 0$. For a strong suppression of the order parameter near SF boundary the critical current of the contact can be presented in the form:

$$j_C^P \approx (T/T_C) B_M^2(T) 2 \sum_{\omega} (\omega^2 - H_e^2)/\{\omega^2 (\omega^2 + H_e^2)^2\} ,$$ (5)

where $B_M(T) = B(T)\pi T_C/\gamma_M$, and in proceeding these relations, we have taken into account that value of Φ_S is small, $\Phi_S \sim \gamma_M^{-1}$. One can see, that if the exchange field is strong enough, namely $H_e >> \pi T_C$ the critical current changes its sign; or, in other words, the crossover of the junction from 0-phase state to the π-phase state takes place.

Antiparallel orientation of the layers magnetization. We have for SFIFS contact with $\gamma_B = 0$ and a strong proximity effect $\gamma_M >> 1$:

$$j_C^A \sim B_M^2(T) T/T_C \sum_{\omega>0} \omega^{-2} \times$$

$$\times \left[(\omega^2 - H_e^2 + B_M^2(T)/\omega^2)^2 + (2\omega H_e)^2 \right]^{-1/2}$$ (6)

One can see that if $0 < H_e \le \sqrt{B_M^2(T) - \omega^4}/\omega$ the critical current increases at low temperature, i.e. there is the region where $j_C^{AP}(H_e) > j_C^{AP}(0)$. As a result, for some values of H_e one can obtain the enhancement of the tunnel current in contrast to its suppression by the magnetic moments aligned in parallel.

On Fig. 1 we show the results of numerical calculations of the amplitude of the Josephson current for the case of weak, $\gamma_M \ll 1$, proximity effect and different quality of the S and F metals electrical contact versus the exchange field strength for parallel (solid curves) and antiparallel (dashed curves) mutual orientation of the electrodes' magnetizations.

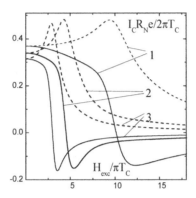

Figure 1. Critical current of an SFIFS tunnel junction for $T \ll T_C$ versus exchange energy at high S/F interface transparency, $\gamma_B = 0.1$ and weak proximity effect $\gamma_M = 0, 0.1,$ and 0.2 (curves 1, 2 and 3, respectively). Solid curves illustrate the case of parallel orientation of F layers magnetization; dashed curves illustrate the case of antiparallel orientation of F layers magnetization; R_N is the junction resistance in normal state.

Another feature of the SF boundary is that the gap $\Delta_S(x)$ is suppressed near the interface more strongly than in the SN case. This is not surprising, since one would expect that induced ferromagnetism suppresses superconducting order parameter at some distance into the S layer in excess of those for nonmagnetic normal layer. The curves on Fig. 2 illustrate the spatial dependencies of the induced exchange correlation in the superconductor for the case of a vanishing interface resistance.

The Green's functions for the S layer $G_{S\uparrow\uparrow}(\omega, x)$ and $G_{S\downarrow\downarrow}(\omega, x)$ for both spin subbands can be obtained using the solutions for the functions $\Phi_S(\omega, x)$ with $\tilde{\omega} = \omega + iH_e$ and $\tilde{\omega} = \omega - iH_e$, respectively. Performing the analytical continuation to the complex plane by substitution $\omega \to -i\varepsilon$ we calculate the spatial dependence of quasiparticle DOS for each spin subband $N_{S\uparrow}(\varepsilon, x) = ReG_{S\uparrow\uparrow}(\omega, x)$ and $N_{S\downarrow}(\varepsilon, x) = ReG_{S\downarrow\downarrow}(\omega, x)$. On Fig. 3 representative $N_{S\uparrow}(\varepsilon, x)$ dependences at different distances from the

SF interface are shown. The spin-splitting decreases with an increase of the distance from the boundary and vanishes in the bulk of the S layer (see curve 4 on Fig. 3) [17,22].

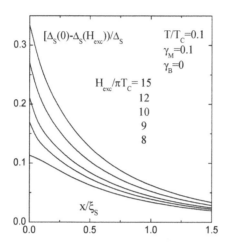

Figure 2. The difference of the superconducting order parameter in the S layer versus distance from the interface for SN and SF structures with the same boundary parameters ($\gamma_M = 0.1$, $\gamma_B = 0$), and different ferromagnetic field energy $H_{exc}/\pi T_C = 8, 9, 10, 12$ and 15.

Other important features, shown on Fig. 3, are the local states that appear inside the energy gap at the distances up to a few ξ_S from the SF boundary [17]. These intergap states are absent far from the SF interface, and also if $H_e = 0$. The local states are due to Cooper pairs breaking in the superconductor by the exchange-induced magnetic correlation. The region of their existence increases with the increasing of H_e, or increasing of pair breaking effects. In the absence of spin-flip (e.g., spin-orbit) scattering, the subgap bands accommodate quasiparticles with a definite ('up' or 'down') spin direction.

The influence of a ferromagnet on a superconductor is reflected in spin-splitting of the densities of states for spin-up and spin-down electrons, $N_{S\uparrow}(\varepsilon,x)$ and $N_{S\downarrow}(\varepsilon,x)$. This DOS splitting causes inhomogeneous quasiparticle spin density in the S metal, i.e., an effective magnetization $M_S(x)$ of the S layer. Fig. 4 illustrates the mechanism of proximity induced magnetization of the S layer. For $T \ll T_C$ all states below the Fermi level, ε_F, are filled, while all states above Fermi energy are empty. One can directly see from the figure that the S layer acquires a nonzero magnetic moment. This suggestion is confirmed by numerical calculations of $M_S(x)$ [17,18].

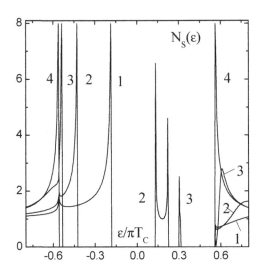

Figure 3. Normalized density of state for spin 'up' quasiparticles in the S layer of the SF sandwich for $\gamma_M = 0.1$, $\gamma_B = 0$ and $H_{exc} = 5\pi T_C$, and various distances from the SF interface: $x/\xi_S = $ 0, 1, 5, and 30 (curves 1, 2, 3, and 4, respectively).

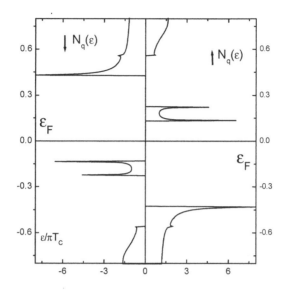

Figure 4. Quasiparticles DOS spin-splitting near Fermi level;
$H_e = 5\pi T_C$, $\gamma_M = 0.1$, $\gamma_B = 0.1$ and $x = \xi_S$

The induced magnetism of the S layer allows to conclude that the 'effective' thickness of a magnetic layer in a layered system may significantly differ from its 'true' (physical) thickness value. Indeed, recent experiments (see, e.g., [23]) revealed that the effective thickness of the magnetic layer in SF hybrid structures is usually much larger than its physical thickness. This is to be contrasted with the case of nonmagnetic spacer layers, where these two lengths are comparable.

4. CONCLUSION

We study the magnetic correlations acquired by a superconductor at the SF interface due to a proximity effect and find that an equilibrium exchange of electrons between the F and S metals results not only in proximity induced superconductivity of the F metal, as was widely discussed earlier, but in proximity induced magnetism of the S metal, too. The existence of the magnetic properties of the S metal is quite important for SF nanoscale structures and should be taken into account while comparing theoretical results with experimental data. To tackle the physics, we consider the limit of very thin magnetic layer. However, it is obvious that for proximity coupled S and F metals the leakage of magnetic correlation into a superconductor is a general phenomena and does not depend on thickness of a ferromagnetic metal. For proximity coupled SF hybrid structures both phenomena - induced superconductivity of the F metal and induced magnetism of the S counterpart - take place simultaneously and should be taken into account in self-consistent calculations.

The author would like to thank V. V. Ryazanov, A. Golubov, M. Kupriyanov, A. Buzdin, L. Tagirov and M. A. Belogolovskii for helpful discussions of some questions discussed in the report. E. A. Koshina is acknowledged for collaboration and performance of numerical calculations.

REFERENCES

[1] Golubov A.A., Kupriyanov M. Yu. (1989) *Josephson effect in SNINS and SNIS tunnel structures with finite SN boundary transparency.* Sov. Phys. JETP 69, 805-818.

[2] Golubov A.A., Houwman E. P., Gisbertsen J. G., Krasnov M., Flokstra J., H. Rogalla, and M. Yu. Kuprijanov (1995). *Proximity effect in superconductor-insulator-superconductor Josephson junctions: Theory and experiment.* Phys. Rev. B **51**, 1073-1089.

[3] Golubov A.A., Kupriyanov M. Yu.and Il'ichev E. *Current-phase relation in Josephson junctions.* to appear.

[4] Buzdin A. I., Bulaevskii L. N., and Panyukov S. V.,(1982) *Critical current oscillations vs exchange field and thickness of a ferromagnetic metal in Josephson S-F-S contact.* Sov. Phys. JETP Lett. **35**, 178-182.

[5] Buzdin A. I., Kupriyanov M. Yu. (1991). *Josephson contact with ferromagnetic layer.* Sov.Phys.JETP Lett. **53**, 321-325.

[6] Ryazanov V. V., Oboznov V. A., Rusanov A. Yu. ,Veretennikov A. V., Golubov A. A., and Aarts J. (2001) *Coupling of two superconductors through a ferromagnet: evidence for a π -junction.* Phys. Rev. Lett. **86**, 2427-2430..

[7] Ryazanov V. V., Oboznov V. A. , Veretennikov A. V., and Rusanov A. Yu., (2002). *Intrinsically frustrated superconducting array of superconductor-ferromagnet-superconductor junction.* Phys. Rev. B **65**, 020501(R).

[8] Blum Y., Tsukernik A., Karpovski M., and Palevski A., (2002). *Oscillations of the superconducting critical current in Nb-Cu-Ni-Cu-Nb junctions.* Phys. Rev. Lett. **89**, 187004..

[9] Kontos T., Aprili M., Lesueur J., Genet F., Stephanidis B., and Boursier R., (2002) *Josephson junction through a thin ferromagnetic layer: negative coupling.* Phys. Rev. Lett. **89**, 137007.

[10] Kontos T., Aprili M.,. Lesueur J., and Grison X, (2001) *Inhomogeneous superconductivity induced in a ferromagnet by proximity effect.* Phys. Rev. Lett. **86**, 304-307.

[11] Mühge Th., Garif'yanov N. N., Guryunov Yu. V., Khaliulin G. G. , Tagirov L.R., Westerholt K., Garifullin I. A. , and Zabel H. (1996). *Possible origin for oscillatory superconducting transition temperature in superconductor/ferromagnet multilayers.* Phys. Rev. Lett. **77**, 1857–1860.

[12] Izumov Yu. A., Proshin, and M. G. Khusainov (2002). *Competition between superconductivity and magnetism in ferromagnet/superconductor heterostructures*. Usp. Fiz. Nauk, **45**, 109 -154.

[13] Bergeret F. S., Volkov A. F., and . Efetov K. E. (2001). *Enhancement of the Josephson current by an exchange field in superconductor-ferromagnet structures*. Phys. Rev. Lett. **86**, 3140-3143.

[14] Krivoruchko V. N. and Koshina E. A..(2001). *From inversion to enhancement of the dc Josephson current in S/F-I-F/S tunnel structures*. Phys. Rev. B **64**, 172511.

[15] Golubov A. A., Kupriyanov M. Yu., and Fominov Ya. V., (2002). *Critical current in SFIFS junctions*. JETP Lett. **75**, 190-194.

[16] Koshina E. A., Krivoruchko V. N.,.(2003) *Critical current of nonsymmetrical SFIFS tunnel structures*. Low. Temp. Phys. **29**, #8 (2003).

[17] Krivoruchko V. N. and Koshina E. A..(2002). *Inhomogeneous magnetism induced in a superconductor at a superconductor-ferromagnetic interface*. Phys. Rev. B **66**, 014521.

[18] Bergeret F. S., Volkov A. F., and Efetov K. E. (2003). *Induced ferromagnetism due to superconductivity in superconductor-ferromanet structures*. arXiv:cond-mat/0307468.

[19] Koshina E. A. and Krivoruchko V. N.(2000). *Spin polarization of quasiparticle states in S/F structures with a finite transparency of the SF interface*. Low. Tem. Phys. **26**, 115-120.

[20] Kuprijanov M. Yu..and Lukichev V. F., (1988). *Effect of boundary transparency on critical current in dirty SS$^{\prime}$S structures*. Sov. Phys. JETP **67**, 1163-1172.

[21] Koshina E. and Krivoruchko V.(2001). *Spin polarization and π - phase state of the Josephson contact: Critical current of mesoscopic SFIFS and SFIS junctions*. Phys. Rev. B **63**, 224515.

[22] Fazio R. and Lucheroni C. (1999). *Local density of states in superconductor-ferromagnetic hybrid systems*. Europhys. Lett. **45**, 707-713.

[23] Ogrin F. Y., Lee S. L., Hillier A. D, Mitchell A, and Shen T.-H.,(2000). *Interplay between magnetism and superconductivity in Nb/Co multilayers*. Phys. Rev. B **62**, 6021-6026.

RECENT EXPERIMENTAL RESULTS ON THE SUPERCONDUCTOR/FERROMAGNET PROXIMITY EFFECT

I.A. Garifullin[a], M.Z. Fattakhov[a], N.N. Garif'yanov[a], S.Ya. Khlebnikov[a], D.A. Tikhonov[a], L.R. Tagirov[b,c], K. Theis-Bröhl[d], K. Westerholt[d], H. Zabel[d]

[a] *Zavoisky Physical-Technical Institute, Russian Academy of Sciences, 420029 Kazan, Russian Federation. E-mail: ilgiz_garifullin@mail.ru*
[b] *Kazan State University, 420008 Kazan, Russian Federation*
[c] *Gebze Institute of Technology, 41400 Gebze-Kocaeli, Turkey*
[d] *Institut für Experimentalphysik/Festkörperphysik, Ruhr-Universität Bochum, 44780 Bochum, Germany*

Abstract: Interaction of ferromagnetism and superconductivity in the superconductor/ferromagnet layered heterostructures is discussed on the basis of the own recent experimental results. These results are: *(i)* the observation of a re-entrant behavior of the superconducting state in Fe/V/Fe trilayers when plotting the superconducting transition temperature T_c as a function of the Fe layer thickness; *(ii)* modification of the ferromagnetic state under the influence of superconductivity in epitaxial V/Pd$_{1-x}$Fe$_x$ bilayers.

Key words: superconductivity, ferromagnetism, proximity effects, multilayers

1. INTRODUCTION

The mutual interaction of two antagonistic phenomena like magnetism and superconductivity in bulk materials remains an exciting topic in the physics of superconductivity (see, e.g., [1] and review [2]). With the advance of nanofabricated heterostructures and multilayers the interaction of superconducting (S) and ferromagnetic (F) layers (the superconductor/ferromagnet proximity effect) raises new questions and

B. Aktaş et al. (eds.), Nanostructured Magnetic Materials and their Applications, 121–144.

opens up a miriad of possibilities for the combination of both materials. Although the basic superconducting properties of S/F systems have been studied for quite some time [3] and in spite of recent activities in this field (see, e.g., reviews [4,5] and Refs. [6,7,8,9,10]), until now the S/F proximity effect is far from being quantitatively understood. The superconducting transition temperature T_c of S/F layered sample should be strongly suppressed because of the penetration of Cooper pairs into the F-layer. The exchange field in the F-layer, H_{ex}, will try to align spins of electrons in a Cooper pair, thus leading to a pair-breaking effect. In order to get a feeling of how strong this suppression is we will consider for simplicity the F/S bilayer in the Cooper limit, which is sufficient for a rough estimate. The Cooper limit means that the thicknesses of the S- and F-layers (d_S and d_F, respectively) are smaller than the corresponding coherence lengths (ξ_S and ξ_F). In case of a perfect S/F interface transparency electrons of both sub-layers are intermixed, and the effective exchange field acting on the Cooper pairs is given by its value averaged over the total thickness of a S/F bilayer:

$$H_{ex}^{eff} = H_{ex} \frac{d_F}{d_S + d_F}. \tag{1}$$

Then we consider the Clogston limit for a spin-split conduction band, i.e. the field $H \sim \Delta/g\mu_B$ (where g is the g-value, μ_B is the Bohr magneton and $\Delta = 1.76 k_B T_c$ is the superconducting energy gap), which completely quenches superconductivity. Taking, for example, $H_{ex} \sim 1$ eV (F=Fe) and $T_c \sim 10$ K (S=Pb or Nb) we obtain $d_F \sim 2 \cdot 10^{-3} d_S$. Consequently, one monolayer of Fe is already enough to suppress T_c to zero of an S-layer with the thickness of about 700 Å. Thus, superconductivity should be strongly suppressed by ferromagnetism.

In fact, this is not the case. The T_c suppression is much weaker. This happens due to two reasons.

First, in some systems, e.g., in Fe/Nb [11] system, an intermediate layer between the S- and F-layers exists at the interface. This layer appears due to interdiffusion. It prevents the direct contact between superconductivity and ferromagnetism.

Second reason is the restriction of quantum mechanical transparency of the S/F interface due to the exchange splitting of the conduction band of the F-layer [8]. This is the classical quantum mechanical problem of reflection of electrons from the interface of two metals due to the Fermi momentum mismatch. It is obvious that two electrons with opposite spins constituting a

Cooper pair never match simultaneously the Fermi momenta of exchange-split sub-bands of a ferromagnet.

Another interesting feature of the S/F proximity effect is the oscillation of the superconducting transition temperature as a function of the F-layer thickness.

One of the reasons for the oscillation of the superconducting transition temperature is the so-called Josephson π-coupling in S/F/S structures. Due to the oscillation of the superconducting pairing function in the F-layer the superconducting pairing function from both sides of the F-layer may have opposite phases at certain F-layer thickness, i.e., the phase difference between neighboring S-layers may be equal to π [12,13,14,15]. Radović *et al.* [13] concluded from their calculations that T_c for the system with π-coupling most probably is higher than that for the system with no phase difference. However, the π-coupling is not the only reason for an oscillation of T_c.

Another reason is the interdiffusion at the interface [11]. Properties of this layer change as the F-layer thickness increases. Thus, its influence on the superconductivity is non-monotonic.

The third reason for the oscillations of T_c is the realization of the so-called Larkin-Ovchinnikov-Fulde-Ferrel (LOFF) state [16]. We believe that we observed the oscillation of T_c in the Pb/Fe system due to this reason [8].

Why does T_c oscillate due to the realization of the LOFF state? One of the best qualitative descriptions of the physics behind this effect has been presented by Demler *et al.* [17]. Imagine a Cooper pair being transported across the S/F interface. In the F-region we have the exchange-split conduction band. Therefore the spin-up electron of the Cooper pair will occupy spin-up sub-band of ferromagnet, while the spin-down electron will occupy spin-down sub-band. So, the spin-up electron in the pair lowers its potential energy by exchange field energy E_{ex} in the F-layer, while the spin-down electron raises its potential energy by the same amount. In order for each electron to conserve its total energy, the spin-up electron must increase its kinetic energy, while the spin-down electron must decrease its kinetic energy. This means that absolute values of momenta of the electrons in pair become unequal. As a result, a pair entering from the S-region into the F-region will change its total momentum from zero to a finite value. Therefore the amplitude of the pairing function in the F-layer will oscillate in space. In the normal metal the pairing function decreases exponentially. In the ferromagnetic metal, in addition, it oscillates in space. It is theoretically explained [13,18,19] that due to these oscillations of the pairing function the $T_c(d_F)$-dependence is non-monotonous, and may have an oscillatory character with an oscillation period on the length scale of the magnetic stiffness length $\xi_I = v_F/E_{ex}$ (where v_F is the Fermi velocity). The oscillatory character of $T_c(d_F)$ can qualitatively be traced back to the propagating character of the superconducting pairing wave function in the ferromagnet. If the thickness of the F-layer is smaller than the penetration depth, the pairing wave function transmitted through the S/F interface into the F-layer will interfere with the wave reflected from the opposite surface of the ferromagnet. As a result, the flux of the pairing wave function which crosses the S/F interface, is modulated as the thickness of the F-layer d_F is changed. Therefore the coupling of the electron systems of ferromagnet and superconductor will be modulated and T_c will oscillate with d_F. If the interference at the S/F interface is essentially constructive (this corresponds to a minimal jump of the pairing function amplitude at the S/F interface), the coupling is weak, and one expects T_c to be maximal. When the interference is destructive, the coupling is maximized and $T_c(d_F)$ should be minimal. For a small enough thickness of the S-layer, $T_c(d_F)$-behavior may be re-entrant, i.e. superconductivity can vanish for a certain range of d_F and then revive for larger d_F-values. Actually, this re-entrant behavior is difficult to realize experimentally and until now it has not been observed.

Using the theoretical model calculations [19] as a guideline, one can extract the important parameters enabling an observation of the re-entrant behavior. A system is required with a large electron mean free path in the F- as well as in the S-layer. A high quantum-mechanical transparency of the S/F interface is important, and the interface should be flat without introducing much diffuse scattering of the electrons. Combining these requirements, a system with high structural quality and mono-atomically sharp interfaces would be optimal. This may be realized in the Fe/V system. This system can be grown with nearly perfect interfaces [20]. In this article we present the first experimental observation of the re-entrant behavior of the superconducting transition temperature in the $T_c(d_{Fe})$-curve of Fe/V/Fe trilayers. Preliminary results have been published in [21].

Apart from the influence of ferromagnetism on superconductivity one can expect the realization of the S/F proximity in the backward direction, i.e., a modification of the F-state under the influence of superconductivity. In bulk magnetic alloys and intermetallic compounds an additional antiferromagnetic term in the Ruderman-Kittel-Kasuya-Yosida interaction between magnetic ions arises from the superconducting correlations in the Cooper condensate. This was predicted theoretically [22] and later on observed experimentally in dilute magnetic alloys by electron spin resonance [23]. In several intermetallic compounds at low temperatures this interaction leads to a domain-like magnetic structure which may coexist with the superconducting state (see, e.g., review [2]). As to the layered S/F systems, the possibility of the formation of a domain-like magnetic structure in a system consisting of a bulk superconductor with a thin ferromagnetic metallic film on its surface was considered theoretically for the first time by Buzdin and Bulaevskii [24]. They calculated the energy of two systems, DS/S and F/S, and concluded that the domain state (DS) will be the ground state for the magnetic layer thicknesses smaller than a certain critical thickness. The tendency for a reconstruction of the ferromagnetic order was observed experimentally in epitaxial Fe/Nb bilayers upon transition to the superconducting state [25]. However, estimates using the theory by Buzdin and Bulaevskii showed that the effect in the Fe/Nb system should occur at the Fe-layer thickness by order of magnitude smaller than observed experimentally. Later on Bergeret *et al.* [26] studied theoretically the possibility of a non-homogeneous magnetic order (cryptoferromagnetic state) due to superconductivity in heterostructures consisting of a bulk superconductor and a ferromagnetic thin layer. They derived a phase diagram which distinguishes the cryptoferromagnetic and ferromagnetic states and discussed the possibility of an experimental observation of the cryptoferromagnetic state in different materials. In particular, they concluded that because of the large magnetic stiffness constant and strong internal exchange field in pure iron the cryptoferromagnetic state is hardly possible in the Fe/Nb structure. Thus, it may be concluded that the tendency for a reconstruction of the ferromagnetic order in the Fe layer observed experimentally in the Fe/Nb films [25], might be caused by discontinuous Fe layers in the thickness range where this effect was observed. Estimates performed by Bergeret *et al.* [26] show that the transition from the ferromagnetic to the cryptoferromagnetic state should be observable, if the exchange field and the magnetic stiffness constant would be an order of magnitude smaller than in pure Fe. A possible system for the experimental observation of the cryptoferromagnetic state in S/F multilayers is the $V/Pd_{1-x}Fe_x$ system due to its low and tunable Curie temperature. Ferromagnetic resonance (FMR) measurements of bulk single crystals of $Pd_{1-x}Fe_x$ [27] suggest that these alloys are convenient systems for FMR studies because of their narrow resonance lines. In addition, it is well known (see, e.g., the review [28]) that at any Fe concentration $Pd_{1-x}Fe_x$ alloys order ferromagnetically. In this article we present our first results on FMR measurements for $V/Pd_{1-x}Fe_x$ single crystalline epitaxial bilayers. Preliminary results have been published in [29]. We show an example where the saturation magnetization determined from the FMR spectra decreases as the temperature lowers below T_c. As a possible explanation for this decrease we suggest the formation of the cryptoferromagnetic state in the $Pd_{1-x}Fe_x$ layer due to S/F proximity effect.

The paper is organized as follows: in Section 2 we present our experimental results concerning the new effect in the influence of ferromagnetism on superconductivity, namely the re-entrant superconductivity in the Fe/V/Fe trilayers. In Section 3 we describe our results on modification of the F-state under the influence of superconductivity in the $V/Pd_{1-x}Fe_x$ epitaxial bilayers. The main results are summarized in Section 4.

2. RE-ENTRANT SUPERCONDUCTIVITY IN THE SUPERCONDUCTOR/FERROMAGNET LAYERED SYSTEM

2.1 Sample Preparation and Characterization

2.1.1 Sample Preparation

First we prepared a series of Fe/V/Fe trilayers with fixed Fe thickness $d_{Fe} = 50$ Å and with a variable V thickness between 200 and 1200 Å (series 1 in Table 1) by rf sputtering.

Table 1. Parameters of the Fe/V/Fe samples. Given are the thickness of the V layer d_V and of the Fe layer d_{Fe}, the residual resistivity ratio *RRR*, the mean free path of conduction electrons in the superconducting layer l_S, the superconducting coherence length in the V layer ξ_S, and the ratio between mean free path of conduction electrons in a F-layer and magnetic stiffness length l_F/ξ_I.

Series	$d_V(Å)$	$d_{Fe}(Å)$	RRR	$l_S(Å)$	$\xi_S(Å)$	l_F/ξ_I
1	200-1200	50	3-6			
2	312	1-28	4.0	40	75	1.0
3	292	1-28	4.1	40	75	1.0
4	308	1-28	4.7	50	80	1.2
5	290	1-28	5.0	50	80	1.2
6	339	2-29	10	120	125	1.2

Pure Ar (99.999 %) at pressures of 5×10^{-3} mbar was used as a sputter gas. Pure V (99.99 %) and Fe (99.99 %) targets were used for the metallic layers deposition. The deposition rates of 0.3 Å/sec for Fe and of 0.5 Å/sec for V were found to be optimal for the structural properties of the films. Then we prepared four series of Fe/V/Fe films with different Fe thickness and fixed V thickness on Al_2O_3 $(11\bar{2}0)$ substrates (series 2-5 in Table 1). Series 2 and 3 were prepared at a substrate temperature of 300° C, whereas for series 4 and 5 we used a slightly lower substrate temperature of $260\,^{o}$ C. The series 6 of Fe/V/Fe trilayers was prepared on high quality MgO (001). This film series was deposited under better vacuum conditions and after a much longer pre-sputtering process to remove any gaseous impurities from the V-target.

In order to get reliable dependencies $T_c(d_{Fe})$ or $T_c(d_V)$, we prepared series of 10 or 16 Fe/V/Fe trilayers with different d_{Fe} (d_V) at constant d_V (d_{Fe}) within each run in our experimental setup. During the deposition of the V-layer in the series with $d_V = const$, the substrates were arranged on the periphery of a circularly shaped substrate holder. During deposition the substrate holder was rotated 4 times by $90\,^{o}$ to ensure a homogeneous V-thickness within each series. As checked by small angle X-ray reflectivity with calibration samples, the thickness variation of V within the series is below 1%. The variation of the Fe-thickness d_{Fe} within the series 2-6 (see Table 1) was achieved by placing the circle of substrates in an asymmetric position relative to the center of the Fe-target and using the natural gradient of the sputtering rate in the discharge. In a final step all samples were covered by a protective Al_2O_3 cap layer of 50 Å thickness.

2.1.2   X-Ray Characterization

For a structural characterization of the samples we used small-angle X-ray reflectivity and out-of-plane Bragg scans. A fit of the reflectivity curves using the Parratt formalism [30] gave an interface roughness of about 10 Å for series 2 and 3 and definitely smaller values of about 6 Å for the series 4-6. The Bragg scans revealed a good crystalline structure with pure (110) texture of V and Fe on a-plane Al_2O_3 substrates and (001) texture on MgO substrates. The rocking width of the V (110) reflection was determined to be about $3\,^{o}$ and for the V (001) peak it was about $2\,^{o}$.

2.1.3   Electrical Resistivity

The analysis of the electrical resistivity showed that the residual resistivity ratio RRR of our samples from series 2-5 varied between 4 and 5 (see Table 1). For the sample series 6 (on MgO) we obtained a definitely higher RRR-ratio of about 10. This reflects the better structural quality of this series with less grain boundaries and impurities and better interface quality. With the BCS coherence length of V ($\xi_0 = 440$ Å) and estimating the mean free path of the conduction electrons l_S in the V-layer from the RRR-values using the Pippard relations [31], we calculated the superconducting coherence length $\xi_S = \sqrt{\xi_0 l_S / 3.4}$. The values are listed in Table 1.

2.1.4   Magnetization

Measurements of the ferromagnetic saturation magnetization by a SQUID magnetometer have been performed at T=20 K for all samples. Assuming that the saturation magnetization of the Fe layer does not depend on d_{Fe}, we calculated the thickness of the Fe layers. The thicknesses thus determined are slightly smaller than those obtained from a fit of the X-ray reflectivity scans, indicating the loss of about 2 atomic monolayers of Fe, probably because of interdiffusion of Fe and V atoms at the interface and oxidation of the top Fe-layer during cooling of the substrates down to room temperature before depositing the Al$_2$O$_3$ cap layer. The thickness d_{Fe} given in the figures below is that derived from the saturation magnetization.

2.1.5   Superconducting Transition Temperature

The superconducting transition temperature was measured resistively in a standard four-terminal configuration using a He^4 cryostat for the temperature range above 1.3 K. For the samples with lower transition temperature a He^3 or He^3/He^4 cryostat was employed. In addition, AC magnetic susceptibility measurements were used to determine T_c for all samples.

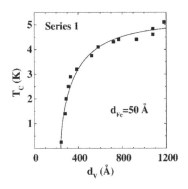

Figure 1. Superconducting transition temperature vs V thickness at fixed iron thickness $d_{Fe} = 50 \mathring{A}$ for series 1 of Fe/V/Fe trilayers. The drawn curve is a theoretical fit with the following parameters: $\xi_{BCS} = 440 \mathring{A}$, $\xi_S = 75 \mathring{A}$, $T_m = 1.6$, and $N_F v_F / N_S v_S = 0.27$.

The dependence of $T_c(d_V)$ at fixed iron thickness, d_{Fe} =50 Å is shown in Fig. 1. The shape of the $T_c(d_V)$ curve is conventional: with decreasing V-thickness, T_c decreases slowly at large d_V and then drops sharply to zero when d_V approaches the critical thickness $d_V^{crit} \approx 240$ Å. Similar curves have been observed for this system by Koorevaar *et al.* [32] before. We can fit these data (drawn solid line in Fig. 1) using the theory developed in [19] (see below).

The essential results of the present paper are represented in Figs. 2-4, where we have plotted $T_c(d_{Fe})$ for five different trilayer series with variable d_{Fe}.

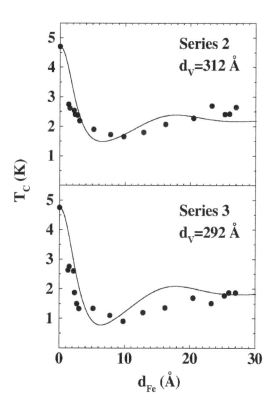

Figure 2. Superconducting transition temperature vs Fe thickness at two fixed V thicknesses for the Fe/V/Fe series 2 and 3. The drawn line is a theoretical curve with the parameters given in Table 1 and in the text.

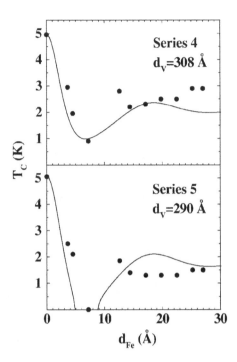

Figure 3. Superconducting transition temperature vs Fe thickness at two fixed V thicknesses for the Fe/V/Fe series 4 and 5. The drawn line is a theoretical curve with the parameters given in Table 1 and in the text.

2.1.6   Analysis and discussion

We begin the discussion with the series 2-5 prepared on Al_2O_3. A common feature for all these curves is a sharp initial drop in $T_c(d_{Fe})$, a minimum at about $d_{Fe} \simeq 10$ Å and a saturation at higher d_{Fe}. The depth of the minimum increases with decreasing d_V and with decreasing the interface roughness. Actually, for the sample with $d_{Fe} = 7$ Å from the series 5 we do not observe a superconducting transition in the temperature range down to 30 mK. Assuming that below this temperature superconductivity does not appear, we may classify the $T_c(d_{Fe})$-behavior in series 5 as a re-entrant superconductivity. For the Fe/V/Fe series 6 we observe definitely clear re-entrant behavior of T_c (Fig. 4). These measurements were performed in a He^3 cryostat down to 0.35 K.

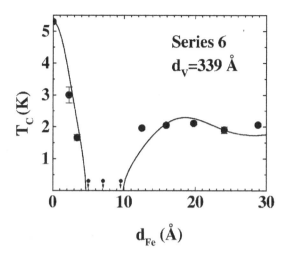

Figure 4. Superconducting transition temperature vs Fe thickness at fixed V thickness for the Fe/V/Fe series 6. The drown line is a theoretical curve with the parameters given in Table 1 and in the text.

We have fitted the $T_c(d_V)$ and $T_c(d_{Fe})$ dependence by the theory [19]. The fit of the $T_c(d_{Fe})$-curve has been performed in a similar way as in [8] and is shown in Figs. 2 and 3 by the solid lines. The fit parameters used in these figures are $\xi_S = 75 - 80$ Å, as estimated from our resistivity data (see Table 1); the transparency parameter $T_m = 1.6$ and the ratio $N_F v_F / N_S v_S = 0.27$ (N_F and N_S denote the electronic density of states at the Fermi energy, v_F and v_S are the Fermi velocities in the F- and S-layers) were fixed by a fit to the data in Fig. 1 (see details in Ref. [8]). It is necessary to note that the $T_c(d_V)$-curve is not sensitive to the ratio l_F/ξ_I, because the thickness $d_{Fe} = 50$ Å physically corresponds to an infinite thickness of the F-layer. The magnetic stiffness length $\xi_I = 6.8$ Å was obtained from a fit of Figs. 2 and 3. The fitting procedure indicated that the shape of the $T_c(d_{Fe})$-curve is qualitatively unaffected by slightly different ξ_S and that the increased l_F/ξ_I ratio is essential for the observation of the re-entrant behavior in Figs. 3 and 4.

We stress that a rough interface of the F-layer has the tendency to suppress a possible re-entrant behavior in two ways. First, because a smaller electron mean free path in the ferromagnet, l_F, leads to a faster decay of the superconducting pair density in the ferromagnet and, second, because the interface roughness will introduce random phase shifts in the interfering waves and thus will smear out the interference pattern and the oscillation amplitude in $T_c(d_{Fe})$.

Principally, the $T_c(d_{Fe})$-behavior for the Fe/V/Fe-series 5 can be characterized as re-entrant, but the agreement with the theoretical curve is not completely satisfactory. In accordance with the theory [19] there are two possibilities to improve the conditions for observing the re-entrant superconductivity further. First, one can try to increase the mean free path l_F. However, this seems not very promising because at $d_{Fe} \sim 10$ Å, where the superconductivity disappears, l_F will be strongly limited by the roughness and thickness of the F-layer. The second possibility, namely an increase of the mean free path of the conduction electrons in the S-layer with a corresponding increase of the superconducting coherence length ξ_S, seems more promising. This requirement is just met by our series 6 (see Table 1) prepared on MgO with a definitely better structural quality and higher *RRR*-value. For this set of samples we obtained clear re-entrant superconductivity (Fig. 4). From the *RRR*-value for this series we estimate a mean free path of the conduction electrons in the S-layer $l_S \sim 120$ Å and a corresponding coherence length $\xi_S \sim 125$ Å. For the theoretical fit of the $T_c(d_{Fe})$-curve in Fig. 4 we used the parameters shown in Table 1 and the same values of T_m, ξ_I and $N_F v_F / N_S v_S$ as for series 2-5.

From the transparency parameter $T_m \simeq 1.6$, derived from our fit, the average quantum mechanical transmission coefficient $\overline{T}$ [8] can be estimated as $\overline{T} = T_m / (1 + T_m) \simeq 0.6$. This value of $\overline{T}$ is about twice as large as the $\overline{T}$-value for the Pb/Fe interface obtained earlier [8]. This relatively high transparency of the Fe/V interface is also essential for observing re-entrance behavior, since it ensures a strong interaction of the superconducting and ferromagnetic electron systems. A highly transparent interface between a superconductor and a ferromagnet is difficult to achieve, since matching of the Fermi momentum at the S/F interface can occur at best for one spin direction in the ferromagnet.

3.   POSSIBLE RECONSTRUCTION OF THE FERROMAGNETIC STATE UNDER THE INFLUENCE OF SUPERCONDUCTIVITY IN EPITAXIAL V/Pd$_{1-x}$ Fe$_x$ BILAYERS

3.1   Sample Preparation and Characterization

3.1.1   Sample Preparation

The Pd$_{1-x}$ Fe$_x$/V and V/Pd$_{1-x}$ Fe$_x$ bilayers were grown on MgO (001) substrates in a molecular beam epitaxy (MBE) system (base pressure ~ $5 \cdot 10^{-11}$ mbar). During the evaporation the background pressure was below 10^{-9} mbar. The MgO (001) substrates were annealed in the growth chamber at 1000° C for 0.5 h prior to the evaporation in order to desorb impurities and to create a well ordered surface. Then the substrates were cooled to the desired temperature. For the Pd$_{1-x}$ Fe$_x$/V/MgO (001) samples (samples 1-4) the substrate temperature during the evaporation of the first V layer was 550° C. The evaporation rates of 0.1 Å/s were found to be optimal for the growth of high quality single crystalline V (001) films. Pd$_{1-x}$ Fe$_x$ (001) films were prepared using two sources. Fe was evaporated from an effusion cell providing a flux of high stability, while Pd was evaporated by an electron beam gun. The substrate temperature during the preparation of the Pd$_{1-x}$ Fe$_x$ layer was 200° C. The deposition rate of Pd was 0.27 Å/s and the Fe deposition rate varied between 0.027 Å/s and 0.009 Å/s. For the V/Pd$_{1-x}$ Fe$_x$/Pd/MgO (001) samples (samples 5 and 6) a buffer Pd (001) layer of about 100 nm was grown on MgO. The substrate temperature during the preparation of the Pd buffer layer was 450° C for the sample 5, and 400° C for the sample 6. The substrate temperature during the evaporation of the Pd$_{1-x}$ Fe$_x$ (001) layer was 450° C and 400° C for the samples 5 and 6, respectively. The V layer was deposited using a substrate temperature of 300° C and the same evaporation rate given above. In both cases the bilayers (except the sample 1) were covered by a protective layer of Pd with a thickness of a few tens of angstroms. The substrate temperature during the growth of the protective layer was 200° C. The growth rates were measured by a quartz-crystal monitor. The final thickness of the Pd$_{1-x}$ Fe$_x$ layer was determined by the evaporation time. The quality of the substrate and of each layer was always controlled by in-situ reflection high-energy electron diffraction characterization to ensure a high-quality growth of our samples. Each sample was prepared separately, but with identical growth conditions. The sample holder was rotated during the deposition to ensure a homogeneous film thickness. Six samples with the experimental parameters summarized in Table 2 have been investigated by FMR.

Table 2. Experimental parameters of the studied samples. Given are the thickness of the V layer d_V and of the Pd and $Pd_{1-x}Fe_x$ layers $d_{Pd} + d_{Pd-Fe}$ obtained from the fit of the small angle reflectivity scans, the Curie temperature T_{Curie} and the thickness of the magnetic $Pd_{1-x}Fe_x$ layer obtained from the SQUID magnetization measurements, the residual resistivity ratio *RRR*, the superconducting transition temperature T_c, $4\pi M_{eff}$ and K_1/M values obtained from FMR measurements.

Sample	d_V	$d_{Pd} + d_1$	T_{Curie}	d_{Pd-Fe}	RRR	T_c	$4\pi M_{eff}$	K_1/M
	(Å)	(Å)	(K)	(Å)		(K)	(kG)	(Oe)
1	372	44	250	44	4.7	4.0	3.9	15.8
2	400	62	100	12	5.0	4.2	2.6	65.0
3	393	62	100	3.0	4.0	3.8	1.7	80.0
4	470	68	90	8.0	4.6	4.0	-	-
5	410	1030	120	9.0	4.0	3.7	-	-
6	370	1010	100	10	4.5	3.1	1.8	72.0

3.1.2   X-ray Characterization

X-ray reflectivity measurements were performed ex-situ using the 1.5 kW X-ray generator with a Mo anode (λ = 7.09 Å) and a Si (111) monochromator. In reflectivity scans well resolved oscillations were clearly seen. Fits gave an interface roughness of less than 4 Å, indicating the high interfacial quality of our samples. The film thicknesses, as obtained from the fits to the X-ray data are given in Table 2. The thicknesses d_{Pd} of the top Pd layer in case of $Pd/Pd_{1-x}Fe_x/V/MgO$ samples and of the Pd buffer layer in case of $V/Pd_{1-x}Fe_x/Pd/MgO$ samples are shown in Table 2 together with d_{Pd-Fe} as $d_{Pd} + d_{Pd-Fe}$. A typical radial Bragg scan covering the angle range of the V (002) and $Pd_{1-x}Fe_x$ (002) peaks reveals the (001) texture of both, the V layer and $Pd_{1-x}Fe_x$ layer for all studied samples.

3.1.3   Superconducting Transition Temperature and Upper Critical Magnetic Field

The superconducting transition temperature T_c and the upper critical field H_{c2} were measured resistively using a standard four-terminal configuration with the current and voltage leads attached to the samples with the silver paint. T_c and H_{c2} were defined as the midpoint of the superconducting transition. The T_c values are presented in Table 2. For all

samples the superconducting transition is very sharp with a transition width
of the order of 0.1 K.

3.1.4   Electrical Resistivity

The residual resistivity ratio RRR=R(300K)/R(T_c) for the samples listed
in Table 1 varies between 4 and 5. The corresponding residual resistivity
values (from 4.6 to 6.1 $\mu\Omega \cdot cm$) allow us to estimate the electron mean-free
path l_S for V with the Pippard relations [31]. This relations permit an
estimate of l_S from the low temperature resistivity ρ and the coefficient of
the electronic specific heat γ. For V, using $\gamma = 9 mJ/moleK^2$ and
$v_F = 3 \cdot 10^7$ cm/s [33], we find l_S-values between 40 and 50 Å.

3.1.5   Magnetization

Magnetization measurements using a SQUID magnetometer were
performed with the film surface parallel to the direction of the magnetic
field. The temperature dependence of the magnetization have been measured
at a small magnetic field of H = 10 *Oe* for all samples. For the precise
determination of the ferromagnetic magnetization the correction of the
substrate contribution to the magnetic moment of the samples is very
important. The measurements of the temperature dependence of the
magnetization of the MgO substrates used in the present study showed that
at temperatures below 4 K the magnetic susceptibility of the substrates starts
to increase strongly. The temperature dependencies of the magnetization in
the ferromagnetic state after subtraction the substrate contribution are shown
in Fig. 5 for two samples. They are typical for bulk $Pd_{1-x} Fe_x$ alloys (see,
e.g., [34]).

The values obtained for the ferromagnetic Curie temperature T_{Curie} are
presented in Table 2. These values of T_{Curie} allow us to estimate the Fe
concentration in $Pd_{1-x} Fe_x$ alloy layers of our samples using the data for
$T_{Curie} = f(x)$ from the review by Nieuwehuys [28]. Values of x determined
in such a way lie between 0.03 and 0.1. Magnetization values for T=0 K
corresponding to these concentrations, give the F-layer thickness d_{Pd-Fe}
from our SQUID magnetization data. These values are also shown in Table
2. The difference observed between the film thicknesses determined by X-
rays and the magnetization measurements are due to the fact that for X-rays
there is basically no electron density contrast between Pd and $Pd_{1-x} Fe_x$.
Therefore with X-ray reflectivity measurements the sum of both layer
thicknesses is determined.

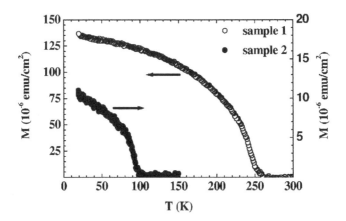

Figure 5. Saturation magnetization per unit surface for the samples 1 and 2 vs. temperature measured by SQUID magnetometer in the magnetic field of H = 10 Oe parallel to the film plane.

3.1.6   FMR Results and Analysis

FMR experiments were carried out at 9.4 GHz in the temperature range from 1.6 K to 250 K using the ESR spectrometer B-ER 418S (Bruker AG). In the normal state of the samples ($T \leq 4.2K$) the FMR signals were observed for four of our samples: 1, 2, 3 and 6; for the other two samples the resonance lines were not found. The angular dependence of the spectra was studied in the in-plane geometry, i.e. with both the dc magnetic field and the high frequency field lying in the film plane. The (001) surface of our thin films contains two principal magnetic axes of the bulk $Pd_{1-x}Fe_x$ crystal ([100] and [110] axes). The observed angular dependence of the resonance exhibits a four-fold anisotropy typical for cubic crystals in the (001)-plane. As an example the angular dependence of the resonance field H_0 of the FMR signal for the samples 1 and 2 is shown in Fig. 6. A qualitatively similar behavior of H_0 was observed for the samples 3 and 6 as well.

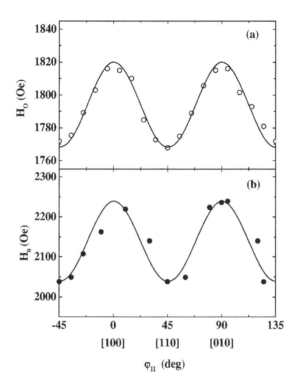

Figure 6. In-plane angular dependences of the resonance field H_0 for the sample 1 (a) and for the sample 2 (b) at T =30 K. The solid lines are the calculated resonance field values with parameters: K_1/M =15.8 Oe, $4\pi M_{eff}$ =3.9 kG (for the sample 1) and K_1/M =65 Oe, $4\pi M_{eff}$ =2.6 kG (for the sample 2).

One can see from Fig. 6 that the [110] axis is the magnetically easy axis for our samples. In the superconducting state we were able to study the behavior of the FMR line parameters for the samples 1 and 2 only. This is due to a drastic increase of the intensity of the electron paramagnetic resonance of non-controlled paramagnetic impurities in the MgO substrate at temperatures below 4 K. This background signal prevented the observation of the FMR signal from $Pd_{1-x}Fe_x$ layers for the samples 3 and 6 in the superconducting state. Examples of FMR lines of the sample 2 in the normal and superconducting states after subtraction of the background signal are shown in Fig. 7 for the dc magnetic field along the magnetically easy axis.

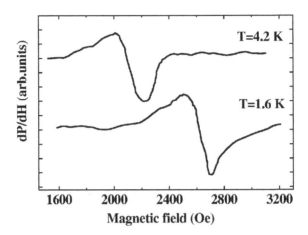

Figure 7. FMR spectra for the sample 2 at T =4.2 K (normal state) and
1.6 K (superconducting state) with dc magnetic field along the [110] axis
of the $Pd_{1-x}Fe_x$ layer.

The FMR results are analyzed using a coordinate system in which the
magnetization **M** of the $Pd_{1-x}Fe_x$ layer makes an angle θ with respect to
the film normal (z direction) and an angle α with respect to the x axis in
the film plane (xy plane). The external magnetic field **H** is applied at an
angle θ_H with respect to the film normal and at an angle α_H with respect to
the x axis. We define the x axis to be parallel to the [100] axis of the
$Pd_{1-x}Fe_x$ layer. In our experiments θ_H was equal to $\pi/2$.

In general, thin films of materials with a cubic structure grown along the
[001] crystallographic axis have a tetragonal symmetry as a result of the in-
plane strain due to epitaxial mismatch. This leads to the corresponding out-
of-plane Poisson distortion. Therefore, the contribution to the free energy
due to the crystal anisotropy should contain the four-fold in-plane anisotropy
constants, which differ from the fourth-order constant for the direction
perpendicular to the film plane. In addition, a non-zero second-order uniaxial
anisotropy term appears, due to the vertical lattice distortion and the broken
symmetry of the crystal field acting on the interface atomic layer. The
corresponding energy term has the form $F_u = -K_u \cos^2\theta$. Since our
experiments were performed in the in-plane geometry only, we will use the
crystal anisotropy energy for cubic instead of tetragonal symmetry. Uniaxial
perpendicular anisotropy and the contribution of the dipolar interaction
(demagnetizing field $4\pi M$) enter the free energy of a system in an additive
way, and therefore one can introduce an effective demagnetizing field

$$4\pi M_{eff} = 4\pi M - \frac{2K_u}{M} \tag{2}$$

in order to account for the second-order perpendicular uniaxial anisotropy. Thus the total magnetic free energy density function appropriate for a (001)-oriented film is written in the form:

$$F = -MH\sin\theta\cos(\phi - \phi_H) + 2\pi M_{eff}^2 \cos^2\theta +$$
$$+ \frac{1}{4}K_1(\sin^2 2\theta + \sin^4\theta\sin^2\phi) \tag{3}$$

Here K_1 is the fourth-order cubic anisotropy constant.

The equilibrium position of M is given by the zeros of the first angle derivatives of F. In our experimental situation the out-of plane equilibrium angle is $\theta_0 = \pi/2$ and the equilibrium in plane angle ϕ_0 is given by the solution of equation

$$H\sin(\phi_0 - \phi_H) = -\frac{K_1}{2M}\sin(4\phi_0). \tag{4}$$

With the general ferromagnetic resonance condition given by Suhl [35] we obtain the second equation:

$$(\omega/\gamma_0)^2 = [H\cos(\phi_0 - \phi_H) + (2K_1/M)\cos(4\phi_0)] \times$$
$$\times [H\cos(\phi_0 - \phi_H) + 4\pi M_{eff} + (K_1/2M)(3 + \cos(4\phi_0)] \tag{5}$$

Here $\gamma_0 = g\mu_B/\hbar$, g is the spectroscopic g-value and $\hbar$ is the Planck constant. The expression (5) together with the condition for equilibrium (4) determine the resonance field position H_0 as a function of the angle ϕ_H, of the effective magnetization $4\pi M_{eff}$, and of the anisotropy constant K_1.

We analyzed numerically the influence of all parameters in Eqs. (4) and (5) on the angular dependencies of the resonance field and obtained the following features: The increase of the $4\pi M_{eff}$ value leads to a total shift of the resonance field for all angles to lower values; the increase of K_1/M leads to an increase of the amplitude of the angular variation of H_0.

To fit the angular dependencies of the resonance field H_0, we solved numerically Eqs. (4) and (5). Typical results with $g = 2.09$ are shown in Fig. 6. They are in a good agreement with our experimental data. These fits gave us K_1 and $4\pi M_{eff}$ values listed in Table 2.

In order to determine the temperature dependence $4\pi M_{eff}(T)$, we used the temperature dependence of the resonance field measured with **H** along the magnetically easy [110] axis of the $Pd_{1-x}Fe_x$ layer. With Eqs. (4) and (5) at $\phi_H = \pi/4$ from the temperature dependence of H_0, we obtain the temperature dependencies of $4\pi M_{eff}$ for the samples 1 and 2 in a wide temperature range as shown in Fig. 8. Our study of the resonance field at fixed ϕ_H-value for the sample 2 clearly reveals a shift of the resonance field to higher values when decreasing the temperature below the superconducting transition temperature T_c (see Fig. 7). The latter fact definitely shows that the observed temperature dependence of H_0 is caused by a decrease of the effective magnetization $4\pi M_{eff}$. The low-temperature part of $4\pi M_{eff}(T)$ derived from that of H_0 is shown in Fig. 9. For comparison we also show the data obtained for the sample 1, where such behavior is not observed. Here we use the low-temperature values of K_1/M and assume that they do not change noticeably in the temperature range from 10 K to 1.6 K.

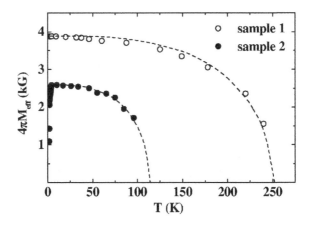

Figure 8. $4\pi M_{eff}$ vs. temperature for the samples 1 and 2 as revealed by FMR measurements. Dashed lines are the theoretical curves for $S = 1/2$ on the molecular field theory.

3.1.7   Discussion

The most interesting result of the present study is the decrease of the effective magnetization $4\pi M_{eff}$ below the superconducting transition for the sample 2 (Fig. 9). At the same time $4\pi M_{eff}$ for the sample 1 does not change in this temperature region. In accordance with Eq. (2), a decrease of $4\pi M_{eff}$ can be caused by a decrease of the saturation magnetization M or by an increase of the perpendicular uniaxial anisotropy constant K_u. One can expect that the uniaxial anisotropy, which appears usually due to the broken symmetry of crystal field acting on the interface atomic layer, is proportional to the reciprocal thickness of the F-layer. Comparison of the values of $4\pi M_{eff}$ for the samples 3 and 6 (Table 2) with nearly the same Fe content (according to the same values of T_{Curie}) but different thicknesses indicates that the second-order perpendicular uniaxial anisotropy is negligible in the thickness range studied here. Thus we have to conclude that the decrease of $4\pi M_{eff}$ is caused by a decrease of the saturation magnetization M.

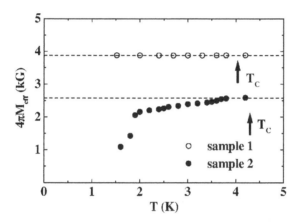

Figure 9. Low-temperature parts of $4\pi M_{eff}(T)$ for the samples 1 and 2. The arrows show the T_c-values at the resonance field H_0. [Please provide a caption for this figure]

We believe that the decrease of the saturation magnetization below T_c, which is observed for the sample 2, is caused by a reconstruction of homogeneous ferromagnetic order in the $Pd_{1-x}Fe_x$ magnetic layer due to the proximity effect with the S-layer. It is well known (see, e.g., [5]) that the superconducting order parameter is strongly suppressed near the S/F interface. This is due to a penetration of the Cooper pairs into the ferromagnet where they are subjected to the strong exchange field. This leads to a T_c suppression. The destructive influence of the exchange field on the superconductivity can considerably be weakened, if a domain [24] (or modulated cryptoferromagnetic [26]) structure on a length scale smaller than the superconducting coherence length appears in the F-layer. In this case the exchange field would be effectively cancelled over the dimension of the Cooper pairs. Bergeret *et al.* [26] presented a microscopic derivation of the phase diagram valid for realistic parameters of the problem involved. They considered a cryptoferromagnetic state with a magnetic moment that rotates in space and concluded that in the absence of a strong anisotropy this state is more favorable than the domain structure [24]. They determined the phase diagram in the vicinity of the superconducting transition for two variables. One of them takes into account the exchange splitting of conduction band in a ferromagnet and another one accounts for the magnetic stiffness of the F-layer. The obtained phase diagram is represented in Fig. 2 of Ref. [26]. The curves are plotted for different values of $\tau = (T_c - T)/T_c$.

We make an estimate for our samples according to the phase diagram by Bergeret *et al.* The magnetic stiffness J is roughly proportional to the Curie temperature. For Fe with $T_{Curie} \sim 1000$ K it is of the oder of 60 K/Å. So, for our sample 2 with $T_{Curie} \sim 100$ K it should be about 6 K/Å. The exchange splitting of the conduction band of ferromagnetic $Pd_{0.97}Fe_{0.03}$ is $h \sim 100$ K, the Fermi velocity $v_{FS} = 3 \cdot 10^7$ cm/s corresponds to the diffusion coefficient $D_S \sim 5$ cm^2/s. Assuming that the Fermi velocities of V and $Pd_{1-x}Fe_x$ are close to each other we obtain $a \sim 1.2$ and $\lambda \sim 1.3 \cdot 10^{-3}$ for our sample 2. In accordance to the phase diagram by Bergeret *et al.* this implies that starting from $\tau \sim 0.2$ ($T \sim 3.2$ K) a transition from the ferromagnetic to the cryptoferromagnetic state should take place, as it is actually observed in our experiments. For the sample 1 with $d_M \sim 44$ Å and $T_{Curie} \sim 250$ K we have $a \sim 20$ and $\lambda \sim 1.4 \cdot 10^{-2}$. With these values of parameters the ferromagnetic state is stable at any temperatures, as it is observed experimentally.

Thus, these estimates support the conclusion that the phase transition from the ferromagnetic to the cryptoferromagnetic state have been observed in our sample 2.

4.   SUMMARY

In summary, we have observed a re-entrant behavior of superconductivity in the $T_c(d_{Fe})$-curve of Fe/V/Fe-trilayers. This behavior gives further evidence of the unconventional superconductivity in S/F thin film systems. This is to the best of our knowledge, the first time that such behavior, which has been predicted theoretically before, has been observed experimentally in an S/F layered system.

FMR measurements of $V/Pd_{1-x}Fe_x$ bilayers prepared by molecular beam epitaxy have been performed over a wide temperature range. We found a decrease of the effective magnetization of the $Pd_{1-x}Fe_x$ magnetic layer below the superconducting transition temperature for the $V/Pd_{1-x}Fe_x$ bilayer system with $x \sim 0.03$ and $d_{Pd-Fe} = 12Å$. We regard this as a clear indication of the formation of the non-homogeneous cryptoferromagnetic state in the $Pd_{1-x}Fe_x$ layer due to S/F proximity effect.

ACKNOWLEDGMENT

The work has been supported by the Deutsche Forschungsgemeinschaft (SFB 491) and by the Russian Fund Fund for Basic Research (Projects No. 02-02-16688 (experiment) and No. 03-02-17656 (theory)).

REFERENCES

[1] S. Rehmann, T. Herrmannsdörfer, and F. Pobell, Phys. Rev. Lett. **78** (1997) 1122.

[2] A. I. Buzdin, L. N. Bulaevskii, M. L. Kulić, S. V. Panyukov, Adv. Phys. **34** (1985) 175-261.

[3] B. Y. Jin, J. B. Ketterson, Adv. Phys. **38** (1989) 189.

[4] C.L. Chien, D. Reich, J.M.M.M. **200** (1999) 83.

[5] I. A. Garifullin, J. Magn. Magn. Mater., **574-579** (2002) 571.

[6] Y. Obi, M. Ikebe, T. Kubo, H. Fujimori, Physica C **317-318** (1999) 149.

[7] M. Vélez, M. C. Cyrille, S. Kim, J. L. Vicent, Ivan K. Schuller, Phys. Rev. B **59** (1999) 14659.

[8] L. Lazar, K. Westerholt, H. Zabel, L.R. Tagirov, Yu.V. Goryunov, N.N. Garif'yanov, and I.A. Garifullin, Phys. Rev. B **61** (2000) 3711.

[9] M. Schöck, C. Sürgers, H.v. Löhneysen, Eur. Phys. J. B **14** (2000) 1.

[10] A. S. Sidorenko, V. I. Zdravkov, A. A. Prepelitsa, C. Helbig, Y. Luo, S. Gsell, M. Schreck, S. Klimm, S. Horn, L. R. Tagirov, and R. Tidecks, Ann. Phys. (Leipzig) **12** (2003) 37.

[11] Th. Mühge, N. N. Garif'yanov, Yu. V. Goryunov, G. G. Khaliullin, L. R. Tagirov, K. Westerholt, I. A. Garifullin, and H. Zabel, Phys. Rev. Lett. **77** (1996) 1857.

[12] A. I. Buzdin, L. N. Bulaevskii, and S. V. Panjukov, JETP Lett. **35** (1982) 178.

[13] Z. Radović, M. Ledvij, L. Dobrosavljević-Grujic, A. I. Buzdin, and J. R. Clem, Phys. Rev. B **44** (1991) 759.

[14] T. Kontos, M. Aprili, J. Lesueur, and S. Grison, Phys. Rev. Lett. **86** (2001) 304.

[15] V. V. Ryazanov, V. A Oboznov, A. Yu. Rusanov, A.V. Veretennikov, A. A. Golubov, and J. Aarts, Phys. Rev. Lett. **86** (2001) 2427.

[16] A. I. Larkin and Yu. N. Ovchinnikov, Zh. Eksp. Teor. Fiz. **47** (1964) 1136 [Sov. Phys. JETP **20** (1965) 762]; P. Fulde and R. A. Ferrell, Phys. Rev. **135** (1964) A550.

[17] E. A. Demler, G. B. Arnold, and Beasley, Phys. Rev. B **55** (1997) 15174.

[18] M. G. Khusainov and Yu. N. Proshin, Phys. Rev. B **56** (1997) 14283; Erratum: Phys. Rev. B **62** (2000) 6832.

[19] L. R. Tagirov, Physica C **307** (1998) 145.

[20] P. Isberg, E. B. Svedberg, B. Hjörvarsson, R. Wäppling, L. Hultman, Vacuum **48** (1997) 483.

[21] I. A. Garifullin, D. A. Tikhonov, N. N. Garif'yanov, L. Lazar, Yu. V. Goryunov, S. Ya. Khlebnikov, L. R. Tagirov, K. Westerholt, and H. Zabel, Phys. Rev. B. **66** (2002) 020505(R).

[22] P. W. Anderson, H. Suhl, Phys. Rev. **116** (1959) 898.

[23] N. E. Alekseevskii, I. A. Garifullin, B. I. Kochelaev, E. G. Kharakhash'yan, Zh. Eksp. Teor. Fiz. **72** (1977) 1523 [Sov. Phys. JETP **45** (1977) 799].

[24] A. I. Buzdin, L. N. Bulaevskii, Zh. Eksp. Teor. Fiz. **94** (1988) 256-261 [Sov. Phys. JETP **67** (1988) 576].

[25] Th. Mühge, N. N. Garif'yanov, Yu. V. Goryunov, K. Theis-Bröhl, K. Westerholt, I. A. Garifullin, H. Zabel, Physica C **296** (1998) 325.

[26] F. S. Bergeret, K. B. Efetov, A. I. Larkin, Phys. Rev. B **62** (2000) 11872.

[27] D. M. S. Bagguley, and J. A. Robertson, J. Phys. F **4** (1974) 2282.

[28] G. J. Nieuwehuys, Adv. Phys. **24** (1975) 515.

[29] I. A. Garifullin, D. A. Tikhonov, N. N. Garif'yanov, M. Z. Fattakhov, K. Theis-Bröhl, K. Westerholt, and H. Zabel, Appl. Magn. Reson. **22** (2002) 439.

[30] L. G. Parratt, Phys. Rev. **95** (1954) 359.

[31] A. B. Pippard, Rep. Prog. Phys. **23** (1960) 176.

[32] P. Koorevaar, Y. Suzuki, R. Coehoorn, and J. Aarts, Phys. Rev. B **49** (1994) 441.

[33] Gschneider K. A.: Solid State Phys. **16** (1964) 275.

[34] Crangle J.: Phil. Mag. **5** (1960) 335.

[35] Suhl H.: Phys. Rev. **97** (1955) 555.

THEORY OF ANDREEV SPECTROSCOPY OF FERROMAGNETS

B.P. Vodopyanov [1,2], L.R. Tagirov [2,3]
[1] *Kazan Physico-technical Institute of RAS, 420029 Kazan, Russian Federation*
[2] *Kazan State University, 420008, Kazan, Russian Federation*
[3] *Gebze Institute of Technology, 41400, Gebze, Kocaeli, Turkey*

Abstract: On the basis of microscopic approach we derive the Eilenberger-type equations of superconductivity for metals with exchange-split conduction band. The equations are valid for arbitrary band splitting and arbitrary spin-dependent mean free paths within the quasiclassic approximation. Next, we deduce general boundary conditions for the above equations. These boundary conditions take into account explicitly the spin-dependence of F/S interface transparency. We apply our theory for the Andreev reflection at F/S interface and derive an original expression for the Andreev conductance. Based on experimental data and our calculations we give estimations of the conduction band spin polarization for series of ferromagnets in contact with superconductors. We consider the superconducting proximity effect for a contact of a strong and clean enough ferromagnet with a dirty superconductor. We show that superconducting T_c of an F/S bilayer oscillates as a function of the F-layer thickness. At small enough superconducting layer thickness the re-entrant behavior of superconductivity is possible. The theory gives also non-monotonic dependence of the superconducting layer critical thickness on the spin-polarization of the ferromagnetic layer. These unconventional and distinctive features of the F/S proximity effect fit well experimental observations.

Key words: Quasiclassic superconductivity theory, Andreev reflection, ferromagnet-superconductor proximity effect

1. INTRODUCTION

At low temperatures, an electric current flows through a normal metal/superconductor (**N/S**) interface as a result of Andreev reflection [1].

B. Aktaş et al. (eds.), Nanostructured Magnetic Materials and their Applications, 145–167.

An electron is reflected from the N/S interface into a subband as a hole with the opposite spin, and the formed Cooper pair moves through the superconductor transferring a charge $2e$. The doubling of the differential conductance of a pure N/S microcontact was demonstrated theoretically in [2] (**BTK**) based on the solution of the Bogoliubov equations. In Ref. [3] an attention was drawn to the fact that the Andreev reflection in ferromagnet/superconductor (**F/S**) contacts is suppressed as the spin polarization of the ferromagnet conduction band sets up. This is associated with the fact that the Andreev reflection efficiency decreases with diminishing the number of conducting channels in the minority spin-subband (the subband with the lower value of the Fermi momentum). In Refs. [4,5,6,7,8,9], it was suggested that the suppression of Andreev reflection in F/S contacts can be used for determining the spin polarization of the conduction band of ferromagnets (Andreev spectroscopy of ferromagnets). Experimental data were interpreted making use of either general phenomenological considerations that the spin-polarized component of the normal current does not pass through a superconductor [4,5,7], or the BTK equations semi-phenomenologically adapted to the F/S contacts [8,9]. More elaborated treatment had been proposed in Ref. [10]. The BTK theory was generalized and applied to F/S point contacts in the theoretical works [11,12,13]. The expressions obtained in those works for Andreev conductance are not consistent with each other. Moreover, the results obtained in [11,12] do not reproduce the Andreev conductance at zero temperature, which follows from physical considerations and previous work, as will be shown below. The number of experiments on Andreev spectroscopy of ferromagnets grows [14,15,16,17,18,19,20], what demands an adequate theoretical understanding and description. The paper summarizes our recent efforts to build a consistent quasiclassical theory of Andreev reflection and proximity effect for the heterogeneous structures with an interface of a superconductor and a strong ferromagnet [21,22,23]. We derive quasiclassical equations of superconductivity for metals with spin-split conduction band, and deduce boundary conditions (**BC**s) for the quasiclassical Green's functions (**GF**s) at the F/S interface. Next, we compute the Andreev conductance of an F/S point contact and give an estimate for the polarization of conduction bands of ferromagnetic metals from a comparison with experiments on Andreev spectroscopy. Finally, we analyze proximity effect in the F/S bilayer structure.

2. EQUATIONS OF SUPERCONDUCTIVITY AND GENERAL BOUNDARY CONDITIONS

2.1 Eilenberger-type equations for a metal with the spin-split conduction band

We start from equations for equilibrium thermodynamic GFs in a matrix form [24], taking into account explicitly the spin splitting of the conduction band,

$$
\left(i\varepsilon_n \tau_z + \frac{1}{2m}\frac{\partial^2}{\partial \mathbf{r}^2} + \hat{\Delta} + \hat{\mu} - U - \hat{\Sigma} \right)\hat{G}(\mathbf{r},\mathbf{r}',\varepsilon_n) = \delta(\mathbf{r}-\mathbf{r}'). \tag{1}
$$

Here, the Green function $\hat{G}$ and the self-energy part $\hat{\Sigma}$ are matrices of the form:

$$
\hat{G} = \begin{pmatrix} G_{\alpha\alpha} & F_{\alpha-\alpha} \\ -\overline{F}_{-\alpha\alpha} & \overline{G}_{-\alpha-\alpha} \end{pmatrix}, \quad \hat{\Sigma} = \begin{pmatrix} \Sigma_{\alpha\alpha} & \Sigma_{\alpha-\alpha} \\ -\overline{\Sigma}_{-\alpha\alpha} & \overline{\Sigma}_{-\alpha-\alpha} \end{pmatrix}. \tag{2}
$$

In addition,

$$
\hat{\Delta} = \begin{pmatrix} 0 & \Delta \\ -\Delta^* & 0 \end{pmatrix}, \quad \hat{\mu} = \frac{1}{2m}\begin{pmatrix} p_{F\uparrow}^2 & 0 \\ 0 & p_{F\downarrow}^2 \end{pmatrix}, \tag{3}
$$

where τ_α are the Pauli matrices, α is the spin index, $\varepsilon_n = (2n+1)\pi T$ is the Matsubara frequency, Δ is the order parameter, p_α^F is the Fermi momentum, U is the electron interaction energy with the electric potential, $\mathbf{r} = (x,\boldsymbol{\rho})$, and $\boldsymbol{\rho} = (y,z)$. Hereafter we use the unit system in which $\hbar = c = 1$, so that we do not distinguish momentum and wave number, for example. We assume that the F/S interface coincides with the plane $x = 0$. Making use the Fourier transformation with respect to the coordinate $(\boldsymbol{\rho} - \boldsymbol{\rho}')$ we obtain the following equation for $\hat{G}(x,x') = \hat{G}(x,x',\boldsymbol{\rho}_c,\mathbf{p}_\parallel,\varepsilon_n)$ ($\mathbf{p}_\parallel$ is the projection of the momentum onto the contact plane, $\boldsymbol{\rho}_c = (\boldsymbol{\rho}-\boldsymbol{\rho}')/2$ is the center of mass coordinate, in that follows the index "c" of $\boldsymbol{\rho}_c$ will be omitted for brevity):

$$
\left(i\varepsilon_n \tau_z + \frac{1}{2m}\frac{\partial^2}{\partial x^2} + i\frac{\mathbf{v}_\parallel}{2}\frac{\partial}{\partial \boldsymbol{\rho}} + \frac{\hat{p}_x^2}{2m} + \hat{\Delta} - U - \hat{\Sigma} \right)\hat{G}(x,x') = \delta(x-x'), \tag{4}
$$

$$\frac{\hat{p}_x^2}{2m} = \hat{\mu} - \frac{\hat{p}_{\parallel}^2}{2m}.$$

For heterogeneous structures with an N/S interface, the quasiclassical equations of superconductivity and BCs for them were derived in [25]. For the $\hat{G}(x,x')$ function, we will use the Zaitsev representation [25] taking into account in a systematic way the spin splitting of the conduction band. The quantities related to the metal on the left (right) side of the interface will be designated by indices 1 (2). For the sake of definiteness, we will assume that the index 1 corresponds to the ferromagnet (F) and the index 2 - to the superconductor (S). Thus, for $x, x' < 0$,

$$\hat{G} = e^{i\hat{p}_{1x}x}\hat{G}_{11}e^{-i\hat{p}_{1x}x'} + e^{-i\hat{p}_{1x}x}\hat{G}_{22}e^{i\hat{p}_{1x}x'} \\ + e^{i\hat{p}_{1x}x}\hat{G}_{12}e^{i\hat{p}_{1x}x'} + e^{-i\hat{p}_{1x}x}\hat{G}_{21}e^{-i\hat{p}_{1x}x'}. \tag{5}$$

Here, $\hat{p}_{1x} = \left[\hat{p}_{1F}^2 - \hat{p}_{\parallel}^2\right]^{1/2}$; for $x, x' > 0$, $\hat{p}_{1x}$ in Eq. (5) should be changed to $\hat{p}_{2x}$. Substituting Eq. (5) into Eq. (4) and neglecting the second derivative with respect to x we obtain the equations for $\hat{G}_{kn}(x,x')$:

$$\left(i\varepsilon_n\tau_z - i(-1)^k\hat{v}_{1x}\frac{\partial}{\partial x} + i\frac{\mathbf{v}_{\parallel}}{2}\frac{\partial}{\partial \boldsymbol{\rho}} + \tilde{\hat{\Delta}}_k - U - \tilde{\hat{\Sigma}}_k\right)\hat{G}_{kn}(x,x') = 0, \quad x \neq x'. \tag{6}$$

Here, $\tilde{\hat{\Delta}}_1 = e^{i(-1)^k\hat{p}_{1x}x}\hat{\Delta}e^{-i(-1)^k\hat{p}_{1x}x}$, and $\tilde{\hat{\Sigma}}_k$ is determined in the same way. For $x, x' > 0$, $\hat{v}_{1x}$ in Eq. (6) should be changed to $\hat{v}_{2x}$. The equation conjugate to Eq. (6) is derived similarly. Let us pass now to the functions $\hat{g} = \hat{g}(x,x',\hat{p}_{jx})$ and $\hat{G} = \hat{G}(x,x',\hat{p}_{jx})$, which depend on the sign of the variable $\hat{p}_{jx}$ and are continuous at the point $x = x'$:

$$\hat{g} = \begin{cases} \hat{g}_> = 2i\sqrt{\hat{v}_{jx}}\hat{A}_1(x)\hat{G}_{11}(x,x')\hat{A}_1^*(x)\sqrt{\hat{v}_{jx}} - \text{sgn}(x-x'), & \hat{p}_{jx} > 0; \\ \hat{g}_< = 2i\sqrt{\hat{v}_{jx}}\hat{A}_2(x)\hat{G}_{22}(x,x')\hat{A}_2^*(x)\sqrt{\hat{v}_{jx}} + \text{sgn}(x-x'), & \hat{p}_{jx} < 0; \end{cases} \tag{7}$$

$$\hat{G} = \begin{cases} \hat{G}_> = 2i\sqrt{\hat{v}_{jx}}\hat{A}_1(x)\hat{G}_{12}(x,x')\hat{A}_2^*(x)\sqrt{\hat{v}_{jx}} - \text{sgn}(x-x'), & \hat{p}_{jx} > 0; \\ \hat{G}_< = 2i\sqrt{\hat{v}_{jx}}\hat{A}_2(x)\hat{G}_{21}(x,x')\hat{A}_1^*(x)\sqrt{\hat{v}_{jx}} + \text{sgn}(x-x'), & \hat{p}_{jx} < 0; \end{cases} \tag{8}$$

In Eq. (5) $\hat{A}_k(x) = \exp\{-i(-1)^k(\hat{p}_{jx} - \hat{\tau}_x\hat{p}_{jx}\hat{\tau}_x)x/2\}$. Let us substitute Eqs. (7) and (8) into Eq. (6) and into the equation conjugate to Eq. (6). Finding the difference (for $n = k$) and the sum (for $n \neq k$) of the resulting equations, we obtain the quasiclassical equations of superconductivity in metals with the spin-split conduction band

$$\text{sgn}(\hat{p}_{xj})\frac{\partial}{\partial x}\hat{g} + \frac{\mathbf{v}_\|}{2}\frac{\partial}{\partial \boldsymbol{\rho}}(\hat{v}_{xj}^{-1}\hat{g} + \hat{g}\hat{v}_{xj}^{-1}) + [\hat{K}, \hat{g}]_- = 0,$$

$$\text{sgn}(\hat{p}_{xj})\frac{\partial}{\partial x}\widehat{G} + \frac{\mathbf{v}_\|}{2}\frac{\partial}{\partial \boldsymbol{\rho}}(\hat{v}_{xj}^{-1}\widehat{G} - \widehat{G}\hat{v}_{xj}^{-1}) + [\hat{K}, \widehat{G}]_+ = 0, \tag{9}$$

$$\hat{K} = -i\hat{v}_{xj}^{-\frac{1}{2}}(i\varepsilon_n\tau_z + \hat{\Delta} - \hat{\Sigma})\hat{v}_{xj}^{-\frac{1}{2}} - i(\hat{p}_{xj} - \hat{\tau}_x\hat{p}_{xj}\hat{\tau}_x)/2,$$

$$[a, b]_\pm = ab \pm ba.$$

The impurity self-energies in Eqs. (9) are

$$\hat{\Sigma}^F = -ic\,|u|^2 \int\frac{d\mathbf{p}_\|}{(2\pi)^2}(\hat{v}_x)^{-\frac{1}{2}}\hat{g}^F(\hat{v}_x)^{-\frac{1}{2}}, \tag{10}$$

$$\hat{\Sigma}^S = -i\frac{1}{2\tau^S}<\hat{g}^S>, \quad \frac{1}{\tau^S} = c\,|u|^2\frac{mp^S}{\pi}, \tag{11}$$

where p^S is the Fermi momentum of a superconductor, u is the potential of interaction of electrons and impurities, c is the concentration of impurities, τ^S is the mean free time of electrons in a superconductor, and brackets mean averaging over the solid angle: $<...>= \oint d\Omega/4\pi$.

In the case of an F/S interface, as well as for an N/S interface, a system of quasiclassical equations arises. In addition to the functions $\hat{g}$, the functions $\widehat{G}$ appear, which describe waves reflected from the interface. The above Eilenberger-type equations for the metal with the exchange-field-split conduction band had been derived for the first time in [21,22]. They are valid for arbitrary band splitting and arbitrary spin-dependent mean free paths within the quasiclassic approximation. The system of Eqs. (9) must be supplemented with boundary conditions at F/S interface.

2.2 General boundary conditions for the Eilenberger equations

We characterize the interface by the transmission coefficient, $\hat{D}$, and the reflection coefficient, $\hat{R} = 1 - \hat{D}$. In the paper we do not consider interactions, which lead to the spin flip of an electron upon its transmission through the interface. Therefore, matrices $\hat{D}$ and $\hat{R}$ have a diagonal form with respect to spin. They have the same matrix structure as $\hat{\mu}$ in Eq. (3). Taking into account the explicit form of the GF given by Eqs. (7) and (8), and matching the quasiclassical functions on both sides of the interface according to the procedure proposed by Zaitsev, we obtain BCs for the quasiclassical equations Eqs. (9). For $p_\| < \min(p_\uparrow^F, p_\downarrow^F, p^S)$ (here, $p_\uparrow^F$ and $p_\downarrow^F$ are the Fermi momenta of spin-subbands of the ferromagnet) it is convenient to represent these conditions in the matrix form

$$
\begin{pmatrix} \hat{a}^* & -\hat{b}^* \\ -\hat{b} & \hat{a} \end{pmatrix} \left(V_x^F \right)^{1/2} \begin{pmatrix} \hat{g}_>^F & \widehat{G}_>^F \\ \widehat{G}_<^F & \hat{g}_<^F \end{pmatrix} \left(V_x^F \right)^{-1/2}
$$
$$
= \left(V_x^S \right)^{1/2} \begin{pmatrix} \hat{g}_>^S & \widehat{G}_>^S \\ \widehat{G}_<^S & \hat{g}_<^S \end{pmatrix} \left(V_x^S \right)^{-1/2} \begin{pmatrix} \hat{a}^* & \hat{b}^* \\ \hat{b} & \hat{a} \end{pmatrix}. \tag{12}
$$

Here $\hat{a} = \hat{d}^{-1}$, $\hat{b} = \hat{r}\hat{d}^{-1}$, $\hat{r}$ and $\hat{d}$ are the scattering amplitudes at the F/S interface [25], and the matrix $\left(V_x^{F(S)} \right)^{1/2}$ is the result of the direct product of the unit matrix and $\hat{v}_x^{F(S)}$. Let us pass in Eq. (12) to the functions $\tilde{g}$ and $\widetilde{G}$ using the relations:

$$
\tilde{g}_>^F = e^{i\frac{\hat{\vartheta}_r}{2}} \hat{g}_>^F e^{-i\frac{\hat{\vartheta}_r}{2}}, \quad \tilde{g}_<^F = e^{-i\frac{\hat{\vartheta}_r}{2}} \hat{g}_<^F e^{i\frac{\hat{\vartheta}_r}{2}},
$$
$$
\widetilde{G}_>^F = e^{i\frac{\hat{\vartheta}_r}{2}} \widehat{G}_>^F e^{i\frac{\hat{\vartheta}_r}{2}}, \quad \widetilde{G}_<^F = e^{-i\frac{\hat{\vartheta}_r}{2}} \widehat{G}_<^F e^{-i\frac{\hat{\vartheta}_r}{2}},
$$

$$
\tilde{g}_>^S = e^{i\frac{\hat{\vartheta}_{rd}}{2}} \hat{g}_>^S e^{-i\frac{\hat{\vartheta}_{rd}}{2}}, \quad \tilde{g}_<^S = e^{-i\frac{\hat{\vartheta}_{rd}}{2}} \hat{g}_<^S e^{i\frac{\hat{\vartheta}_{rd}}{2}}, \tag{13}
$$
$$
\widetilde{G}_>^S = e^{i\frac{\hat{\vartheta}_{rd}}{2}} \widehat{G}_>^S e^{i\frac{\hat{\vartheta}_{rd}}{2}}, \quad \widetilde{G}_<^S = e^{-i\frac{\hat{\vartheta}_{rd}}{2}} \widehat{G}_<^S e^{-i\frac{\hat{\vartheta}_{rd}}{2}},
$$

$$
\hat{\vartheta}_{rd} = \hat{\vartheta}_r/2 - \hat{\vartheta}_d,
$$

where $\hat{\vartheta}_r$ and $\hat{\vartheta}_d$ are the scattering phases associated with the scattering amplitudes $\hat{r}$ and $\hat{d}$ at the F/S interface, respectively. Next we pass to the $\tilde{g}_{s(a)}$ and $\widetilde{G}_{s(a)}$ matrices, symmetric (s) and antisymmetric (a) with respect to the variable p_{jx}:

$$
\tilde{g}_{s(a)} = \frac{1}{2}\left[\tilde{g}_> \pm \tilde{g}_<\right], \quad \widetilde{G}_{s(a)} = \frac{1}{2}\left[\widetilde{G}_> \pm \widetilde{G}_<\right]. \tag{14}
$$

After this transformation the boundary conditions can be solved with respect to $\widehat{G}_{s(a)}$ matrices and take the form:

$$\left(\tilde{\tilde{g}}_a^S\right)_d = \left(\tilde{\tilde{g}}_a^F\right)_d, \quad \left(\tilde{\tilde{G}}_a^S\right)_d = \left(\tilde{\tilde{G}}_a^F\right)_d,$$

$$\left(\sqrt{R_\alpha} - \sqrt{R_{-\alpha}}\right)\left(\tilde{\tilde{G}}_a^+\right)_n = \alpha_3\left(\tilde{\tilde{g}}_a^-\right)_n,$$

$$\left(\sqrt{R_\alpha} - \sqrt{R_{-\alpha}}\right)\left(\tilde{\tilde{G}}_a^-\right)_n = \alpha_4\left(\tilde{\tilde{g}}_a^+\right)_n,$$

$$-\tilde{\tilde{G}}_s^- = \sqrt{R_\alpha}\left(\tilde{\tilde{g}}_s^+\right)_d + \alpha_1\left(\tilde{\tilde{g}}_s^+\right)_n,$$

$$-\tilde{\tilde{G}}_s^+ = 1\sqrt{R_\alpha}\left(\tilde{\tilde{g}}_s^-\right)_d + \alpha_2\left(\tilde{\tilde{g}}_s^-\right)_n,$$

(15)

where $\tilde{\tilde{g}}_{s(a)}^\pm = 1/2\left[\tilde{\tilde{g}}_{s(a)}^S \pm \tilde{\tilde{g}}_{s(a)}^F\right]$, the $\tilde{\tilde{G}}_{s(a)}^\pm$ functions are determined in the same way. Indices n and d denote the diagonal and off-diagonal parts of the matrices: $\hat{T}_{d(n)} = 1/2[\hat{T} \pm \tau_z\hat{T}\tau_z]$. The coefficients α_i equal

$$\alpha_{1(2)} = \frac{1 + \sqrt{R_\alpha R_{-\alpha}} \mp \sqrt{D_\alpha D_{-\alpha}}}{\sqrt{R_\alpha} + \sqrt{R_{-\alpha}}},$$

$$\alpha_{3(4)} = 1 - \sqrt{R_\alpha R_{-\alpha}} \pm \sqrt{D_\alpha D_{-\alpha}}.$$

(16)

If the interference of waves arriving from the neighboring interfaces can be neglected, and $\tilde{\tilde{g}}$ does not depend on ρ, the boundary condition containing only the function $\tilde{\tilde{g}}$ can be obtained:

$$\tilde{\tilde{g}}_a^+\hat{b}_1 + \hat{b}_2\tilde{\tilde{g}}_a^+ + \tilde{\tilde{g}}_a^-\hat{b}_3 + \hat{b}_4\tilde{\tilde{g}}_a^- = \hat{b}_3 - \hat{b}_4,$$

$$\tilde{\tilde{g}}_a^-\hat{b}_1 + \hat{b}_2\tilde{\tilde{g}}_a^- + \tilde{\tilde{g}}_a^+\hat{b}_3 + \hat{b}_4\tilde{\tilde{g}}_a^+ = \hat{b}_1 - \hat{b}_2.$$

(17)

The $\hat{b}_i$ matrices equal

$$\hat{b}_1 = \tilde{\tilde{G}}_s^+\tilde{\tilde{g}}_s^- + \tilde{\tilde{G}}_s^-\tilde{\tilde{g}}_s^+, \quad \hat{b}_2 = \tilde{\tilde{g}}_s^-\tilde{\tilde{G}}_s^+ + \tilde{\tilde{g}}_s^+\tilde{\tilde{G}}_s^-,$$

$$\hat{b}_3 = \tilde{\tilde{G}}_s^+\tilde{\tilde{g}}_s^+ + \tilde{\tilde{G}}_s^-\tilde{\tilde{g}}_s^-, \quad \hat{b}_4 = \tilde{\tilde{g}}_s^-\tilde{\tilde{G}}_s^- + \tilde{\tilde{g}}_s^+\tilde{\tilde{G}}_s^+.$$

(18)

In this case, the system of BCs consists of Eqs. (17) and the first equation of Eqs. (15). Note that only the diagonal part of the functions is continuous at the F/S interface. These BCs take into account explicitly the spin-dependence of F/S interface. Previously quoted boundary conditions for the F/S interface can be obtained (if they are not wrong) as approximations, or limiting cases of our general BCs.

3.   ANDREEV CONDUCTANCE OF AN F/S POINT CONTACT

3.1   Basic formulas for a point contact

We will consider a hole of the radius a in an impenetrable membrane as a model for a point contact. At zero temperature, the expression for Andreev conductance G_A can be written down from the following physical considerations. Let us find a current in the ferromagnet at $p_\downarrow^F < p^S, p_\uparrow^F$. Then, in the case of specular reflection from the interface, $p_\parallel = p_\downarrow^F \sin\theta_\downarrow = p_\uparrow^F \sin\theta_\uparrow = p^S \sin\theta^S$, and the incidence angles of electrons, which can be Andreev-reflected from the interface in the spin-up subband, are determined from the relationship $p_\downarrow^F \sin\theta_\downarrow = p_\uparrow^F \sin\theta_\uparrow$ and depend only on the parameter $\delta = p_\downarrow^F / p_\uparrow^F$. Electrons incident at more slanted trajectories will undergo total internal reflection. The problem becomes equivalent to the problem of finding the conductance of a point contact of normal metals with different Fermi momenta (in this case, these are $p_\downarrow^F$ and $p_\uparrow^F$), when these metals are in direct contact. Using the known solution of this problem by Zaitsev (equation (38') in Ref. [25]), we find

$$G_A(T=0) = G_\downarrow \frac{8\delta(2+\delta)}{3(1+\delta)^2}, \quad G_\downarrow = \frac{e^2\left(p_\downarrow^F\right)^2 A}{4\pi^4}, \tag{19}$$

where A is the contact area. The equations for $G_A(T=0)$ obtained in [11,12] do not coincide with this result. At $\delta = 1$ (non-magnetic metal with equal $p_\downarrow^F$ and $p_\uparrow^F$), the Andreev conductance (19) equals the doubled Sharvin conductance, which corresponds to the doubling of conductance as a result of Andreev reflection. Let us now find an expression for the Andreev conductance in the case of arbitrary transmission coefficients. We start with the equation for current I in the linear approximation with respect to electric field $\mathbf{E} = (E_x, 0, 0)$. The current is calculated on the ferromagnet side of the contact at $x \to 0$:

$$I_x = \frac{e^2}{2m^2} \lim_{r \to r'} \left(\frac{\partial}{\partial x} - \frac{\partial}{\partial x'}\right) Tr\left\{\tau_z \int_{-\infty}^{\infty} \frac{d\varepsilon}{4\pi\cosh^2(\varepsilon/2T)}\right.$$
$$\left. \times \int d\mathbf{r}_1 G^R(\varepsilon, \mathbf{r}, \mathbf{r}_1) E_x(\mathbf{r}_1) \tau_z \frac{\partial}{\partial x_1} G^A(\varepsilon, \mathbf{r}_1, \mathbf{r}')\right\}. \tag{20}$$

Here, $G^{R(A)}$ is the retarded (advanced) GF, which is obtained from the temperature GFs (Eqs. (1) to (6)) by substituting $\varepsilon \pm i\delta$ for $i\varepsilon_n$. Let us substitute representations given by Eqs. (5), (7) and (8) into Eq. (20). After the Fourier transformation with respect to the $\boldsymbol{\rho} - \boldsymbol{\rho}'$ coordinate, we obtain the ballistic conductance $G_{F/S}$ of an F/S point contact:

$$G_{F/S} = \frac{Ae^2}{16\pi} Tr \left\{ \tau_z \int_{-\infty}^{\infty} \frac{d\varepsilon}{\cosh^2(\varepsilon/2T)} \int \frac{d\mathbf{p}_{\parallel}}{(2\pi)^2} \right.$$

$$\left. \times \left[1 - \hat{g}_s^R \tau_z \hat{g}_s^A - \hat{g}_a^R \tau_z \hat{g}_a^A + \hat{G}_s^R \tau_z \hat{G}_s^A - \hat{G}_a^R \tau_z \hat{G}_a^A \right] \right\}. \tag{21}$$

Now, we must solve the first of Eqs. (9) with the BCs given by Eq. (17). When $\hat{g}$ is independent of $\boldsymbol{\rho}$ the solution to Eq. (9) takes the form

$$\hat{g}_j = e^{-\text{sgn}(\hat{p}_{jx})\hat{K}x} \hat{C}_j(\mathbf{p}_{Fj}) e^{\text{sgn}(\hat{p}_{jx})\hat{K}x} + \hat{C}_j. \tag{22}$$

Matrices $\hat{C}_j$ represent the values of GF $\hat{g}_j$ at large distances from the F/S interface and

$$\hat{C}_2 = \begin{pmatrix} g & f \\ -f^+ & -g \end{pmatrix} = \frac{1}{\sqrt{\varepsilon_n^2 + |\Delta|^2}} \begin{pmatrix} \varepsilon_n & -i\Delta \\ i\Delta^* & -\varepsilon_n \end{pmatrix}. \tag{23}$$

In Eq. (22) $\hat{g}_1$ must tend to $\hat{C}_1 = \tau_z \text{sgn}(\varepsilon_n)$ at $x \to -\infty$, and $\hat{g}_2$ must tend to $\hat{C}_2$ at $x \to \infty$. Performing the matrix multiplication in Eq. (22), we find that these conditions are fulfilled if the following relationships are obeyed:

$$\hat{C}_j \hat{C}_j(\mathbf{p}_{Fj}) = -\hat{C}_j(\mathbf{p}_{Fj})\hat{C}_j = \text{sgn}(p_{jx})(-1)^j \hat{C}_j(\mathbf{p}_{Fj}). \tag{24}$$

From these relationships, it follows that

$$\hat{g}_{js} = \hat{C}_j + (-1)^j \hat{C}_j \hat{C}_{ja}, \quad \hat{g}_{ja} = \hat{C}_{ja}. \tag{25}$$

Passing in Eq. (25) to the functions $\hat{g}_s^{\pm}$, substituting them into the system of BCs given by Eqs. (17), and solving the resulting equations in the linear approximation with respect to $\hat{C}_a^{\pm} = 1/2[\hat{C}_a^S \pm \hat{C}_a^F]$, we find,

$$\hat{C}_a^{\pm} = -\frac{f(1 - \sqrt{R_\alpha R_{-\alpha}} \pm \sqrt{D_\alpha D_{-\alpha}})}{2[1 + \sqrt{R_\alpha R_{-\alpha}} + (1 - \sqrt{R_\alpha R_{-\alpha}})g]} \tau_x. \tag{26}$$

Now, from Eq. (21) and making use of Eqs. (15) and (26), we find the Andreev conductance G_A of an F/S point contact

$$G_A = G_{F/S}(V = 0) = Ae^2 \int\limits_0^\Delta \frac{d\varepsilon}{4\pi \cosh^2(\varepsilon/2T)}$$

$$\times \int \frac{d\mathbf{p}_\parallel}{(2\pi)^2} \frac{4|\Delta|^2 D_\downarrow D_\uparrow}{(1 + \sqrt{R_\downarrow R_\uparrow})^2 |\Delta|^2 - 4\sqrt{R_\downarrow R_\uparrow}\, \varepsilon^2}. \tag{27}$$

It depends on the relationship between the Fermi momenta $p_\downarrow^F$, $p_\uparrow^F$, and p^S. For example, at $p_\downarrow^F < p^S < p_\uparrow^F$ the expression for G_A takes the form:

$$G_A = \frac{Ae^2 \left(p_\downarrow^F\right)^2 |\Delta|}{4\pi^2 T} \int\limits_0^1 \frac{dx}{\cosh^2\left(\dfrac{|\Delta|}{2T} x\right)}$$

$$\times \int\limits_0^{\pi/2} d\theta_\downarrow \sin(2\theta_\downarrow) \frac{D_\downarrow D_\uparrow}{(1 + \sqrt{R_\downarrow R_\uparrow})^2 - 4\sqrt{R_\downarrow R_\uparrow}\, x^2}. \tag{28}$$

In the case of a nonmagnetic metal, where $D_\downarrow = D_\uparrow$, the expression for the Andreev conductance obtained in Ref. [25] follows from Eq. (28).

3.2 Discussion of experiments on Andreev spectroscopy

The ratio of G_A to $G_{F/N}$, where $G_{F/N}$ is the conductance of an F/S contact in the normal state, is given in [4,5,6,7,8]. The latter quantity is

$$G_{F/N}(V = 0) = \frac{Ae^2 \left(p_\downarrow^F\right)^2}{8\pi^2} \int\limits_0^{\pi/2} d\theta_\downarrow \sin(2\theta_\downarrow) D_\downarrow$$

$$+ \frac{Ae^2 \left(p^S\right)^2}{8\pi^2} \int\limits_0^{\pi/2} d\theta_N \sin(2\theta_N) D_\downarrow. \tag{29}$$

Equations (27)–(29) are valid for the arbitrary transmission coefficients D_α. For particular calculations we use the model expressions for the transmission coefficients corresponding to the direct contact between the metals

$$D_\uparrow = \frac{4 p_{x\uparrow} p_x^S}{(p_{x\uparrow} + p_x^S)^2}, \quad D_\downarrow = \frac{4 p_{x\downarrow} p_x^S}{(p_{x\downarrow} + p_x^S)^2}. \tag{30}$$

With these transmission coefficients, $G_A(T=0)$ and $G_{F/N}$ can be calculated analytically:

$$G_{F/N} = \frac{A e^2 (p^S)^2}{6\pi^2} \left(\frac{\delta_\uparrow^N (1 + \delta_\uparrow^N)}{(1 + \delta_\uparrow^N)^2} + \frac{(\delta_\uparrow^N)^3 (2 + \delta_\downarrow^N)}{(1 + \delta_\downarrow^N)^2} \right), \tag{31}$$

for $G_A(T=0)$, Eq. (19) is obtained. Here, $\delta_\uparrow^N = p^S/p_\uparrow^F$ and $\delta_\downarrow^N = p_\downarrow^F/p^S$. From Eqs. (19) and (31) it follows that the Andreev conductance at $\delta \le 0.26$ becomes smaller than the conductance of the contact in the normal state. The dependence of the ratio $G_A(V=0)/G_{F/N}$ on the parameter δ is given for various temperatures in Fig. 1. The ratio $\Delta/2T = 5.5$ corresponds to the experimental conditions [5] ($T = 1.6$ K, $\Delta_{Nb} = 1.5$ meV). In order to interpret universally the experimental data obtained in [5] for a series of ferromagnetic materials in contact with superconducting Nb, we fixed the Fermi momentum of the superconducting metal by the equation $(p^S)^2 = \left[(p_\uparrow^F)^2 + (p_\downarrow^F)^2 \right]/2 = \text{const}$. Now, the values of δ (abscissa) can be estimated by the value of the reduced conductance at zero voltage across the contact (ordinate). Emphasize that in this calculation we assumed the absence of an oxide or similar barrier at the F/S interface ($Z_{BTK} = 0$). The estimated results for δ are given in the table.

Table 1.

Material under study [5]	δ	$P_c(\%)$	$P_c(\%)$[5]
NiFe	0.64	42	37±5.0
Co	0.55	52	42±2.0
NiMnSb	0.48	63	58±2.3
LMSO	0.31	83	78±4.0
CrO$_2$	0.18	94	94±3.6

Note that the values $\delta(\text{Ni}) \approx 0.64$ and $\delta(\text{Co}) \approx 0.55$ obtained from the Andreev spectroscopy turned out to be close to the upper estimates for $\delta(\text{Ni}) \approx 0.64$ and $\delta(\text{Co}) \approx 0.57$, which we obtained in [26] from data on the giant magnetoresistance in magnetic point contacts [27].

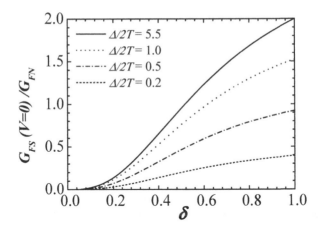

Figure 1. Dependence of the normalized Andreev conductance on the ratio δ of the Fermi momenta of spin subbands of the ferromagnet conduction band ($\delta = p_\downarrow^F / p_\uparrow^F$).

Let us now compare our results with the original estimates of polarization obtained in [5]. The authors argue that the normalized conductance measured in their work depends on polarization as $G_{F/S}/G_n = 2(1 - P_I)$ (Eqs. (4)–(6) in [5]), where $P_I = (I_\uparrow - I_\downarrow)/(I_\uparrow + I_\downarrow)$ and $G_n \simeq G_{F/N}$ is the conductance at high voltages across the contact ($eV \gg \Delta$). In the course of discussion, the authors identified the current polarization P_I with the contact polarization

$$ P_c = (N_\uparrow v_\uparrow^F - N_\downarrow v_\downarrow^F)/(N_\uparrow v_\uparrow^F + N_\downarrow v_\downarrow^F) = (1 - \delta^2)/(1 + \delta^2), $$

where N_α and v_α are the density of states and the Fermi velocity in the α-spin subband of the ferromagnet, respectively. We think that this identification is not quite correct, because it tacitly implies the independence of the total current $I_\uparrow + I_\downarrow$ through the contact in the normal phase on the spin polarization of the ferromagnet. It is evident from Eq. (31) that $G_{F/N}$ essentially depends on δ. As a result, the reduced conductance $G_A(V = 0)/G_{F/N}$ is a nonlinear function of the contact polarization P_c (Fig. 2). It is seen in Fig. 2 that the identification of P_I with P_c leads to a systematic underestimation of the estimate for P_c (compare the third and the fourth columns of the table). Note here that the numerical calculations of the conductance at zero voltage performed in [13] (see Fig. 4 of the work for $Z = 0$, $T/T_c = 0.2$) fit well the linear dependence on the contact polarization proposed in [5] (dash line in our Fig. 2). From this observation we conclude that the calculations made in [13] also give underestimated values of the contact polarization taken from the conductance at zero voltage. Our theory allows to estimate the polarization parameter δ of the ferromagnet conduction band, through which the polarization of the density of states P_{DOS}, the tunneling polarization P_T, and the contact polarization P_c are expressed. Our analysis of experiments on Andreev spectroscopy leads to values of P_c that are systematically higher than those estimated previously.

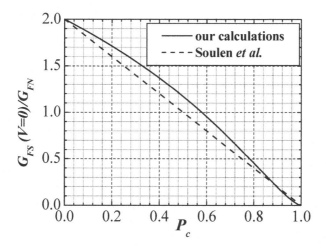

Figure 2. Dependence of the normalized Andreev conductance on the contact polarization P_c.

4.   THE SUPERCONDUCTOR - STRONG FERROMAGNET PROXIMITY EFFECT

The ferromagnet-superconductor contacts are interesting not only as a tool to measure conduction-band polarization of ferromagnetic metals, but also as a unique combination of materials to build π-contacts [28] and superconducting logic circuits [29,30]. The physics behind of the π-contacts is the unconventional Larkin-Ovchinnikov-Fulde-Ferrell (LOFF) pairing in ferromagnetic superconductors [31,32], which manifests itself in F/S contacts by oscillations of the superconducting transition temperature [33,34], tunneling density of states [35], Josephson current [36,37,38] as a function of F-layer thickness or temperature. The realization of the LOFF-like pairing in weak ferromagnets (ferromagnetic alloys having low Curie temperatures ~ 100 K) is certainly proved [30,35,36,37,38]. The case of the contacts of superconductors with strong ferromagnets, like Fe, Co, Ni, is still questionable. In fact, there are calculations of the local tunneling density of states [39,40] and the superconducting transition temperature [40], which do not predict oscillations of the above mentioned quantities, if the ferromagnet is clean (with mean free path much longer than the pairing function oscillation length). Direct experimental verification is very difficult, because one needs ultra-thin, few monolayers thick, ferromagnetic films of excellent quality - a big problem. In the section below we show, that in the case of clean enough ferromagnets the critical temperature still oscillates as a function of the F-layer thickness.

4.1   Boundary conditions for the "dirty" superconductor - strong ferromagnet bilayer

To find quasiclassic Green functions at the F/S interface one needs to solve the Eilenberger-type equations (9) for the every metal in a contact. Near the transition temperature the equations can be linearized with respect to the anomalous Green function $\hat{f}^{S(F)}$:

$$\hat{g}^{S(F)} = \frac{\varepsilon_n}{|\varepsilon_n|}\tau_z + \hat{f}^{S(F)}, \, \hat{f}^{S(F)} = \begin{pmatrix} 0 & f_{\uparrow\downarrow}^{S(F)} \\ -\overline{f}_{\downarrow\uparrow}^{S(F)} & 0 \end{pmatrix}. \tag{32}$$

Then, the first of equations in (9), applied to the superconductor homogeneous in the plane of the contact, reads

$$\text{sgn}(p_x^S)\frac{\varepsilon_n}{|\varepsilon_n|}\tau_z l_x^S \frac{\partial}{\partial x}\hat{f}^S + \lambda^S \hat{f}^S = <\hat{f}^S> -2i\tau^S\hat{\Delta}, \tag{33}$$

$$\lambda^S = 1 + 2|\varepsilon_n|\tau^S.$$

Here $l_x^S = |v_x^S|\tau^S$ is the component of the mean free path along the normal to the contact plane. Passing to the symmetric, $\hat{f}_s^S$, and antisymmetric, $\hat{f}_a^S$, GFs according to Eq. (14) we obtain the differential equations with respect to $\hat{f}_s^S$ and $\hat{f}_a^S$:

$$l_x^2 \frac{\partial^2}{\partial x^2}\hat{f}_s^S - (\lambda^S)^2 \hat{f}_s^S + \lambda^S <\hat{f}_s^S> = 2i\lambda^S\tau^S\hat{\Delta},$$

$$l_x \frac{\partial \hat{f}_s^S}{\partial x} = -\lambda^S \text{sgn}(\varepsilon_n)\tau_z \hat{f}_a^S. \tag{34}$$

Analogously, we deduce the equations to find GF $\hat{f}_s^F$ and $\hat{f}_a^F$ for the ferromagnet:

$$\left(\frac{2l_{x\uparrow}l_{x\downarrow}}{l_{x\uparrow}+l_{x\downarrow}}\right)^2 \frac{\partial^2}{\partial x^2}\hat{f}_s^F - (\lambda^F)^2\hat{f}_s^F = -2\frac{\sqrt{v_{x\uparrow}v_{x\downarrow}}}{l_{x\uparrow}+l_{x\downarrow}}\tau_\uparrow\tau_\downarrow\lambda^F\left\langle\frac{1}{\tau_{\uparrow\downarrow}}\hat{f}_s^F\right\rangle,$$

$$\hat{f}_a^F = -\mathrm{sgn}(\varepsilon_n)\frac{2l_{x\uparrow}l_{x\downarrow}}{\lambda^F(l_{x\uparrow}+l_{x\downarrow})}\tau_z\frac{\partial}{\partial x}\hat{f}_s^F, \tag{35}$$

where

$$\lambda^F = 1+\frac{2\tau_\uparrow\tau_\downarrow(v_{x\uparrow}+v_{x\downarrow})}{l_{x\uparrow}+l_{x\downarrow}}|\varepsilon_n|-i\frac{2l_{x\uparrow}l_{x\downarrow}}{l_{x\uparrow}+l_{x\downarrow}}(p_{x\uparrow}-p_{x\downarrow})\mathrm{sgn}(\varepsilon_n),$$

$$\left\langle\frac{1}{\tau_{\uparrow\downarrow}}\hat{f}_s^F\right\rangle = c|u|^2\int\frac{d\mathbf{p}_\parallel}{(2\pi)^2}\frac{1}{\sqrt{v_{x\uparrow}v_{x\downarrow}}}\hat{f}_s^F. \tag{36}$$

In the above, $\left(v_x^{F(S)}\right)^{1/2}$, τ_α^F is the mean free path in the α-th spin-subband of the ferromagnet. In the second line of (36) the angular integration is constrained [41] to fulfill the specular scattering condition:

$$p_\parallel = p_\downarrow^F\sin\theta_\downarrow = p_\uparrow^F\sin\theta_\uparrow = p^S\sin\theta^S. \tag{37}$$

In Eqs. (35)-(37) and hereafter $p_\uparrow^F$ and $p_\downarrow^F$ are the Fermi momenta of the spin–subbands of the ferromagnet.

Solution of Eq. (33) with the boundary condition $\hat{f}_a^S(x\to\infty)=0$ reads:

$$\hat{f}_s^S(x) = \mathrm{sgn}(\varepsilon_n)\tau_z\,\hat{f}_a^S(x)$$

$$+\frac{1}{l_x^S}\int_x^\infty d\xi\, e^{-\frac{\lambda^S(\xi-x)}{l_x^S}}[<\hat{f}_s^S(\xi)>-2i\tau^S\hat{\Delta}(\xi)]. \tag{38}$$

In the equation (38) the integrand in the square parentheses has the spatial range δ , where $D^S = v^S l^S/3$ is the diffusion coefficient of electrons in a superconductor. Expanding the slow varying function around the point $\xi = x$ and taking it out the integral we obtain $\hat{f}_s^S(x)$:

$$\hat{f}_s^S(x) = \mathrm{sgn}(\varepsilon_n)\tau_z\,\hat{f}_a^S(x)+\frac{1}{\lambda^S}\left(1+\frac{l_x^S}{\lambda^S}\frac{d}{dx}\right)<\hat{f}_s^S(x)>. \tag{39}$$

Solution for the ferromagnet is sought in the form:

$$\hat{f}_s^F(x) = C^F(\theta_\downarrow)\cosh\left[\kappa^F(\theta_\downarrow)(x + d^F)\right], \tag{40}$$

where

$$\kappa^F(\theta_\downarrow) = \kappa_1^F(\theta_\downarrow) + i\,\mathrm{sgn}(\varepsilon_n)\kappa_2^F(\theta_\downarrow),$$

$$\kappa_1^F(\theta_\downarrow) = (1 - \eta_1)\frac{l_{x\uparrow}^F + l_{x\downarrow}^F}{2l_{x\uparrow}^F l_{x\downarrow}^F} + \frac{v_{x\uparrow} + v_{x\downarrow}}{2v_{x\uparrow}v_{x\downarrow}}|\varepsilon_n|, \tag{41}$$

$$\kappa_2^F(\theta_\downarrow) = |\,p_{x\uparrow} - p_{x\downarrow}\,| + \eta_2\frac{l_\uparrow^F + l_\downarrow^F}{l_\uparrow^F l_\downarrow^F}.$$

In the equations (41), the quantities η_1 and η_2 do not depend on the angles $i\varepsilon_n$ and $\theta_\uparrow$. As in the above Sections we assume $p_\uparrow^F > p_\downarrow^F$. Substituting the solution (40) into the equation (35) one obtains the integral equation to find η_1 and η_2:

$$\frac{2l_\uparrow}{p_\downarrow^F p_\uparrow^F}\int\frac{d\mathbf{p}_\parallel}{2\pi}\frac{\lambda^F(l_{x\uparrow}^F + l_{x\downarrow}^F)}{(\lambda^F)^2(l_{x\uparrow}^F + l_{x\downarrow}^F)^2 - (2l_{x\uparrow}^F l_{x\downarrow}^F)^2(\kappa^F)^2} = 1. \tag{42}$$

For the strong ferromagnet the solution of the above equation is (the relative accuracy is $\left(p_\uparrow^F l_\uparrow^F\right)^{-1} \ll 1$):

$$\eta_1 = \frac{l_\uparrow^F}{p_\downarrow^F p_\uparrow^F}\int\frac{d\mathbf{p}_\parallel}{2\pi}\frac{1}{l_{x\uparrow}^F + l_{x\downarrow}^F}, \quad \eta_2 = 0. \tag{43}$$

Now we express symmetric GF $\hat{f}_s^F(x)$ via the antisymmetric one, $\hat{f}_a^F(x)$ at $x = 0$:

$$\hat{f}_s^F(0) = -\mathrm{sgn}(\varepsilon_n)\frac{\lambda^F(l_{x\uparrow}^F + l_{x\downarrow}^F)}{2l_{x\uparrow}^F l_{x\downarrow}^F \kappa^F \tanh(\kappa^F d^F)}\tau_z\hat{f}_a^F(0), \tag{44}$$

combine $\hat{f}_s^S(x)$ (39) and $\hat{f}_s^F$ (44) in $\hat{f}_s^\pm$ at $x = 0$, satisfy the boundary conditions (17) and obtain $\hat{f}_a^S(0)$ and $\hat{f}_a^F(0)$:

$$\hat{f}_a^S(0) = -\operatorname{sgn}(\varepsilon_n)\tau_z \frac{B}{\lambda^S}\left(1 + \frac{l_x^S}{\lambda^S}\frac{d}{dx}\right) < \hat{f}_s^S(x) >,$$

$$\hat{f}_a^F(0) = -2\operatorname{sgn}(\varepsilon_n)\tau_z \frac{\sqrt{D_\uparrow D_\downarrow}}{\lambda^S \Gamma}\left(1 + \frac{l_x^S}{\lambda^S}\frac{d}{dx}\right) < \hat{f}_s^S(x) >,$$

(45)

where

$$B = \frac{\Gamma^S}{\Gamma},$$

$$\Gamma^S = D_\uparrow + D_\downarrow + (\sqrt{R_\uparrow} - \sqrt{R_\downarrow})^2 v^F,$$

$$\Gamma = 2\left[1 + \sqrt{R_\uparrow R_\downarrow} + (1 - \sqrt{R_\uparrow R_\downarrow})v^F\right],$$

(46)

$$v^F = \frac{\lambda^F(l_{x\uparrow}^F + l_{x\downarrow}^F)}{2 l_{x\uparrow}^F l_{x\downarrow}^F \kappa^F \tanh(\kappa^F d^F)}.$$

In deriving the above equations we neglected spin-dependence of the phases of the scattering amplitudes in the general boundary conditions Eqs. (15).

To formulate particular BC for the contact of a strong ferromagnet with a dirty superconductor we use the Ansatz proposed in Ref. [42]: at distances of the order of the mean free path l^S in a superconductor, when the terms proportional to ε_n and Δ can be neglected, one may write down

$$\left\langle \cos(\theta^S)\hat{f}^S \right\rangle = \frac{1}{2}\int_0^{\pi/2} d\theta^S \sin(2\theta^S)\hat{f}_a^S = \hat{C}^S,$$

(47)

where $\hat{C}^S$ is constant. This constant can be found substituting into Eq. (47) the antisymmetric combination

$$\hat{f}_a^S = -\hat{\tau}_z \operatorname{sgn}(\varepsilon_n) l_x^S \frac{d}{dx} < \hat{f}_s^S >,$$

which corresponds to the solution of the Usadel equation [43] for the "dirty" superconductor far away from the F/S interface. The result is:

$$\hat{C}^S = -\hat{\tau}_z \operatorname{sgn}(\varepsilon_n)\frac{1}{3}l^S \frac{d}{dx} < \hat{f}_s^S >.$$

(48)

Now, we calculate the same constant with the use of the function $\hat{f}_a^S(0)$ taken from the equation (45) above, and obtain the boundary condition for the averaged over the solid angle GF, $\hat{F}_s^S(x) = <\hat{f}_s^S>$:

$$l^S \frac{d}{dx}\hat{F}_s^S(x) = \gamma \hat{F}_s^S(x), \quad \gamma = \frac{\gamma_1}{1-\gamma_2},$$

$$\gamma_1 = \frac{3}{2}\int_0^\varphi d\theta^S \sin(2\theta^S)B, \quad \gamma_2 = \frac{3}{2}\int_0^\varphi d\theta^S \cos(\theta^S)\sin(2\theta^S)B. \tag{49}$$

The upper limit in Eq. (49) depends on the relation between the Fermi momenta of contacting metals, and is determined from conservation of the parallel component of the transferred momentum (37). If the Fermi momentum p^S is the smallest of three, then $\varphi = \pi/2$. The quantity B is determined in Eq. (46), it is a function of the angles θ^S and $\theta_\downarrow$, which obey the scattering specularity condition (37). The boundary condition (49) is valid for the dirty superconductor - strong ferromagnet interface at arbitrary transparency.

4.2   Critical temperature of F/S bilayer

To find the superconducting transition temperature T_c of the bilayer we solve the linearized Usadel equation at temperatures close to T_c :

$$D^S \frac{d^2}{dx^2}\hat{F}_s^S - 2|\varepsilon_n|\hat{F}_s^S = 2i\hat{\Delta}, \tag{50}$$

and satisfy with this solution the boundary condition (49). The problem is easily solved in the single-mode approximation [44], which is valid at intermediate suppressions of T_c against unperturbed transition temperature of the isolated superconducting film, T_{c0} :

$$\hat{F}_s^S = -\frac{2i\hat{\Delta}}{|\varepsilon_n|+D^S(\kappa^S)^2}, \quad \hat{\Delta} = \hat{\Delta}_0 \cos[\kappa^S(x-d^S)], \tag{51}$$

where κ^S is determined from BC (49),

$$l^S \kappa^S \tan(\kappa^S d^S) = \gamma. \tag{52}$$

Substituting (51) into the self-consistency equation,

$$\hat{\Delta}\ln(t_c) = \pi T \sum_{n=-\infty}^{\infty}\left(i\hat{F}_s^S - \frac{\hat{\Delta}}{|\varepsilon_n|}\right); \quad t_c = \frac{T_c}{T_{c0}}, \tag{53}$$

one finally gets the equation for finding the transition temperature of the dirty superconductor - strong ferromagnet bilayer:

$$\ln(t_c) = \Psi\left(\frac{1}{2}\right) - Re\,\Psi\left(\frac{1}{2}+\frac{\rho}{t_c}\right), \tag{54}$$

$$\rho = \frac{D^S(\kappa^S)^2}{4\pi T_{c0}}.$$

4.3   Results and discussion of proximity effect

Upon solution of the equation (54) we will neglect dependence of κ^S on ε_n because we consider strong ferromagnet with the energy of the exchange splitting of conduction band which is much larger than the thermal energy. As it was done above, in the discussion of experiments on Andreev spectroscopy, we fix the Fermi momentum of the superconductor by the relation, $\left(p^S\right)^2 = \left[\left(p_\uparrow^F\right)^2+\left(p_\downarrow^F\right)^2\right]/2 = \text{const}$, and use the above formulas (30) to evaluate the interface transmission coefficients.

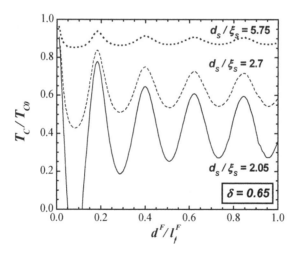

Figure 3. Dependence of the superconducting critical temperature of F/S bilayer on the thickness of the ferromagnetic layer at $\delta = 0.65$. Values of other parameters are given in the text.

The results of calculations for the set of parameters: $\delta = p_\downarrow^F / p_\uparrow^F = 0.65$, $p_\uparrow^F l_\uparrow^F = 40.0$, $l_\downarrow^F / l_\uparrow^F = 2.5$, $\xi^S / \xi_{BCS}^S = 0.25$ [$\xi^S = (D^S / 2\pi T_{c0})^{1/2}$ - is the coherence length of dirty superconductor], are displayed in Fig. 3. The parameters approximately correspond to the contact of nickel with niobium or vanadium. The figure shows damped oscillations of transition temperature as a function of the ferromagnetic layer thickness. As the superconducting layer becomes thin enough, the re-entrant behavior of the superconducting transition temperature is possible (the lower, solid curve in Fig. 3), which has been observed in the experiment [45].

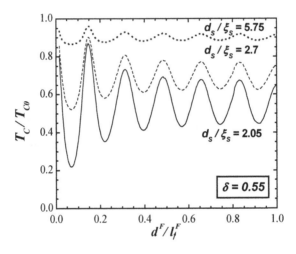

Figure 4. Dependence of the superconducting critical temperature of F/S bilayer on the thickness of the ferromagnetic layer at $\delta = 0.55$. Values of other parameters are the same as in Fig. 3.

The results for another calculation with only the exchange splitting parameter is changed, $\delta = p_\downarrow^F / p_\uparrow^F = 0.55$ (this corresponds approximately to cobalt), are displayed in Fig. 4. Comparison with the previous figure shows that the superconducting T_c suppression weakened in the contact with the stronger ferromagnet ($\delta(\text{Co}) < \delta(\text{Ni})$), which seems to contradict expectation. However, one should keep in mind that when the conduction band polarization growths, the interface transparency decreases as a result of increasing mismatch of Fermi momenta of the superconductor and the ferromagnet. The growing isolation of S and F layers dominates the increase of depairing influence of the exchange field. This scenario has been realized in the layered system $Fe_x V_{1-x} / V$ [46]. With increasing the iron content x in the ferromagnetic alloy $Fe_x V_{1-x}$ the non-monotonic behavior of the superconductor critical thickness was observed. The pure iron layer suppressed T_c weaker than the alloy with the iron concentration $x \simeq 0.6$.

Our calculations take into account explicitly dependence of the interface transparency on the conduction-band exchange splitting, giving the theoretical basis for the extensive discussions of the F/S interface transparency based on the experimental data [41,45,46,47]. Our results do not contradict conclusions by Bergeret *et al.* [40]: figures 3 and 4 show that at small T_c suppression, when $T_{c0} - T_c \ll T_{c0}$, and for $d^F > l_\uparrow^F$, the oscillations amplitude is considerably smaller than the asymptotic value of the suppression, $\delta T_c = T_{c0} - T_c (d^F \to \infty)$ (see upper curves on Figs. 3 and 4). Thus, the oscillations of T_c are beyond the approximation adopted in Ref. [40].

It interesting to note, that at certain thickness of the F layer the decrement of oscillations decay does not increase when decreasing the mean free paths l_α^F, at it could be expected, but decreases. This unusual behavior is explained by exclusion of slanted trajectories, on which the path of electrons inside the ferromagnetic film exceeds their mean free path. In the certain range of the thickness d^F, closing of the cone of effective trajectories, which couple F and S layers, dominates over their decay. As the thickness $\hat{g}_J$ approaches mean free paths, the cone of effective trajectories, those interference gives rise geometric oscillations of T_c, collapses to the film normal, and the solution of the problem in our approach becomes single-exponential [41].

CONCLUSION

In this paper we summarized results of our theoretical studies of the contacts of superconductors with strong ferromagnets. On the basis of microscopic approach we derived for the first time the Eilenberger-type equations of superconductivity in metals with exchange-split conduction band. The equations are valid for arbitrary band splitting and arbitrary spin-dependent mean free paths within the quasiclassic approximation. As a next step we deduced general boundary conditions for the above equations. These BCs take into account explicitly the spin-dependence of F/S interface transparency. All other formulations of the boundary conditions for the F/S interface (if they are correct) can be obtained as approximations, or limiting cases of our general BC.

First, we applied our theory for the Andreev reflection at F/S interface and derived an original expression for the Andreev conductance. Our expression takes into account explicitly the spin dependence of the interface transparency and the spin-dependent conservation laws at scattering on the F/S interface. Based on the experimental data and our calculations we give estimations of the conduction band spin-polarization for series of ferromagnets in contact with superconductors.

Second, we considered the superconducting proximity for contact of strong and clean enough ferromagnet (mean free paths much longer than the oscillation period of the pairing function) with dirty (short mean free path) superconductor. We showed that superconducting T_c of the F/S bilayer oscillates as a function of the F-layer thickness. At small enough superconducting layer thickness the re-entrant behavior of superconductivity is possible. The theory takes into account explicitly the spin dependence of the interface transparency, which results in non-monotonic dependence of the superconducting layer critical thickness on the spin-polarization of the ferromagnetic layer. These unconventional and distinctive features of the F/S proximity effect fit qualitatively experimental observations.

Finally, the set of Eilenberger-type equations and the boundary conditions for them can be used for calculations and re-examinations of variety of properties and physical effects in F/S contacts like local densities of states, Josephson effect, magnetic moment distribution, *etc.*

ACKNOWLEDGMENT

The work was supported by the Russian Foundation for Basic Research (grants no. 03-02-17432 and no. 03-02-17656) and the INTAS project no. 03-5489.

REFERENCES

[1] A. F. Andreev, Zh. Eksp. Teor. Fiz. **46**, 1823 (1964) [Sov. Phys. JETP **19**, 1228 (1964)].
[2] G. E. Blonder, M. Tinkham, and T. M. Klapwijk, Phys. Rev. B **25**, 4515 (1982).
[3] M. J. M. de Jong and C. W. J. Beenakker, Phys. Rev. Lett. **74**, 1657 (1995).
[4] S. K. Upadhyay, A. Palanisami, R. N. Louie, and R. A. Buhrman, Phys. Rev. Lett. **81**, 3247 (1998).
[5] R. J. Soulen, J. M. Byers, M. S. Osofsky, *et al.*, Science **282**, 85 (1998); J. Appl. Phys. **85**, 4589 (1999).
[6] M. S. Osofsky, B. Nadgorny, R. J. Soulen *et al.* J. Appl. Phys. **85**, 5567 (1999).
[7] B. Nadgorny, R. J. Soulen, M. S. Osofsky *et al.*, Phys. Rev. B **61**, 3788(R) (2000).
[8] Y. Ji, G. J. Strijkers, F. Y. Yang, et al., Phys. Rev. Lett. **86**, 5585 (2001).
[9] G. J. Strijkers, Y. Ji, F. Y. Yang, and C. L. Chien, Phys. Rev. B **63**, 104510 (2001).
[10] I. I. Mazin, A. A. Golubov, and B. Nadgorny, J. Appl. Phys. **89** 7576 (2001).
[11] S. Kashiwaya, Y. Tanaka, N. Yoshida, and M. R. Beasley, Phys. Rev. B **60**, 3572 (1999).
[12] A. A. Golubov, Physica C (Amsterdam) **326–327**, 46 (1999).
[13] K. Kikuchi, H. Imamura, S. Takanashi, and S. Maekawa, Phys. Rev. B **65**, 20508 (2001).
[14] M. S. Osofsky, R. J. Soulen, B. E. Nadgorny *et al.*, Mat. Scien. Eng. **84**, 49 (2001).

[15] B. Nadgorny, I. Mazin, M. Osofsky *et al.*, Phys. Rev. B **63**, 184433 (2001).

[16] Y. Ji, C. L. Chien, Y. Tomioka, and Y. Tokura, Phys. Rev. B **66**, 012410 (2002).

[17] C. H. Kant, O. Kurnosikov, A.T. Filip *et al.*, Phys. Rev. B **66**, 212403 (2002).

[18] B. Nadgorny, M. S. Osofsky, D. J. Singh *et al.*, Appl. Phys. Lett. **82**, 427 (2003).

[19] P. Raychaudhuri, A. P. Mackenzie, J. W. Reiner, and M. R. Beasley, Phys. Rev. B **67**, 020411 (2003).

[20] N. Auth, G. Jakob, T. Block, and C. Felser, Phys. Rev. B **68**, 024403 (2003).

[21] B. P. Vodopyanov and L.R. Tagirov, Physica B **284-288**, 509 (2000).

[22] B. P. Vodopyanov and L.R. Tagirov, Pisma Zh. Eksp. Teor. Fiz. **77**, 153 (2003) [JETP Letters **77**, 126 (2003)].

[23] B. P. Vodopyanov and L.R. Tagirov, Pisma Zh. Eksp. Teor. Fiz. **78**, 1043 (2003) [JETP Letters **78**, (2003)].

[24] A. I. Larkin and Yu. N. Ovchinnikov, J. Low Temp. Phys. **10**, 401 (1973).

[25] A. V. Zaitsev, Zh. Eksp. Teor. Fiz. **86**, 1742 (1984) [Sov. Phys. - JETP **59**, 1015 (1984)].

[26] L. R. Tagirov, B. P. Vodopyanov, and K. B. Efetov, Phys. Rev. B **65**, 214419 (2002).

[27] N. García, M. Muñoz, and Y.-W. Zhao, Phys. Rev. Lett. **82**, 2923 (1999); G. Tatara, Y.-W. Zhao, M. Muñoz, and N. García, Phys. Rev. Lett. **83**, 2030 (1999).

[28] Z. Radovic, M. Ledvij, L. Dobrosavljević-Grujić *et al.*, Phys. Rev. B **44**, 759 (1991).

[29] L. R. Tagirov, Phys. Rev. Lett. **83**, 2058 (1999).

[30] V. V. Ryazanov, V. A . Oboznov, A. V. Veretennikov, and A. Yu. Rusanov, Phys. Rev. **65**, 020501(R) (2001).

[31] A. I. Larkin and Yu. .N. Ovchinnikov, Zh. Eksp. Teor. Fiz. **47**, 1136 (1964) [JETP]

[32] P. Fulde and R. Ferrell, Phys. Rev. **135A**, 1550 (1964).

[33] J. S. Jiang, D. Davidović, D. H. Reich, and C. L. Chien, Phys. Rev. Lett. **74**, 314 (1995).

[34] Th. Mühge, N. N. Garifyanov, Yu. V. Goryunov *et al.*, Phys. Rev. Lett. **77**, 1857 (1996).

[35] T. Kontos, M. Aprili, J. Lesueur, and X. Grison, Phys. Rev. Lett. **86**, 304 (2001).

[36] V. V. Ryazanov, V. A . Oboznov, A. Yu. Rusanov *et al.*, Phys. Rev. Lett. **86**, 2427 (2001).

[37] T. Kontos, M. Aprili, J. Lesueur *et al.*, Phys. Rev. Lett. **89**, 137007 (2002).

[38] H. Sellier, C. Baraduc, F. Lefloch, and R. Calemczuk, Phys. Rev. B **68**, 054531 (2003).

[39] I. Baladie and A. I. Buzdin, Phys. Rev. **64**, 224514 (2001).

[40] F. S. Bergeret, A. F. Volkov, and K. B. Efetov, Phys. Rev. **65**, 134505 (2002).

[41] L. R. Tagirov, Physica C **307**, 145 (1998).

[42] M. Yu. Kupriyanov, V. F. Likichev, Zh. Eksp. Teor. Fiz. **94**, 139 (1988) [JETP].

[43] K. D. Usadel, Phys. Rev. Lett. **25**, 507 (1970).

[44] Z. Radović, L. Dobrosavlijević-Grujić, A.I. Buzdin, J.R. Clem, Phys. Rev. B **38**, 2388 (1988).

[45] I. A. Garifullin, D. A. Tikhonov, N. N. Garifyanov *et al.*, Phys. Rev. B **66**, 020505(R) (2002).

[46] J. Aarts, J. M. E. Geers, E. Brück *et al.*, Phys. Rev. B **56**, 2779 (1997).

[47] L. Lazar, K.Westerholt, H. Zabel *et al.*, Rhys. Rev. B **61**, 3711 (2000).

4

PART 4: MAGNETIC RESONANCE IN NANOSTRUCTURED MATERIALS

INVESTIGATION OF MAGNETIC COUPLING BY FERROMAGNETIC RESONANCE

A. Layadi
Département de Physique, Université Ferhat Abbas, Sétif 19000, Algeria

Abstract: Ferromagnetic Resonance (FMR) is a powerful method for study the magnetic coupling in multilayers. In the present work, a trilayer system is assumed to be characterized by the bilinear J_1, biquadratic J_2 coupling, and in-plane uniaxial magneto-crystalline anisotropies with anisotropy axis directions in the two layers making a δ angle. A detailed discussion of the effect of J_2 and δ on the mode characteristics (resonant frequency and the mode intensity) will be given. In certain situations, (for instance when $J_1 < 0$, and the magnetizations are parallel) only the acoustic mode will appear with constant mode position and intensity for all J_2 values, making it difficult to detect any additional biquadratic coupling. Also, it will be shown that, for small J_2 and δ values, the effect of the misalignment of the easy axes in the two layers, δ, on the FMR modes can be similar to that of J_2, making it hard to distinguish the effect of δ from that of the biquadratic coupling strength J_2 from an FMR experiment. A relation establishing the equivalence between J_2 and δ in the small (J_2, δ) region, is derived. Thus precautions have to be taken to avoid confusion which may lead to incorrect interpretations of the FMR experimental data and consequently to erroneous values of J_1 and J_2. All these effects will be described and discussed.

Key words: Ferromagnetic resonance, magnetic anisotropy, magnetic coupling, multilayer films.

1.   INTRODUCTION

The present work is devoted to the use of Ferromagnetic resonance (FMR) for the study of magnetic coupling in multilayers. Ferromagnetic

B. Aktaş et al. (eds.), Nanostructured Magnetic Materials and their Applications, 171–186.

Resonance is a powerful method for investigation of magnetic properties of single and coupled-layer thin films. It will be shown, however, that sometimes the interpretation of FMR spectra might not be straightforward, and precautions have to be taken to avoid incorrect analysis of experimental data. This fact will be shown on a particular example, and it is the objective of the present paper. The article is divided as follows. Different types of coupling (bilinear and biquadratic with coupling parameters J_1 and J_2 respectively) will be discussed in Section 2. A brief description of the FMR method will be given in Section 3. The particular situation studied here consists of coupled layers with the in-plane uniaxial anisotropy axes making an arbitrary angle δ between them (Section 4). The resonant frequency and FMR intensity of such a system are derived in Section 5. Detailed discussions of the effect of an additional biquadratic coupling J_2 and of the angle δ on the mode position and intensity will be investigated in Sections 6 and 7, respectively. It will be shown that, in some cases, an additional biquadratic coupling J_2 cannot be detected. Also, a small J_2 may lead to the same spectrum as a small δ value, making it difficult to distinguish the effect of δ from that of the coupling strength J_2. A relation establishing the equivalence between J_2 and δ in the small (J_2, δ) region, is derived in Section 8. Concluding remarks are compiled in Section 9.

2.   MAGNETIC COUPLING

The interlayer magnetic coupling is an interaction between the magnetizations of two ferromagnetic layers separated by a non magnetic interlayer (see Fig. 1). It was discovered [1,2] in 1986 and has been intensively studied since then [1-9,15-29]. It was observed that the two magnetizations (labeled as $\mathbf{M_A}$ and $\mathbf{M_B}$) prefer to be parallel to each other (Fig. 1a) or antiparallel (Fig. 1b). In the first case the coupling is said to be ferromagnetic and in the latter one, the coupling is antiferromagnetic. This interaction can be described by the following bilinear term in the energy

$$E_{AB} = -J_1 \frac{\mathbf{M}_A \cdot \mathbf{M}_B}{M_A \cdot M_B} \tag{1}$$

J_1 is the bilinear coupling parameter. The nature and the strength of the coupling are described by the sign and the magnitude of J_1. When J_1 is positive, the energy [Eq. (1)] is minimal when M_A and M_B are parallel (ferromagnetic coupling), while if it is negative, then the lowest energy is achieved when M_A and M_B are antiparallel (antiferromagnetic coupling). T

RKKY interaction could explain this kind of coupling; in this case the coupling is due to indirect exchange through conduction electrons of the intermediate layer.

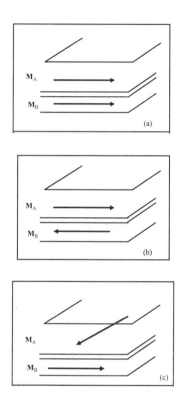

Figure 1. The configurations of the magnetization for (a) Ferromagnetic coupling ($J_1 > 0$) (b) Antiferromagnetic coupling ($J_1 < 0$) and (c) Biquadratic (the 90°-type coupling) coupling ($J_2 < 0$).

In 1991, another type of coupling, called biquadratic coupling [3], was observed. This coupling may favor a perpendicular alignement of the magnetizations (see Fig. 1c). It can be described by the following biquadratic term in the energy

$$E_{AB} = -J_2 \left(\frac{\mathbf{M}_A \cdot \mathbf{M}_B}{M_A \cdot M_B} \right)^2 \qquad (2)$$

J_2 is the biquadratic coupling parameter. If, J_2 dominates and is negative (which was experimentally observed), then the minimum energy [Eq. (2)] occurs when the magnetizations are oriented perpendicular to each other (the 90°-type coupling). The biquadratic exchange could arise from the variation of the interlayer thickness [4,5]. Magnetostatic coupling is also believed to be the source of this kind of coupling in certain systems [6].

Several methods have been used to detect and measure the magnetic coupling. Among them, one may cite the hysteresis curve, e.g [7], the torque method [8,9] and the Ferromagnetic resonance (FMR) [15-29]. The latter one will be described in that follows.

3.   FERROMAGNETIC RESONANCE

Ferromagnetic Resonance (FMR) refers to the the the absorption of power by a ferromagnetic material subject to a DC magnetic field H and excited by an alternating magnetic field h with a (angular) frequency ω. Generally, the frequency of the alternating magnetic field is in the microwave range. The absorbed energy ultimately appears as heat (the vibration of the lattice), since, by spin-lattice coupling, the energy is transferred from the spin system to the lattice. There are two ways to perform an FMR experiment. One may fix the frequency (using a fixed frequency resonant cavity spectrometer) and a variable DC field. At a certain field, called the resonant field H_R, the required relation between the frequency ω and H is satisfied and absorption of microwave power is observed. The resonance relation between ω and H involves the magnetization, the various magnetic anisotropy field contributions and the gyromagnetic ratio of the material. Alternatively one may use a variable frequency set up (frequency sweeper) and a fixed DC field, in this case the frequency corresponding to the absorption is the resonant frequency.

In single thin films, FMR has been used successfully to analyze the magnetic properties [10-14]. From the mode position, one may obtain, under appropriate circumstances, the anisotropy fields and the gyromagnetic ratio, γ, or equivalently the g-value. The linewidth is another fingerprint of the magnetic material. The linewidth is defined as the full width at half amplitude of the absorption curve. There are two main contributions to the linewidth [10-13]: intrinsic broadening, which relates to damping (thus one can measure the damping constant) and inhomogeneous broadening, which is associated with inhomogeneity [13] and the anisotropy dispersion in the film. Also, in single thin films and under certain circumstances, the number of peaks in the FMR spectrum will correspond to the number of magnetically different regions in the films [14]; since the different regions resonate at different applied DC fields (or different frequencies). Moreover, the intensity (area under the curve) is, in some cases, proportional to the product of the volume of the region and its magnetization. Hence the magnetic properties and the relative volume of each region in the film can be found. Thus FMR can single out different regions with different magnetic properties [14]. This is indeed an advantage of FMR over other magnetic characterization methods such as torque magnetometer or the vibrating sample magnetometer which give a volume average of the magnetic properties.

In a coupled layer system, FMR has been used since the discovery of the magnetic coupling [15-29]. It was found that the number, the position, the linewidth and the intensity of the FMR peaks are related to the magnetic coupling parameters, J_1 and J_2. Therefore, FMR can be used to detect and measure the magnetic coupling parameters, J_1 and J_2. However, it will be shown that in some situations some precautions have to be taken to avoid incorrect interpretation of the FMR data, which may lead to erroneous values of J_1 and J_2. One of these situations is illustrated in the following section.

4. SYSTEM OF COUPLED LAYERS WITH ARIBITRARY IN-PLANE ANISOTROPY AXES

It is assumed that the two thin film layers, denoted as A and B, lie in the *x-y* plane, with the *z*-axis normal to the film plane (see Fig. 2). They are coupled to each other through a non-magnetic layer. The magnetization M_A of layer A is defined, in spherical coordinates, by the angles θ_A and ϕ_A; and similarly M_B (layer B) by the angles θ_B and ϕ_B. The layers are supposed to have in-plane uniaxial magnetocrystalline anisotropies. For the layer A, the easy axis of such an anisotropy is taken to be along the *x*-axis, the corresponding energy then is

$$E_{Ain} = K_A \sin^2\theta_A \cos^2\phi_A \tag{3}$$

For the layer B, the easy axis has an arbitrary direction within the plane, it makes a δ angle with the *x*-axis. Following the analysis in ref. [18], the magnetocrystalline energy can be written as

$$E_{Bin} = -K_B \sin^2\theta_B \cos^2(\phi_B - \delta) \tag{4}$$

Here K_A (K_B) is the in-plane magnetocrystalline anisotropy constant for the layer A (layer B); K_A and K_B are positive since it is assumed that the anisotropy axes are easy directions. The external applied magnetic field H is taken to be in the plane of the films, making an α angle with the x-axis (the general formula have been derived for an arbitrary α value, however in the following analysis, α is set equal to zero, see Fig. 2). The microwave field h is along the y-axis. With all these considerations, the total free energy of the system per unit area can be explicitly written as

$$
\begin{aligned}
E = t_A & \left\{ \begin{array}{l} -M_A H \sin\theta_A \cos(\alpha - \varphi_A) + \\ + K_{ueffA} \sin^2\theta_A - K_A \sin^2\theta_A \cos^2\varphi_A \end{array} \right\} + \\
+ t_B & \left\{ \begin{array}{l} -M_B H \sin\theta_B \cos(\alpha - \varphi_B) + K_{ueffB} \sin^2\theta_B - \\ -K_B \sin^2\theta_B \cos^2(\varphi_B - \delta) \end{array} \right\} - \\
& -J_1 \left\{ \sin\theta_A \sin\theta_B \cos(\varphi_A - \varphi_B) + \cos\theta_A \cos\theta_B \right\} - \\
& -J_2 \left\{ \sin\theta_A \sin\theta_B \cos(\varphi_A - \varphi_B) + \cos\theta_A \cos\theta_B \right\}^2
\end{aligned}
\tag{5}
$$

In the two first terms of Eq. (5), t_A and t_B are the thicknesses of the layers A and B respectively. The total energy for the layer A (the first term) consists of (i) the Zeemann energy (interaction of the magnetization M_A with the external magnetic field), (ii) the shape and any out-of-plane uniaxial anisotropy characterized by the constant $K_{ueffA} = K_{uA} - 2\pi M_A^2$, K_{uA} is the out-of-plane magnetocrystalline anisotropy constant for layer A and (iii) the in-plane magnetocrystalline anisotropy [Eq. (3)]. Similarly for layer B (the second term in Eq. (5)), the same relations hold by changing the subscript A by B; the in-plane anisotropy is, however, given by Eq. (4). The inter-layer coupling energy is given by the two last terms. Note that at equilibrium, the magnetizations M_A and M_B must lie in the film plane, i.e. $\theta_A = \theta_B = 90°$, due to the strong demagnetizing field of the thin films and to the fact that the applied magnetic field lies in the plane of the films.

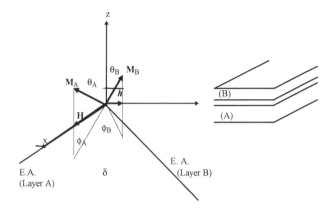

Figure 2. The coupled layers with the coordinate system. E.A.: Easy Axis. M_A [M_B] : magnetization of layer (A)[(B)]. H : DC applied magnetic field. h : rf magnetic field. δ : angle between the easy axes of layers (A) and (B).

5. FMR MODE POSITION AND INTENSITY

5.1 Mode position

The normal modes of the system can be found by the well-known energy method [18,19]. In this case the equations coupling $\Delta\theta_i$, $\Delta\phi_i$ (i = A, B), the excursions during oscillations about the equilibrium position can be written in a matrix form [18,19]. The matrix elements consist of the second derivatives of the energy E with respect to θ_i and ϕ_i (i = A,B). A solution of the form exp($i\omega t$) will be taken, ω is the (angular) frequency of precession. The solutions (normal modes) of the system will be found by setting the determinant of the matrix to zero. Using a variable frequency and a fixed DC field set up, one finds the following fourth-order equation in ω (the resonant frequency) [20].

$$\left[\frac{ab}{\gamma_A\gamma_B}\right]^2\omega^4-\left[a^2b^2\left(\frac{H_1^A H_2^A}{\gamma_B^2}+\frac{H_1^B H_2^B}{\gamma_A^2}\right)+abc_1\left(\frac{aH_2^B}{\gamma_A^2}+\frac{bH_2^A}{\gamma_B^2}\right)\right.$$

$$+abc_2\left(\frac{aH_1^B}{\gamma_A^2}+\frac{bH_1^A}{\gamma_B^2}\right)+c_1c_2\left(\frac{a^2}{\gamma_A^2}+\frac{b^2}{\gamma_B^2}\right)+\frac{2c_0c_2ab}{\gamma_A\gamma_B}\right]\omega^2 \qquad (6)$$

$$+\left[abH_2^A H_2^B+c_2(aH_2^A+bH_2^B)\right]\times$$

$$\times\left[abH_1^A H_1^B+c_1(aH_1^A+bH_1^B)+(c_1^2-c_0^2)\right]=0$$

Here $a=t_A M_A$, $b=t_B M_B$, γ_A and γ_B denote the gyromagnetic ratios of layers A and B respectively. The parameters c_j contain the coupling strength : $c_0 = J_1 + 2J_2 \cos(\phi_A-\phi_B)$, $c_1 = J_1 \cos(\phi_A-\phi_B) + 2J_2\cos^2(\phi_A-\phi_B)$ and $c_2 = J_1 \cos(\phi_A-\phi_B) + 2J_2\cos2(\phi_A-\phi_B)$. The fields H_j^i are $H_1^A = H\cos(\alpha-\phi_A) - H_{KeffA} + H_{KA}\cos^2\phi_A$, $H_2^A = H\cos(\alpha-\phi_A) + H_{KA}\cos2\phi_A$, $H_1^B = H\cos(\alpha-\phi_B) - H_{KeffB} + H_{KB}\cos^2(\phi_B-\delta)$ and $H_2^B = H\cos(\alpha-\phi_B) + H_{KB}\cos2(\phi_B-\delta)$. Finally $H_{KeffA} = 2K_{ueffA}/M_A$ $(H_{KeffB} = 2K_{ueffB}/M_B)$ and $H_{KA} = 2K_A/M_A$ $(H_{KB} = 2K_B/M_B)$ are respectively the effective uniaxial and the planar anisotropy fields for layer A (layer B).

5.2   FMR intensity

The corresponding mode intensities are given by [20]

$$I=\frac{2ab\omega^2(aq+b)^2}{ab\omega^2(aq^2+b)+b\gamma_A^2[qaH_2^A+c_2(q-1)]^2+a\gamma_B^2[bH_2^B-c_2(q-1)]^2} \qquad (7)$$

where

$$q=\frac{\gamma_A^2 c_2 b(aH_1^A+c_1)+\gamma_A\gamma_B c_0 a(bH_2^B+c_2)}{\gamma_A^2 b(aH_1^A+c_1)(aH_2^A+c_2)+\gamma_A\gamma_B c_0 c_2 a-a^2 b\omega^2}$$

When the layers are uncoupled, Eq. (7) should reduce to the expressions giving the intensity of two individual thin films (layer A and B). Indeed, for the uncoupled layers, $c_j = 0$, thus $q = 0$. The intensity of layer A is found to be:

$$I_A=\frac{2M_A t_A(H+H_{KA}-H_{KeffA})}{(2H+2H_{KA}-H_{KeffA})} \qquad (8)$$

Equation (8) is in agreement with the formula of the intensity for a thin film [21], with specific anisotropy (in the present case, the in-plane uniaxial anisotropy).

In the following, Eqs. (6) and (7) will be solved. The coupling is assumed to be antiferromagnetic ($J_1 < 0$). The purpose of the following analysis is to study the effect on the resonance modes, (i) of an additional biquadratic coupling (superposed to the bilinear negative J_1) (ii) of the relative directions of the in-plane anisotropy axes (given by the angle δ).

6. EFFECT OF AN ADDITIONAL BIQUADRATIC COUPLING

For a fixed coupling ($J_1 = -2$ ergs/cm^2), the problem is to see the effect of the additional biquadratic coupling (J_2 from 0 to -0.5 erg/cm^2) on the FMR mode position and intensity. In this (AF) coupling situation, the higher (lower) frequency mode is the acoustic (optical) mode, which is the opposite of the (F) coupling case [20] (and also opposite to the resonant field analysis, Ref. [15]). Two situations can arise. When the applied field H is greater than the saturation field [20], H$_{sat}$, the magnetizations are parallel, this case is displayed in Fig. 3 (resonant frequency and mode intensity vs the biquadratic coupling $|J_2|$). One can see that Eq. (6) predicts two modes. The position of the high frequency (the acoustic) mode does not vary with increasing $|J_2|$ (see Fig. 3) while that of the low frequency (the optical) one decreases almost linearly. However, the latter mode has zero intensity [Eq. (7)] and the acoustic mode has a constant intensity (see Fig. 3, for the intensity). Thus, in practice, only one peak will appear in the FMR spectrum. Since its position and intensity are constant with J_2, it is hard to detect the biquadratic coupling in this case.

On the other hand, and for the same parameters, when the magnetizations are antiparallel, the mode behavior is different. In Fig. 4, the applied field, $H = 2$ kOe, is less than the critical field H$_{crit}$ [20], the magnetizations will align antiferromagnetically under the strong bilinear J_1 coupling (assuming $t_A M_A > t_B M_B$, one will have $\phi_A = 0°$ and $\phi_B = 180°$). Two modes are expected. The variations of the mode positions and intensities as a function of $|J_2|$ are shown in Fig. 4. The resonant frequencies of the the lower mode (the optical mode) and the upper (acoustic) mode decrease as $|J_2|$ increases. Note that, contrary to the first case, the intensities of both modes are non-zero and do vary with J_2. The intensity of the acoustic (optical) will increase (decrease) with J_2. Thus, in this case from the mode position and intensity, one can detect and measure the biquadratic coupling strength.

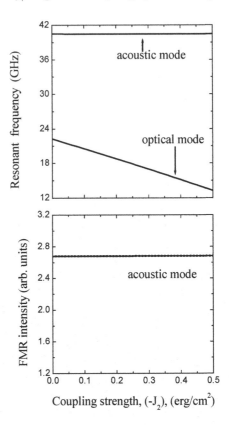

Figure 3. Resonant frequency and FMR intensity vs. the biquadratic coupling $|J_2|$. (J_1 = - 2 erg/cm^2) H = 10 kOe > H_{sat}. Other parameters: Layer A : $4\pi M_A$ = 10kG, H_{keffA} = -10kOe, H_{KA} = 0.5 kOe, $\gamma_A/2\pi$ = 2.8 GHz/kOe, t_A =200Å. Layer B : $4\pi M_B$ = 6 kG, H_{keffB} = -6 kOe, H_{KB} = 0.5 kOe, $\gamma_B/2\pi$ = 2.9 GHz/kOe, t_B = 100 Å.

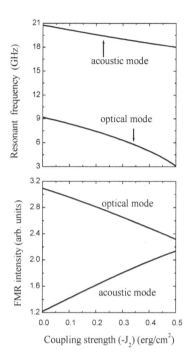

Figure 4. Resonant frequency and FMR intensity vs. the biquadratic coupling $|J_2|$. ($J_1 = -2$ erg/cm^2) $H = 2$ kOe $< H_{crit}$. Other parameters are as in Fig. 3.

7.   EFFECT OF ANISOTROPY AXES MISALIGNMENT

In the following, the effect of the δ angle on the resonant frequency and mode intensity will be explicitly examined. The fourth-order equation in ω [Eq. (6)] has to be solved for different δ values. Once again, the coupling is assumed antiferromagnetic (AF) (J_1 is negative). Also, depending on the applied magnetic field value, two situations have to be separately studied. When H is large (greater than the saturation field [20]), the magnetizations are parallel to the direction of H. In this case, it is found that the intensity of the optical mode is almost zero; thus practically, only one mode (the acoustic one) will be seen in the FMR spectra. As δ increases from 0° to 90°, the resonant frequency of this acoustic mode decreases slightly for all J_1, J_2 values, while the intensity remains practically constant with δ. Thus, the δ value will not greatly affect the FMR modes and, within the uncertainty of the experimental measurements, the effect of δ can hardly be detected in this situation [22].

When H is less than a critical value [20], the magnetizations will be antiparallel, with M_A along H ($\phi_A = 0$) and M_B in the opposite direction ($\phi_B = \pi$) (assuming $t_A M_A > t_B M_B$). Two modes are expected for this antiparallel alignment of the magnetizations [22]. When δ varies from 0° to 90°, the resonant frequency of both modes decreases. However, the variation of δ seems to affect more the optical than the acoustic mode positions for all J_1, J_2 values. An additional negative biquadratic coupling J_2 will only accentuate this effect; as an example, for the case where $J_1 = -2$ and $J_2 = -0.5$ erg/cm^2, the positions of the optical and acoustic modes shift downward by 6 and 0.4 GHz respectively [22]. As for the FMR intensities, it is found [22] that when δ is increased from 0° to 90°, the acoustic mode intensity, I_{ac}, increases by a large amount while the optical mode intensity, I_{op}, decreases slightly, with I_{op} remaining always greater than I_{ac} for the set of coupling values used here (usual experimental values [2]).

The behavior of the FMR modes as a function of δ [the decrease of the resonant frequency of both modes with δ and the increase (decrease) of I_{ac} (I_{op})] is qualitatively similar to that of the FMR modes vs. $|J_2|$ [20]. To illustrate this fact, the resonant frequency and the FMR intensity are plotted against δ (solid line) and against the biquadratic coupling $|J_2|$ (dashed line) in the same figure (Fig. 5) and under the same conditions ($J_1 = -2$ erg/cm^2 and $H = 0.90$ kOe, less than the critical field). Note, that in Fig. 5, for the variation with δ, it is assumed that there is no biquadratic coupling ($J_2 = 0$) and for the variation with ($-J_2$) it is assumed that $\delta = 0°$ (the easy axes are parallel), so that one can see the effect of one or the other. For certain ranges of δ (between 0° and 30°) and of $|J_2|$ (less than 0.15 erg/cm^2), one notices that the variation of δ and J_2 will give exactly the same optical mode resonant frequencies and practically the same intensity (see Fig. 5). For very low δ and J_2 values, even the corresponding frequencies and intensities of the acoustic mode are close enough to, experimentally, measure the same value. Therefore in this latter situation, it is hard to distinguish the effects of δ and J_2 from a FMR experiment; a difference in the easy axis directions could be easily taken for the existence of a biquadratic coupling. For instance, a 10° off-alignment of the easy axes in the two layers ($\delta = 10°$) will produce the same FMR spectrum as an additional biquadratic coupling $J_2 = -0.01$ erg/cm^2. This value ($J_2 = -0.01$ erg/cm^2) is a non-negligible one. Heinrich and co-workers have derived J_2 values as low as -0.0022, -0.015 and -0.027 erg/cm^2 in Fe/Cu/Fe/GaAs(100) [23], Fe/Cu/AgCu/Fe [24] and Ag/Fe/Cu/Fe [4] systems respectively. Also, theoretically, fluctuations of dipole interactions may give rise to a biquadratic coupling of the order of -0.01 erg/cm^2 [5]. Therefore, it is shown here that, if FMR data are based on the mode position, which is usually done, then a small difference in the in-plane easy axis directions may be incorrectly interpreted as a biquadratic coupling. Even the study of the mode intensity will not lift the degeneracy.

As δ and J_2 increase further, the difference between the FMR peaks (position and intensity) become more apparent. In this region, the shift to low frequency induced by J_2 is more important than the one produced by δ, for the acoustic mode (compare in Fig. 5, the solid and dotted lines). Moreover, the increase of the intensity ratio, I_{ac}/I_{op} , induced by J_2 is larger than the one due to δ.

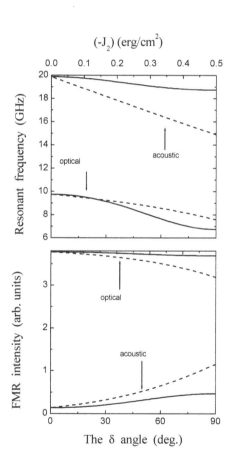

Figure 5. Mode position and intensity vs. the δ angle (solid line) and vs. the biquadratic coupling strength (- J_2) (dashed line). $J_1 = -2$ erg/cm^2, $H = 0.90$ kOe (less than the critical field, antiparallel magnetizations). For the solid line $J_2 = 0$, for the dashed line $\delta = 0°$. Other parameters are as for Fig. 3.

8.   EQUIVALENCE BETWEEN J₂ AND δ

In fact, going back to low δ and J_2 values, it is noted that the degeneracy does not arise only in the FMR experiments. It may also occur when deriving J_2 from a fit to the critical field H_{crit}, the field below which the magnetizations become antiparallel under a negative J_1. It will be shown in the following that low δ and J_2 values will lead to the same H_{crit}. The critical field might be found by the minimization of the total energy [Eq. (5)]. It is assumed here that $H_{KA} = H_{KB} = H_K$, i.e. the anisotropy fields are the same, only the relative directions of the easy axes in the two layers (the δ angle) is of interest. When $\delta = 0°$ and J_2 is small compared to J_1, the critical field is found to be:

$$H_{crit} = \frac{1}{2}\left[H_0 - \frac{J_1}{d_-}\right] + J_2\left[\frac{1}{d_-} + \frac{1}{H_0}\left(\frac{2H_K}{d_+} - \frac{J_1}{d_-^2}\right)\right]$$ (9)

Here

$$H_0 = \sqrt{\frac{J_1^2}{d_-^2} - \frac{4H_K J_1}{d_+} + 4H_K^2}\,,$$

and

$$\frac{1}{d_\pm} = \frac{1}{t_B M_B} \pm \frac{1}{t_A M_A}\,.$$

On the other hand, when $J_2 = 0$, and δ is a small angle, then H_{crit} is given by the following relation :

$$H_{crit} = \frac{1}{2}\left[H_0 - \frac{J_1}{d_-}\right] + \delta^2\left[-H_K + \frac{1}{H_0}\left(\frac{J_1 H_K}{d_+} - 2H_K^2\right)\right]$$ (10)

where δ is expressed in rad. Comparing Eqs. (9) and (10), the following relation between δ and J_2 is derived

$$J_2 = \frac{-H_0 H_K + \dfrac{J_1 H_K}{d_+} - 2H_K^2}{\dfrac{H_0}{d_-} + \dfrac{2H_K}{d_+} - \dfrac{J_1}{d_-^2}} \delta^2 \tag{11}$$

Eq. (11) establishes an equivalence between the biquadratic coupling J_2 and the misalignment of the in-plane easy axes. Numerically, using Eq. (11), $\delta = 10°$ will lead to $J_2 = -0.012$ erg/cm^2, a value very close to the one derived by considering FMR mode positions and intensities. Thus, this relationship between J_2 and δ [Eq. (11)], which was derived from the critical field, accounts also for data based on FMR.

9. CONCLUSION

The effect of the biquadratric coupling J_2 and the relative direction of the in-plane uniaxial anisotropy axes in the two coupled layers (given the angle δ) on the mode caracteristics are studied. The bilinear coupling is assumed to favor the antiferromagnetic (AF) alignement, i.e. $J_1 < 0$.

(i) If the magnetizations are parallel ($H > H_{sat}$) only the acoustic mode will appear with constant mode position and intensity for all J_2 values, making it difficult to detect any additional biquadratic coupling. On the other hand, and for the same parameters, if the magnetizations are antiparallel ($H < H_{crit}$), then two modes are predicted; as $|J_2|$ increases the intensity of the acoustic (optical) mode will increase (decrease) while the resonant frequency of both modes decrease. The additional biquadratic coupling can then be detected.

(ii) When the magnetizations are antiparallel ($J_1 < 0$ and $H < H_{crit}$), a coupled system with a small δ value will produce the same FMR spectrum (position and intensity) as one with a small $|J_2|$ value, and even the critical field, H_{crit}, will be the same. Hence, a slight deviation of the easy axis direction of one layer with respect to the one of the second layer may be interpreted as an existence of a small biquadratic coupling.

(iii) A relation establishing the equivalence between J_2 and δ in the region of small (J_2, δ), is derived, a relationship which is consistent with the data based on FMR and on the critical fields.

REFERENCES

[1] P. Grünberg, R. Schreiber, Y. Pang, M.B. Brodsky and H. Sowers, Phys. Rev. Lett. 57, 2442, (1986).

[2] P. Grünberg, Acta mater. 48, 239 (2000).

[3] M. Rührig, R. Schäfer, A. Hubert, R. Mosler, J.A. Wolf, S. Demokritov and P. Grünberg, Phys. Stat. Sol. (a) 125, 635 (1991).

[4] B. Heinrich, Can. J. of Phys. 78 (3), 161 (2000).

[5] J.C. Slonczewski, J. Magn. Magn. Mat. 150, 13 (1995).

[6] U. Rüker, S. Demokritov, B. Tsymbal, P. Grünberg and W. Zinn, J. Appl. Phys. 78, 387 (1995).

[7] H.J. Elmers, G. Liu, H. Fritzshe, and U. Gradmann, Phys. Rev. B 52, R696 (1995).

[8] C. Williams, J.J. Krebs, F. J. Rachford, G. A. Prinz, A. Chaiken, J. Magn. Magn. Mater. 110, 61 (1992).

[9] A. Layadi, J. Magn. Magn. Mater. 192, 353 (1999).

[10] A. Layadi, J. Appl. Phys. 87, 1429 (2000).

[11] A. Layadi, J. Appl. Phys. 86, 1625 (1999).

[12] B. Heinrich, S.T. Purcell, J.R. Dutcher, K.B. Urquhart, J.F. Cochran and A.S. Arrot, Phys. Rev. B 38, 12879, (1988).

[13] B. Heinrich, J.F. Cochran, M. Kowelewski, J. Kirschner, Z. Celinski, A.S. Arrot and K. Myrtle, Phys. Rev. B 44 (17), 9348 (1991).

[14] J. O. Artman , J. Appl. Phys. 61, 3137 (1987).

[15] B. Heinrich, In Ultrathin magnetic structures II. Edited by B. Heinrich and J.A.C. Bland, Springer-Verlag, Berlin, 1994.

[16] B. Heinrich and J.F. Cochran, Adv. Phys. 42, 523 (1993).

[17] A. Layadi and J.O. Artman, J. Magn. Magn. Mater. 92, 143 (1990), J. Magn. Magn. Mater.176, 175 (1997).

[18] A. Layadi, Phys. Rev. B 63, 174410, (2001).

[19] Z. Zhang, L. Zhou, P.E. Wigen and K. Ounadjela, Phys. Rev. B50, 6094 (1994).

[20] A. Layadi, Phys. Rev. B 65, 104422, (2002).

[21] B. Heinrich et al., Phys. Rev. Lett. 59, 1756, (1987).

[22] A. Layadi, J. Magn. Magn. Mat. 266, 282 (2003).

[23] T.L. Monchesky, B. Heinrich, R. Urban, K. Myrtle, M. Klaua and J. Kirchner, Phys. Rev. B 60 (14), 10242 (1999).

[24] M. Kowalewski, B. Heinrich, J. F. Cochran and P. Schurer, J. Appl. Phys. 81 (8), 3904 (1997).

[25] M.E. Filipkowski, C.J. Gutierrez, J.J. Krebs and G.A. Prinz, J. Appl. Phys. 73, 5963 (1993).

[26] A. Layadi, J. Appl. Phys. 83, 3738, (1998). (Errata) : , J. Appl. Phys. 85, 7483, (1999).

[27] A. Azevedo, C. Chesman, S.M. Rezende, A.F. de Aguiar, X. Bian and S.S.P. Parkin, Phys. Rev. Lett. 76, 4837 (1996).

[28] S.M. Rezende, C. Chesman, M.A. Lucena, A. Azevedo, F.M. de Aguiar and S.S.P. Parkin, J. Appl. Phys. 84, 958, (1998).

[29] S.M. Rezende, C. Chesman, M.A. Lucena, A. Azevedo, F.M. de Aguiar and S.S.P. Parkin, J. Magn. Magn. Mat. 177-181, 1213 (1998).

MAGNETIC PROPERTIES OF AN INTERCALATE
$Mn_{0.86}PS_3(ET)_{0.46}$: AN ESR STUDY

Y. Köseoğlu [a], F. Yıldız [b] and B. Aktaş [b]

[a] *Fatih University, Physics Department, Istanbul, Turkey*
[b] *Gebze Institute of Technology, Kocaeli, Turkey*

Abstract: $MnPS_3$ intercalated with BEDT-TTF (ET) has been studied by Electron Spin Resonance (ESR) technique in the temperature range of 10-300K. A single and symmetric ESR peak was observed. The peak is much narrower than the expected from the paramagnetic Mn^{2+} ion. The ESR signal intensity, line width and effective g-value were observed to change dramatically at about 45 K, indicating a magnetic phase transition. That is, while the intensity is monotonically increased until a critical temperature T_c=45 K , both the line width and g-value remain almost constant above Tc. However, a very sharp decrease in ESR signal intensity, accompanied by line broadening below T_c, indicates a very strong spin frustration at lower temperatures. Some small and relatively smooth changes also occurred at about 30 K, 100 K and 170 K. However, below 30 K, the magnetization however starts to increase again and the line-width decreases with decreasing temperature. This points to the onset of a second transition toward ferromagnetic phase.

Key words: Molecular magnet, Intercalation, Electron Spin Resonance (ESR).

1.   INTRODUCTION

The intercalated transition metal phosphorous trichalcogenides (hexathiohypodiphosphates) MPS_3, (M= Mn, Co, Fe, Ni, Zn and Cd, etc.) have attracted considerable attention recently due to their interesting electronic and magnetic properties. The compounds MPS_3 crystallize in layered structures (monoclinic, space group C_2/m) and order antiferromagnetically below a critical temperature, viz. 78K for $MnPS_3$, 120K for $FePS_3$. Above these temperatures, they are paramagnets with

B. Aktaş et al. (eds.), Nanostructured Magnetic Materials and their Applications, 187–197.
© 2004 *Kluwer Academic Publishers. Printed in the Netherlands.*

antiferromagnetic coupling parameter, J [1-2]. MnPS$_3$ is a quasi-2D S=5/2 antiferromagnet with inter-planar coupling $J' = J/405$ in which the Mn^{2+} moments form honeycomb layers in the ab-planes. Below T$_N$=78.6 K the moments align perpendicular to the planes. The isotropic exchange coupling parameters are J_1=0.77 meV, J_2=0.07 meV and J_3=0.18 meV for the nearest, second- and third-nearest neighbours, respectively [3]. Unlike MnPS$_3$, the magnetic structure of FePS$_3$ and NiPS$_3$ consists of parallel ferromagnetic chains coupled to each other antiferromagnetically but the Mn^{2+} ions in MnPS$_3$ do not form one dimensional magnetic chains [4].

The intercalation compounds show spontaneous magnetization at low temperatures. For intercalates Mn$_{0.80}$PS$_3$ (α-aminopyridine)$_{0.40}$ and Mn$_{0.86}$PS$_3$(bipy)$_{0.56}$, the antiferromagnetic transition (characteristic of pure MnPS$_3$) no longer exists. Instead, an abrupt increase occurs in magnetic susceptibility at about 40K, which indicates a spontaneous magnetization [5]. Magnetic susceptibility measurements and electrochemical impedance spectroscopy show that the antiferromagnetic interactions present in the phases MPS$_3$ (M= Cd, Fe, Mn) get attenuated in the mixed phases M$_{0.5}$In$_{0.33}$PS$_3$. This is attributed to the larger separation of the M^{2+} ions in the mixed phase and, therefore, a decrease of magnetic interaction. Also, Fe$_{0.5}$In$_{0.33}$PS$_3$ shows an electrical conductivity of σ = 3.0 x10^{-8} S/cm [6].

Investigation of the temperature/field/composition magnetic phase diagram of Mn$_{1-x}$Mg$_x$PS$_3$ (x=0, 0.05, 0.075 and 0.10) indicates that these samples are also quasi-2D layered antiferromagnets with an out-of-plane Ising anisotropy and, therefore, they undergo a spin-flop transition, except in the case of x=0.10 where no spin-flop was found [7]. The successful insertion of polyaniline in MnPS$_3$ leads to a paramagnetic state, which is different from antiferromagnetic interaction of the pure host. This might be the result of disorder of metallic vacancies [8]. In the case of Me-Mepepy intercalated into thin films of MPS$_3$ (M=Mn, Cd) irradiation causes trans to cis isomerization as shown by photoisomerization. Irradiation also modifies the interactions between the chromophores, which cause a modification in the value of the spontaneous magnetization of the material [9]. Intercalation of potassium and indium causes a decrease in the antiferromagnetic interactions and, in addition, produces a significant increase of the magnetic susceptibility below 40 K, which supports the conjecture that the trivalent cation penetrates into the lamellae of the host matrix [10]. Neutron scattering experiments on single crystal MnPS$_3$ reveal temperature-dependent scattering well above the Neel temperature, due to two-dimensional critical magnetic fluctuations, not Bragg rods [11]. The S-K and P-K polarized absorption spectra measurements on the layered MPS$_3$ (M=Mn, Fe, Ni, Zn, Mg) single crystals indicate that the S-p$_z$ state is largely mixed with the d orbitales of transition metal ion. Also, the P-p states and S-p states are overlapped each other in the transition metal thiophosphates [12].

EPR studies on intercalation ferromagnets $Mn_{0.79}PS_3(4,4'\text{-bipy})_{0.42}$ and $Mn_{0.84}PS_3(1,10\text{-phen})_{0.64}$ [13] and $Mn_{0.86}PS_3(bipy)_{0.56}$ [14] indicate that the nature of organic guest cation appears to have a strong influence on the magnetic behaviour of this inorganic-organic hybrid system. The continued appearance of the exchange-narrowed sharp line, characteristic of the paramagnetic phase, even at 15 K suggests that the bulk of the intercalate ferromagnets in this system do not attain a completely ordered state [13]. In addition, one observes the presence of regions of magnetic frustration at 70 K, just below the antiferromagnetic ordering temperature of the parent phase $MnPS_3$, fluctuations in the region 35-20 K, which is just below the Curie temperature of the intercalate, and the existence of two inequivalent sites of Mn in the ferromagnetic state [14].

Hybrid compounds in which organic species are interleaved with inorganic planar lattices offer an opportunity to design new architectures with novel electric/magnetic properties [15]. For instance, hybrid intercalates $M2(OH)3X$, where M=Co or Cu, and X is an exchangeable organic anion, spontaneous magnetization is known to occur (Tc = up to 21K), even for interlayer spacing d =40 Angstroms, even though the parent inorganic layered hydroxide is not ferromagnetic. This has been explained to occur due to dipolar through-space interactions, which promote super spins within the magnetic layers to yield ferromagnetic coupling [16].

The hexathiohypodiphosphate antiferromagnet $MnPS_3$ (TN=78K), which forms lamellar compounds with unique reactivity, since they are able to exchange a fraction of their Mn2+ cations with a great variety of cationic species which can occupy guest sites in the interlayer galleries. The observation of spontaneous magnetization in MnPS3 intercalates has been suggested to result from the imbalance of spins on the two magnetic sub-lattices of the parent antiferromagnetic (TN=78K) caused by intercalation of organic cation guest molecules which bring about interlayer vacancies of Mn2+ cations, and in turn a drastic modification of the long range magnetic behaviour of $MnPS_3$. It is, therefore, of interest to try to intercalate different organic cation hosts into the $MnPS_3$ host lattice, which should lead to alteration of $MnPS_3$ sandwiches and Van der Waals gaps along the stacking c-direction. The presence of empty and accessible sites in the VdW gaps made it possible for us to insert, for the first time, the large cation molecules BEDT-TTF, which are known to be excellent 'donor' molecules in the charge-transfer conductors and organic superconductors. As expected, this has resulted in the observation of a rich variety of magnetic features in the

BEDT-TTF intercalated $MnPS_3$, including a magnetic transition at 45 K and spin frustration effects at low temperatures.

In this study, we have reported the preliminary results of high frequency magnetic properties by using Electron Spin Resonance technique in the temperature range of 10K-300K.

2.   EXPERIMENTAL

BEDT-TTF intercalated $MnPS_3$ sample was obtained by treating preintercalate, phen-$MnPS_3$ with $(ET)_2Ix$, followed by filtration and then dried in air. The preintercalate phen-$MnPS_3$ was obtained from the reaction of $MnPS_3$ with 1,10-phenanthroline monohydrate in the presence of HOAc in ethanol for several days. The occurrence and completion of the intercalation reaction were ascertained by the X-ray powder diffraction of the final product. Elemental analysis led to the stoichiometry of the intercalate to be $Mn_{0.86}PS_3(ET)_{0.46}$. The X-ray powder diffraction reflections of the intercalate was indexed in a monoclinic unit cell with lattice parameters a =5.977 Å, b = 10.231 Å, c = 23.147 Å, β = 114.63° [18]. Then, the polycrystalline powders of BEDT-TTF-$MnPS_3$ placed in nonmagnetic paraffin just above its melting temperature have been tried to orient in the presence of a strong magnetic field (15 kG). The samples were subsequently cooled down below the melting temperature of paraffin in this field to have magnetic orientation. A sample with dimensions 1.5x2x2.5 mm was cut from this ingot for ESR measurements. A conventional X-band (f $\cong$ 9.7 GHz) Bruker EMX model spectrometer employing an ac magnetic field modulation technique was used to record the first derivative absorption signal. An Oxford continuous helium gas flow cryostat has been used, allowing the X-band microwave cavity to remain at ambient temperature during ESR measurements at low temperatures. The temperature was stabilized by a conventional Lakeshore 340 temperature controller within an accuracy of 1 degree between 10-300 K. A goniometer was used to rotate the sample with respect to the external magnetic field in order to observe angular variations of the ESR spectra

3.   RESULTS AND DISCUSSION

ESR measurements have been carried out at different samples temperature. A strong and narrow microwave resonance absorption signal (with approximately g=2.009, which is close to g-value of the free electron

2.0023, and peak-to-peak line width = 80 G) is observed at room temperature as seen in the Fig. 1. A value of about 2 for g is commonly expected from an S state ion like Mn^{2+}.

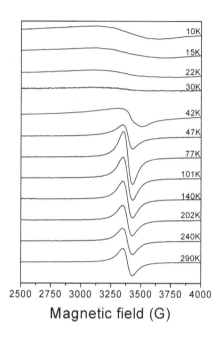

Figure 1. ESR spectra of BEDT-TTF Intercalated $MnPS_3$ in the temperature range of 10-290 K.

As can be seen from this figure, the spectra remain almost unchanged as the temperature is decreased until 50 K. However there are dramatic changes under this temperature. That is the ESR signal is strongly broaden and accordingly its amplitude start to diminish as the temperature is decreased further below this temperature. This may come from the domination of antiferromagnetic interactions.

The peak-to-peak line width values obtained from the experimental spectra as a function of temperature are given in Fig. 2.

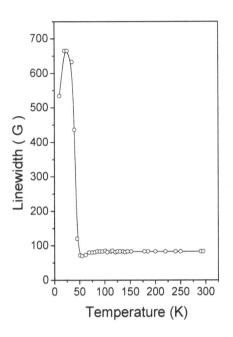

Figure 2. Line width against temperature of BEDT-TTF intercalated MnPS$_3$. Lines are guides to eye.

As can be seen from this figure, the ESR line width remains practically the same down to about 50 K. After this temperature it increases very sharply, indicating a phase transition. Normally one would expect for ESR spectrum originating from the transitions between the magnetic states ($|Ms>$-$|Ms-1>$) to split into 6 hyperfine components separated about 90-110 gausses in field from each other. Since paramagnetic Mn^{2+} spins see their own nuclear spins distributed randomly between 6 nuclear states from one ion to another, then each individual Mn^{2+} electronic spin in the sample would resonate independently from each other. If the intrinsic line width had been larger than this splitting, then the six components would overlap giving broader (at least 600 gauss) single line. However, the present line width is about 80 gauss. This means that, the Mn^{2+} ion does not behave as an isolated paramagnetic ion even at higher (room) temperatures. That is they are coupled together so strongly that they exhibit collective behaviour. If they are strongly exchange coupled, then a cluster of molecules containing Mn^{2+} spins become ferromagnetic (or antiferromagnetic). In this case, due to exchange narrowing the hyperfine splitting is suppressed in the FMR spectra. In other words there is a large exchange narrowing effect even at room temperature.

Below 50 K, ESR line is broadening possibly due to spin disorder (or frustrations,) coming from antiferromagnetic interactions (or spin-fluctuations) between clusters or within a single cluster [21]. However the line width curve passes a maximum at around 30 K and starts to decrease again with decreasing temperature. This decrease can be taken as a sign for ferromagnetic phase inclusion at lower temperatures. This ferromagnetic phase inclusion might be the result of disorder of metallic vacancies [8]. It should be noted that the guest cation in organic intercalated compound is already ferromagnetic [14,18] at low temperatures. These ferromagnetic interactions cause an exchange narrowing in ESR spectra at low temperatures.

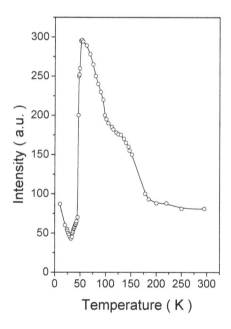

Figure 3. ESR signal intensity versus temperature of BEDT-TTF intercalated MnPS₃.

The intensity value obtained from the second integral of the ESR signal is shown in Fig. 3 as a function of the temperature. In order to study temperature behaviour of the magnetization more closely, also inverse-intensity (1/intensity) values have been plotted in Fig. 4.

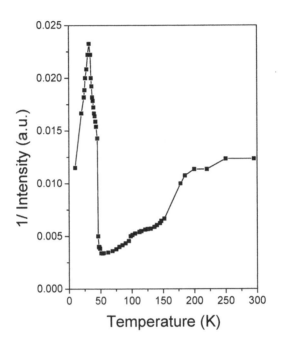

Figure 4. 1/Intensity versus temperature of BEDT-TTF intercalated MnPS3. Lines are guides to eye.

As well known, for a magnetic material in paramagnetic regime obeying Curie law, the temperature dependence of the susceptibility value should be simply proportional to 1/T, and the intersection of inverse susceptibility curve with the horizontal axis should corresponds to the Curie (or Neel) temperature, which is positive for ferromagnetic (or negative for antiferromagnetic) interactions [19]. While the susceptibility (which is proportional linearly to the ESR intensity) above 50K obeys very roughly the Curie law, its behaviour is drastically different below 50 K. That is the sample is not purely paramagnetic. Obviously, there are some anomalies at some certain temperature intervals limited by 30 K, 50 K, 100 K, 150 K.

As one can see from Fig. 3 the intensity (that is proportional to the magnetization value) decreases sharply at the temperatures below 50 K, which is consistent with, spin disorder interpretation given above. It should be noted that the magnetization passes through a minimum at the same temperature (Tc=30 K) corresponding to a maximum for the line width curve and then starts to increase again as the temperature is decreased further. This behavior is also consistent with ferromagnetic phase inclusion below 30 K.

On the other hand the magnetization roughly decreases with the increasing temperature above 50 K. However some critical temperatures still exist for this curve. The monotonic decrease up to 100 K becomes faster at this temperature and the curve retains its previous trend until about 150 K. The curve shows another small and sharp decrease at this temperature value and then almost level of beyond this temperature value. The overall temperature behaviour might be accounted for mainly competition between antiferromagnetic and ferromagnetic interactions.

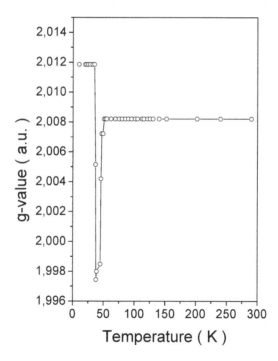

Figure 5. g-value of BEDT-TTF intercalated MnPS₃ against temperature. Lines are guides to eye.

Lastly, the g-values calculated by using the resonance field values obtained from the intersection of ESR line with base line are plotted in Fig.

5. As seen from this figure, effective g-value remains almost constant as the temperature is decreased until 50 K while some insignificant changes appear at the temperatures of 100 and 150 K. However there is a sudden change at about 50 K. Moreover, the increase below 50 K is stabilized at around 30 K.

As a summary, almost all the ESR parameters, g-value, line width, signal intensity and thereby magnetization exhibit remarkable changes at about 150 K, 100 K, 50 K and 30 K. The changes of these parameters occurred at these temperatures could be considered as a sign of mixed magnetic phase for this material. It should be noted that under about 30 K, magnetization starts to increase while the line width decreases with decreasing temperatures. This indicates a second transition toward ferromagnetic phase. These results imply that the magnetic properties of the intercalate are neither similar to the pre-intercalate $Mn_{0.86}PS_3(bipy)_{0.56}$ which exhibits spontaneous magnetization below 40 K [1] nor similar to the pristine $MnPS_3$ which shows antiferromagnetic behaviour below 78 K [20]. The critical temperature of pristine $MnPS_3$ is changed from 78 K to 50 K and additional change in susceptibility curve occurred at 30 K by ET intercalation. The successful insertion of ET in $MnPS_3$ leads to mixed magnetic phases, which is different from antiferromagnetic interaction of the pure host. This might be the result of disorder of metallic vacancies [8]. That is the magnetic structure of the host has been remarkably changed by ET intercalation.

REFERENCES:

[1] Jingui Qin, Chuluo Yang, Kyuya Yakushi, Yasuhiro Nakazawa and Kenji Ichimura, Solid State Commun., Vol. 100, No. 6, (1996) 427-431.

[2] Jingui Qin, Chuluo Yang, Kyuya Yakushi, Yasuhiro Nakazawa and Kenji Ichimura, Daoyu Liu, Synthetic Metals, 85 (1997) 1673-1674.

[3] H. M. Ronnow, A. R. Wildes, S.T. Bramwell, Physica B., 276-278 (2000) 676-677.

[4] Vladlen Zhukov, Santiago Alvarez and Dmitrii Novikov, J. Phys. Chem. Solids Vol. 57, No. 5, (1996) 647-652.

[5] Chuluo Yang, Xingguo Chen, Jingui Qin, Kyuya Yakushi, Yasuhiro Nakazawa and Kenji Ichimura, Yunqi Liu, Synthetic Metals 121 (2001) 1802-1803.

[6] V. Manriquez, P. Barahona, O. Pena, M. Mouallem-Bahout, R.E. Avila, J. Alloys and Compounds 329 (2001) 92-96.

[7] D.J. Goossens, T. J. Hicks, J. Magn. And Magn. Mater. 177-181 (1998) 721-722.

[8] Donghui Zhang, Jingui Qin, Kyuya Yakushi, Yasuhiro Nakazawa and Kenji Ichimura, Material Science and Engineering A, 286 (2000) 183-187.

[9] Anne Leaustic, Angelique Sour, Eric Riviere, Rene Clement, C.R.Acad.Sci.Paris, Chimie/ Chemistry 4 (2001) 91-95.

[10] Victor Manriquez, Antonio Galdamez, Anibal Villanueva, Pilar Aranda, Juan Carlos Galvan and Eduardo Ruiz-Hitzky, Materials research Bulletin, Vol. 34, No. 5, (1999) 673-683.

[11] A.R. Wildes, M.J. Harris, K.W. Godfrey, J. Magn. and Magn. Mater., 177-181 (1998) 143-144.

[12] A. Kamata, S.Nakai, T. Kashiwakura, K. Noguchi, N. Kozuka, T. Yokohama, H. Tezuka, Y. Ohno and O. Matsudo, J. Electron Spectroscopy and Related Phenomena, 79 (1996) 203-206.

[13] M.K. Bhide, R.M. Kadam, M.D. Sastry, Jingui Qin, Chuluo Yang, K. Yakushi, Yasuhiro Nakazawa, Kenji Ichimura, A.K. Sra, J.V. Yakhmi, Mol. Cryst. and Liq. Cryst., Vol. 341, (2000) 119-124.

[14] M.K. Bhide, M.D. Sastry, Jingui Qin, Chuluo Yang, K. Yakushi, Yasuhiro Nakazawa, Kenji Ichimura, A.K. Sra, J.V. Yakhmi, C.R. Acad. Sci.Paris, Chimie/Chemistry 4 (2001) 189-192.

[15] M. Drillon and P. Panissod, JMMM 188 (1998) p. 93.

[16] V. Laget, C. Hornick, P. Rabu, M. Drillon, R. Ziessel, Coord. Chem. Rev. 178-180 (1998) 1533.

[17] L. Ouahab, Coord. Chem. Rev 178-180 (1998) 1501.

[18] Chuluo Yang, Jingui Qin, Kyuya Yakushi, Yasuhiro Nakazawa and Kenji Ichimura, Synthetic Metals 102 (1999) 1482.

[19] C. Kittel, Introduction to Solis State Physics, p.330, John Wiley and Sons, Inc., New York, 1986.

[20] G L. Pascal, R. Clement, K. Nakatani, J. Zyss, I. Ledoux, Science 263 (1994) 658.

[21] Juh-Tzeng Lue, J. Phys. Chem. Solids 62 (2001) 1599.

ELECTRON MAGNETIC RESONANCE STUDY OF SONOCHEMICALLY PREPARED NANOPOWDERS OF La$_{1-X}$Sr$_X$MnO$_3$ (X = 0.16, 0.3) CMR MANGANITES

A. I. Shames [1], E. Rozenberg [1]* and G.Gorodetsky [1], G. Pang [2] and A. Gedanken [2]

[1]*Department of Physics, Ben-Gurion University of the Negev, P.O. Box 653, 84105, Beer-Sheva, Israel*

* evgenyr@bgumail.bgu.ac.il

[2] *Department of Chemistry, Bar-Ilan University, 52900, Ramat-Gan, Israel*

Abstract: Electron magnetic resonance (EMR) spectra, comprising FMR and EPR ones below and above the Curie point T$_c$, were recorded within the interval $100 \leq T \leq 420$ K for nanosized powders of La$_{1-x}$Sr$_x$MnO$_3$ (x = 0.16 and 0.3) with the mean particle size about 20-30 nm. These powders were produced by the ultrasonic irradiation (sonication-assisted coprecipitation). EMR spectra obtained demonstrate relatively low magnetocrystalline anisotropy and high magnetic homogeneity of the studied nanosized powders comparing to crushed single crystals of the same composition. Above data directly confirm that sonochemically prepared nanosized powders may be used as a testing systems for study of intrinsic properties of colossal magnetoresistive doped manganites.

Key words: Ferromagnetic   and   Electron   Paramagnetic   resonance,   colossal magnetoresistance, doped manganites, magnetic homogeneity, sonochemistry

1.   INTRODUCTION

It is widely accepted nowadays that the presence of inherent magnetic inhomogeneities is an important prerequisite of electron/magnetic phase separation (PS) and related phenomena [1] in doped manganites that, in turn, are the core of colossal magnetoresistance (CMR) effect in these compounds. At the same time, magnetic inhomogeneities may originate from

B. Aktaş et al. (eds.), Nanostructured Magnetic Materials and their Applications, 199–204.

technologically induced local chemical/structural defects such as grain boundaries in polycrystalline samples (see, for example, Chaps. 1 and 5 in Ref. [2]); chemical inhomogeneities in bulk crystals grown by fused-salt electrolysis [3,4] etc. It is shown [2-4] that such mesoscopic defects/inhomogeneities modify drastically conductive and resonant properties of CMR manganites. Namely, bulk or bulk-like samples considered demonstrate clear features of PS state (presence of mesoscopic regions with different magnetic orderings). Thus, a general question could be addressed - is the PS-like state observed in such bulk crystalline originated samples the inherent one corresponding to that, widely discussed in theory [1], or it is an extrinsic property caused by used technology of growing/sintering of these objects?

The study of nanosized doped manganites, which are free from above defects may bring us important information regarding the aforenamed problem. In this connection the study of electron magnetic resonance (EMR) were carried out on sonochemically prepared nano-powders of $La_{1-x}Sr_xMnO_3$ (x = 0.16 and 0.3). EMR, comprising ferromagnetic and electron paramagnetic resonance below and above the Curie point (T_C), is known to be highly sensitive to minute magnetic phases, which allows its observation even in considered nano powders.

2.   EXPERIMENTAL

The nano-powders of $La_{1-x}Sr_xMnO_3$ (marked further as LSMO-1 and LSMO-2 for x = 0.16 and 0.3, respectively) with the mean particle size about 20-30 nm were prepared by a sonication-assisted coprecipitation method. The coprecipitation reaction was carried out with ultrasonic radiation. Fully crystallized powders were obtained after the as-prepared mixture was annealed at 900° C for 2 h. The details of samples preparation and characterization are described in [5]. The Curie points of LSMO-1 and LSMO-2 were found to be about 300 and 340 K respectively, which accord pretty well with the known values for the nominal Sr-doping concentrations [1]. EMR spectra of the loose packed samples were recorded using a Bruker EMX-220 X-band (ν = 9.4 GHZ) spectrometer in the temperature (T) range $100 \leq T \leq 420$ K. For a detailed description of the method of measurements and the choice of the loose packed samples - see [6].

3.   OBTAINED RESULTS AND DISCUSSION

EMR spectra recorded at the paramagnetic (PM) and ferromagnetic (FM) regions of T are presented in Figs. 1(a) and 1(b) for LSMO-1 and LSMO-2, respectively. At PM temperatures, well above T_C, EMR spectra of both samples show a singlet Lorentzian line with g ~ 1.98, typical for doped manganites. Below T_C the signal is broadened and shifted towards lower resonant fields. One can note that all considered spectra consist of a singlet line. Only a "knee"-like feature (marked by solid arrow in Fig. 1(b)) is observed on such lines within the narrow interval of temperatures approaching T_C from below.

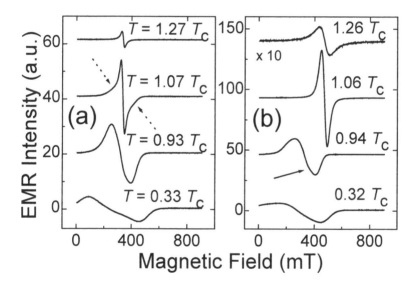

Figure 1. EMR spectra recorded for LSMO-1 (a) and LSMO-2 (b) nano-powders. Solid arrow marks the traces of PM phase; dots ones mark the contribution of LSMO-2 impurity in LSMO-1 sample (see text).

Figure 2 demonstrates the temperature dependences of resonance fields (H_r) for LSMO-1 and LSMO-2 samples. It is directly seen that the EMR line position shifts from its PM value about 0.34 Tesla to H_r ~ 0.24 and ~ 0.28 Tesla in LSMO-1 and LSMO-2 nano-powders, respectively, upon decreasing T up to 100 K (below T_C) - Fig. 2.

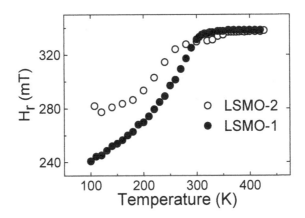

Figure 2. Temperature dependences of EMR signal resonance fields for LSMO-1 and LSMO-2 nano-powders.

Temperature dependences of the linewidths (ΔH_{pp}) show a broad flat minimum in the range $1.1T_C < T < 1.3T_C$, as well as relatively narrow one in the vicinity of $1.1T_C$ for LSMO-1 and LSMO-2 samples, respectively - Fig. 3. The zoom of ΔH_{pp} versus T curve - see insert in Fig. 3, directly demonstrates the presence of some LSMO-2 impurity in LSMO-1 nano-powder (additional local minimum at $T \sim 1.1T_C$, characteristic for LSMO-2 on the background of the broad flat one in LSMO-1). This observation is confirmed also by traces of additional (broader) resonance line, which manifest itself at T slightly above T_C on resonant spectra of LSMO-1 - see dotted arrows in Fig. 1. At PM temperatures ΔH_{pp} increases with T obeying the pseudolinear law [7] for both samples.

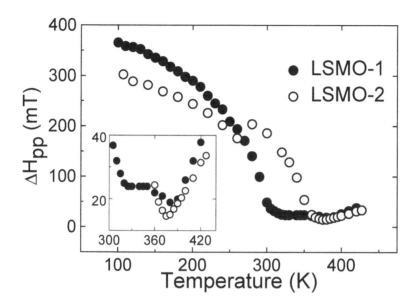

*Figure 3.*Temperature dependences of EMR signal linewidths at PM and FM intervals of temperature for LSMO-1 and LSMO-2 nano-powders. Insert is the zoom of ΔH_{pp} vs. T curves in the region of their minima.

Let us discuss the data obtained. As it was noted yet, EMR spectra of LSMO-1 and LSMO-2 represent a single resonant line below and above their T_C. One may directly compare above spectra with multiline ones obtained on crushed single crystalline [3,4] and polycrystalline [6] doped manganites, showing complex (PS-like) magnetic ordering. This facts allows us suggesting that considered samples do not contain any nanosized regions distinguished by both the type of magnetic ordering and/or different magnetic anisotropies. Moreover, about the same size and shape of individual particles lead to almost symmetrical shape of FMR lines observed for LSMO-1 and LSMO-2 samples (Fig. 1). In contrast, FMR signals of crushed single crystalline [3,4] and polycrystalline [6,8] loose packed samples are strongly asymmetric. Thevalues of resonant field in the FM state point out relatively low magnetocrystalline anisotropy for both LSMO-1 and LSMO-2 - Fig. 2. It should be noted that such anisotropy is some stronger in LSMO-1 nano-powder, which may be connected with complex competition between surface and crystallographic anisotropies (rhombohedral for LSMO-

2 and orthorhombic for LSMO-1 [1]) in considered nano-sized compounds. Extremely small size of individual particles causes also uniform penetration of microwave field within the sample bulk, i.e. the effects of non-resonant surface absorption [8] may be excluded for LSMO-1 and LSMO-2.

In summary, the results obtained in this work unambiguously show that nano-sized powders of $La_{1-x}Sr_xMnO_3$ manganites ($x = 0.16$ and 0.3) free from strains, grain boundaries, twins etc., characteristic for usual bulk single- and polycrystalline doped manganites, do not demonstrate phase separated-like state. Only very weak traces of paramagnetic signals are observed at temperatures closed to the Curie points (Fig. 1). Thus, above phase separated-like states, very often observed in bulk manganites may be considered as extrinsic ones, induced by technology of their growing/sintering.

ACKNOWLEDGEMENTS

This work was supported by the Israeli Science Foundation administered by the Israel Academy of Sciences and Humanities (grant 209/01).

REFERENCES

1. E. Dagotto, T. Hotta, A.Moreo, *Phys. Repts.*, **344**, 1 (2001).
2. C.N.R. Rao and B. Raveau (Eds.) *Colossal Magnetoresistance Charge Ordering and Related Properties of Manganese Oxides*, World Scientific, Singapore, 1998.
3. A. I. Shames, E. Rozenberg, W. H. McCarroll, M. Greenblatt , G. Gorodetsky, *Phys. Rev.* B **64**, 172401 (2001).
4. A.I. Shames, E. Rozenberg, G. Gorodetsky, A.A. Arsenov, D.A. Shulyatev, Ya. M. Mukovskii, A. Gedanken, G. Pang, *J. of Appl. Phys.*, **91**, 7929 (2002).
5. G. Pang, X. Xu, V. Markovich, S. Avivi, O. Palchik, Yu. Koltypin, G.Gorodetsky, Y. Yeshurun, H.P. Buchkremer, A. Gedanken, *Mat. Res. Bulletin*, **38**, 11 (2003).
6. A. I. Shames, E. Rozenberg, V. Markovich, M. Auslender, A. Yakubovsky, A. Maignan, C. Martin, B. Raveau, G. Gorodetsky, *Solid State Comm.*, **126**, 395 (2003).
7. D.L. Huber, G. Alejandro, A. Caniero, M.T. Causa, F. Prado, M. Tovar, S.B. Oseroff, *Phys. Rev.* B **60**, 6324 (1999).
8. A.I. Shames, E. Rozenberg, G. Gorodetsky, J. Pelleg, B.K. Chaudhuri, *Solid State Comm.*, **107**, 91 (1998).

EMR STUDY OF THIN La$_{0.7}$Ca$_{0.3}$MnO$_3$ FILMS EPITAXIALLY GROWN ON SrTiO$_3$

A. I. Shames [1*], E. Rozenberg [1], G. Gorodetsky [1] and M. Ziese [2]

[1]*Department of Physics, Ben-Gurion University of the Negev, P.O. Box 653, 84105, Be'er-Sheva, Israel.*
[*]E-mail: sham@bgumail.bgu.ac.il
[2]*Department of Superconductivity and Magnetism, University of Leipzig, 04103 Leipzig, Germany*

Abstract: The electron magnetic resonance (EMR) study was done at temperatures 130 ≤ *T* ≤ 340 K for two La$_{0.7}$Ca$_{0.3}$MnO$_3$ films –22 nm thickness, Curie temperature T_C = 254 K and 70 nm, T_C = 263 K - epitaxially grown on SrTiO$_3$ under identical conditions using pulsed laser deposition,. At $T < T_C$ both films showed a strong angular dependence of EMR lines due to magnetocrystalline and shape anisotropies. At increasing T, the effective anisotropy decreases and lines converged at $T \sim 1.2T_C$. The EMR line widths demonstrated gradual narrowing at T increase, having a spike-like feature in the vicinity of T_C. On the other hand, EMR spectra of these two films revealed significant differences. The 22 nm film showed enormously broad lines in both orientations. Within the certain temperature range, additional lines in low and high fields were clearly observed. Spin wave resonance (SWR) modes were detected in spectra of the 70 nm film at 230 ≤ T ≤ 280 K. The peculiarities observed may originate from the different structural states of these films: the 70 nm film is a quasi-relaxed and bulk-like one while the 22 nm film has an additional strained phases distinguishing in magnetic characteristics.

Key words: CMR, doped manganite, thin film, EPR, FMR

1.   INTRODUCTION

Thin films of doped rare earth manganites showing colossal magnetoresistance (CMR) in the vicinity of their Curie point (T_C) are considered now as an important class of objects regarding their technological applications (magnetic field sensors, opto- and magneto-electronic devices,

B. Aktaş et al. (eds.), Nanostructured Magnetic Materials and their Applications, 205–212.

solid electrolytes etc.) - see, for instance, reviews [1,2]. Thus, fabrication of high quality single crystalline films of CMR manganites is very important task for both understanding of their physical properties and applications. Some lattice mismatch between the single crystalline perovskite oxides, usually used as substrates, and the epitaxial films, induces an epitaxial strain (tensile or compressive depending on substrate/film lattice parameters ratio) [1-3]. Such strain, in turn, may affect physical properties of above films.

In particular, Refs.[4,5] showed that the anisotropy of transport properties measured perpendicular to plain and in plane was observed in coherently strained $La_{0.67}Ca_{0.33}MnO_3$ thin films grown on $SrTiO_3$ substrates. The strain-induced transition from an orbitally disordered to orbitally ordered (OO) ferromagnetic state was suggested in [4,5] for the explanation of this anisotropy. Electron magnetic resonance (EMR), comprising ferromagnetic and electron paramagnetic resonances below and above T_C, is known to be highly sensitive to minute magnetic phases. The main goal of this work is experimental EMR study of two $La_{0.7}Ca_{0.3}MnO_3$ films (thickness 22 nm and 70 nm), epitaxially grown on $SrTiO_3$ under identical conditions using pulsed laser deposition. The significant difference in the resonance properties of these films was explained by their different structural states - the 70 nm film is a quasi-relaxed and bulk-like one, while the 22 nm film has an additional strained phase, which may be in OO state below T_C [4,5].

2.   EXPERIMENTAL DETAILS

2.1   EMR measurements

EMR spectra were recorded using a Bruker EMX-220 X-band ($\nu = 9.4$ GHz) spectrometer within the temperature range 130 ($\pm$ 0.5) $\leq T \leq$ 340 ($\pm$ 0.5) K. Square shaped 2×2 mm^2 thin film plates were attached to the home built goniometer and centered into the standard rectangular cavity. The EMR spectra were recorded in parallel ($H^{\parallel}$, $\alpha = 0°$) and almost perpendicular ($H^{\perp}$, $\alpha = 87.5°$) orientations of the dc magnetic field with respect to the film plane (here α is angle between the film plane and the H direction). The latter orientation was chosen because of the low temperature EMR line at true perpendicular orientation ($H^{\perp}$, $\alpha = 90°$) appeared at resonance field $H_r^{\perp}$ which is slightly above the upper limit of the dc field available for the spectrometer. This fact did not allow us recording the entire EMR line in perpendicular orientation. EMR spectra processing (integration, differentiation, filtering etc.) was done using a Bruker WIN-EPR Software. Resonance field values were determined from the 2nd derivative of EMR spectra.

2.2   Samples preparation and characterization

La$_{0.7}$Ca$_{0.3}$MnO$_3$ films were fabricated using pulsed laser deposition from a stoichiometric polycrystalline target onto (001) single crystal SrTiO$_3$ substrates. Substrate temperature was 700 ^{o}C and oxygen partial pressure during deposition was 0.13 mbar. An Excimer Laser (Lambda Physik) operating at 308 nm (XeCl) with a pulse energy of 200 mJ was employed at repetition rate of 10 Hz. The fluence on the target was estimated to be about 1 J/cm^2. After deposition the films were slowly cooled (rate of ~ 4 K/min) to room temperature in an oxygen atmosphere of 0.13 mbar. Afterwards the films were transferred to a furnace and annealed for 2 h at 950 ^{o}C in 1 bar of flowing oxygen. Heating rate was ~ 8 K/min and cooling rate ~ 4 K/min. Film thickness was estimated from the deposition times and has an uncertainty of ~ 10%. Two film samples on SrTiO$_3$ were investigated. The following parameters were found for the 22 nm sample: $a_{\|} = 0.3905$ nm, $a_{\perp}$ = 0.3811 nm, $T_C = 254$ K. The same parameters for the 70 nm sample were: $a_{\|} = 0.3875$ nm, $a_{\perp} = 0.3833$ nm, $T_C = 263$ K. More detailed samples characterization as well as results of resistivity and magnetic measurements, obtained on these samples, will be reported elsewhere.

3.   RESULTS AND DISCUSSION

The EMR spectra of the 70 nm film sample, measured in both orientations ($H^{\|}$ and $H^{\perp}$) at different temperatures (T), are shown in Fig. 1(a). At low T for each sample's geometry of measurements the only intensive peak is ascribed to the uniform precession mode. For $H^{\|}$ a nearly symmetric EMR line located at low fields is observed over the entire temperature range except at temperatures near T_C. For out-of-plane geometry the high field EMR signals observed are asymmetrical at $T < T_C$. At approaching T_C from below, several additional weak signals are clearly observed – Fig. 2: a higher field in-plane signal (marked by asterisk) and two lower field satellites of the out-of-plane signal (marked by arrows). On increasing T, the resonance field positions H_r for the in-plane and out-of-plane signals converge towards $g = 1.99$ in paramagnetic (PM) phase. It should be noted here that $H_r^{\|}$ and $H_r^{\perp}$

are distinguishably separated up to $T \sim 1.2T_C$ – Fig. 3(a). On heating, the line widths $\Delta H_{pp}^{\parallel}$ and $\Delta H_{pp}^{\perp}$ decrease, passing through a minimum at $T \sim 245$ K, then exhibiting sharp spike-like features at $T_p = 270$ K $(\Delta H_{pp}^{\parallel})$ and $T_p = 285$ K $(\Delta H_{pp}^{\perp})$ – Fig. 3(c).

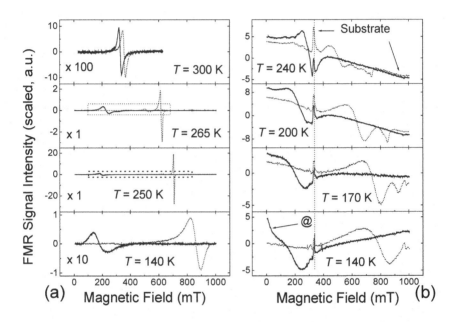

Figure 2. Temperature dependences of EMR spectra of two film samples: (a) - 70 nm and (b) - 22 nm; $\nu = 9.43 \div 9.44$ GHz. Solid lines – in-plane $(\alpha = 0^0)$ signals, dashed lines – out-of-plane $(\alpha = 87.5^{\,0})$ signals. Spectra (a) are represented in comparable scales with the multiplication coefficient marked in the left corner. Symbol @ in (b) points out the intensive broad low field line. Dotted rectangles in (a) show the zoom region - see Fig. 2.

EMR spectra of the 22 nm film are shown in Fig. 1(b). It is clearly seen, that both in-plane and out-of-plane EMR lines of this sample are significantly broader, than the corresponding lines of the 70 nm sample, within the entire $T \leq T_C$ range. Two high field satellites were observed for the out-of-plane line. The temperature dependences of resonance fields and line widths for the 22 nm sample are represented in Figs. 3(b) and 3(d), respectively. In general, the temperature behavior of $H_r^{\parallel}$, $H_r^{\perp}$, the orientation driven line separation $\Delta H_r = H_r^{\parallel} - H_r^{\perp}$, and the line widths $\Delta H_{pp}^{\parallel}$ and $\Delta H_{pp}^{\perp}$ look similar for both films. However, the 22 nm films' $\Delta H_{pp}(T)$ plots do not show the characteristic spike-like feature in the vicinity of T_C that was observed in the 70 nm sample (compare Figs. 3(c) and 3(d)). Moreover, at $T < 150$ K another intensive broad lines were observed in EMR spectra at all orientations. These lines overlap with the relatively strong background EMR signal, originated from the crystalline $SrTiO_3$ substrate (multiple angular dependent sharp peaks and the low-to-high field negative slope, clearly seen in Fig. 1(b)), as well as more intensive "main" EMR lines, which complicates detailed study of their temperature and angular dependences. Nevertheless, in the EMR spectra at $T = 140$ K and $\alpha = 0°$ this signal is well observed as a negative part of some broad intensive EMR line, centered in zero field (marked by the symbol @ in Fig. 1(b)). Here the "main" in-plane EMR signal is relatively narrow one located at $H_r^{\parallel} \sim 200$ mT.

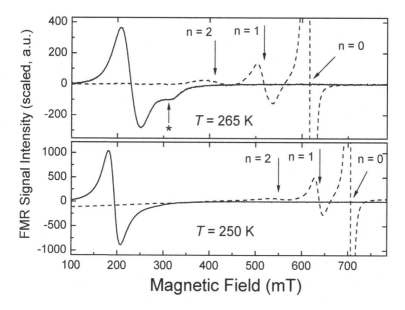

Figure 2. Zoom of the EMR spectra of the 70 nm film at T approaching T_C. Asterisk points out additional in-plane line. Arrows point out the position of a uniform precession mode (n = 0) and two non-uniform modes with n = 1, 2.

Let us discuss the EMR data obtained. Temperature dependences of resonance fields evidence that the ferromagnetic (FM) phase appears in these films at temperatures $T \sim 1.2T_C$ (here T_C was determined from the upper inflection points of the magnetization measured in a field of 100 mT). Moreover, the magnetic phase appeared is quite homogeneous since no mixture of signals, belonging to other FM and PM phases, was observed. This result significantly differs from the data, obtained on more thick $(La_{0.7}Ca_{0.3})_{1-x}Mn_{1+x}O_3$ (LCMO), ($x = 0$, 0.1, 0.2) films by FMR [6]. In principle, the presence of the FM phase (and, correspondingly, the coexistence of FM and PM signals) at $T_C < T < 1.2T_C$ has been already observed in bulk (single- and polycrystalline samples). It was attributed to the internal nanoscale sample's non-homogeneities characterized by different local Curie temperatures [7]. One can conclude, that the considered thin films are free of such type of non-homogeneities: the only FM phase appears at certain temperature and then progressively develops down to T, determined as T_C by magnetic measurements.

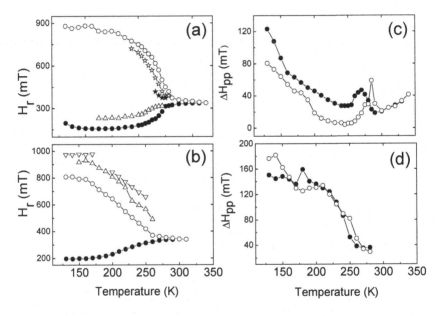

Figure 3. Temperature dependences of EMR spectra parameters of two film samples: (a) 70 nm film, solid circles - $H_r^{\parallel}$, main line; open circles - $H_r^{\perp}$, n = 0 mode; triangles up - $H_r^{\parallel}$, additional line; open and solid stars - $H_r^{\perp}$, n = 1 and n = 2 modes, respectively; (b) 22 nm film, solid circles - $H_r^{\parallel}$, main line; open circles - $H_r^{\perp}$, main line; triangles up, down - $H_r^{\perp}$, additional lines; (c) 70 nm film, solid circles – $\Delta H_{pp}^{\parallel}$, main line; open circles - $\Delta H_{pp}^{\perp}$, n = 0 mode; (c) 22 nm film, solid circles – $\Delta H_{pp}^{\parallel}$, main line; open circles - $\Delta H_{pp}^{\perp}$, main line. Lines are guide for eyes.

The maximum values of ΔH_r occur at low temperatures and are determined by the values of magnetic anisotropy and demagnetization fields. With decreasing the film thickness, ΔH_r slightly decreases – Figs. 3(a), (b). In general, H_r (T) dependences agree with those, obtained on thicker manganite films [6]. Thus, one can conclude, that in both films the demagnetization field exceeds the sample's shape brings the main contribution to the magnetic anisotropy observed.

Likewise the resonance fields, the line width vs. T dependences for the 70 nm film resemble those reported earlier for the stoichiometric ($x = 0$) LCMO film [6]. For both geometries ΔH_{pp} exhibit a sharp maximum at T_p, in the vicinity of PM-FM transition – Fig. 3(c). Following [6] it may be concluded, that sharp ΔH_{pp} peak in the vicinity of the transition temperature is a typical feature of manganite films. On the other hand, this feature was not observed in the thin 22 nm film – Fig. 3(d). We could also emphasize that such a peak was not observed neither for single crystals nor for polycrystalline manganite samples – see, for example [7].

There could be various mechanisms responsible for the broadening of the EMR line at $T < T_C$ and a ΔH_{pp} anomaly near the transition. Atsarkin and co-workers convincingly showed that no signs of critical acceleration of the longitudinal EMR relaxation time T_1 were observed in CMR manganites [8]. They also supposed the broadening, observed in some manganites near T_C is in fact non-homogeneous and caused by non-uniform magnetization in the presence of strong FM correlations. The further line narrowing, occurred in the 70 nm sample at 245 K $< T < T_p$, may be explained by the reduction in the internal field non-homogeneities on moving off T_C area.

Taking into account that the line widths ΔH_{pp} in the 22 nm film exceed the same in the 70 nm film - Fig. 3(d), one can assume an additional mechanism for the line broadening in this sample. Following [9] this line broadening may be attributed to the spin disorder originating from the interface (substrate, outer surface) anisotropy, which, for thinner films, influences all the spins at interior region.

Additional weak EMR signals revealed in spectra of two films (Figs. 1, 2) also evidence magnetic non-homogeneities of these samples. The angular dependent weak in-plane signal in the 70 nm film as well as the similar out-of-plane high-field signals in the 22 nm film may be attributed to some regions distinguishing by magnetic parameters (both anisotropy and magnetization). These regions could originate from the interface areas and their relative contribution increases on the decrease of the film' thickness.

Two weak signals, observed on the low-field shoulder of the out-of-plane EMR signal in the 70 nm film (Fig. 2) correspond to spin wave resonance (SWR) modes. These modes become observable within the certain temperature range where the EMR line is narrow enough. Two SWR modes (designated to n = 1, 2) were resolved. Unfortunately, the fact that just two SWR modes were resolved does not allow determination of the origin of this SWR spectra (one- or two-sided pinning and/or type of non-homogeneity), which bases on the analysis of line separation and intensities. No SWR modes were observed in EMR spectra of the 22 nm film. Like the above-mentioned additional line broadening, the absence of SWR modes in thinner film is consistent with the model, proposed in Ref. [9]. The increasing contribution of interfaces modifies pinning conditions and affects both line width and SWR spectrum of the prevalent magnetic phase.

Let us discuss the broad intensive signals, which appear in EMR spectra of the 22 nm film at $T < 150$ K – see lower spectrum in Fig. 1(b). These signals also demonstrate strong angular dependence and, correspondingly, belong to some new FM ordered phase. Considering only the well-resolved low field in-plane EMR signal, it may be supposed, that this new phase is characterized by another parameters of magnetic anisotropy. Appearance of these additional signals correlates well with the results of the magnetization measurements (not shown here), which, just in the vicinity of $T = 150$ K, indicates the presence of two different strain-states in the 22 nm film: a quasi-relaxed phase with a bulk-like Curie temperature and a strained phase with a strongly reduced transition temperature. Above results are in good correlation with the model by Klei et. al [4,5] on the OO strain-induced transition in manganite films.

ACKNOWLEDGEMENTS

This work was supported by the Israeli Science Foundation administered by the Israel Academy of Sciences and Humanities (grant 209/01); and by the DFG under Contracts No. DFG ES 86/ 6-1 and No. DFG ES 86/ 7-1

REFERENCES

1. W. Prellier, Ph. Lecoeur, B. Mercey, *J.Phys.: Cond. Matter* **13**, R915 (2001).
2. A-M. Haghiri-Gosnet, J-P. Renard, *J.Phys. D: Appl. Phys.*, **36**, R127 (2003).
3. H.-U. Habermeier, *Physica B*, **321**, 9 (2002).
4. J. Klein, J.B. Philipp, G. Carbone, A. Vigliante, L. Alff, R. Gross, *Phys. Rev.* B **66**, 052414 (2002).
5. J. Klein, J.B. Philipp, D. Reisinger, M. Opel, A. Marx, A. Erb, L. Alff, R. Gross, *J. Appl. Phys.*, **93**, 7373 (2003).
6. V. Dyakonov, V. Shapovalov, E. Zubov, P. Aleshkevych, A. Klimov, V. Varyukhin, V. Pashchenko, V. Kamenev, V. Mikhailov, K. Dyakonov, V. Popov, S. J. Lewandowsky, M. Berkowsky, R. Zuberek, A. Szewczyck, H. Szymczak, *J. Appl. Phys.*, **93**, 2100 (2003).
7. A. I. Shames, E. Rozenberg, and G. Gorodetsky, A. A. Arsenov, D. A. Shulyatev, Ya. M. Mukovskii, A. Gedanken and G. Pang, *J. Appl. Phys.*, **91**, 7929 (2002).
8. V. A. Atsarkin, V. V. Demidov, F. Simon, R. Gaal, Y. Moritomo, K. Conder, A. Jánossy, L. Forró, *J. Magn. Magn. Mater.*, **258-259**, 256 (2003).
9. S. Budak, F. Yildiz, M. Özdemir, B. Aktaş, *J. Magn. Magn. Mater.*, **258-259**, 423 (2003).

MAGNETIC ANISOTROPIES IN THE ULTRA-THIN IRON FILM GROWN ON THE GaAs SUBSTRATE

B. Aktaş [a], B. Heinrich [b], G. Woltersdorf [b], R. Urban [b], L. R. Tagirov [a,c],
F. Yıldız [a], K. Özdoğan [a], M. Özdemir [a], O. Yalçın [a]

[a] *Gebze Institute of Technology, 41400, Çayırova, Gebze, Turkey*
[b] *Simon Fraser University, Burnaby, BC, V5A 1S6, Canada*
[c] *Kazan State University, Kazan 420008, Russia*

Abstract: We study magnetic anisotropies of epitaxial, ultra-thin iron films grown on GaAs substrate (Au/Fe/GaAs(001)). The ferromagnetic resonance (FMR) technique has been explored to determine magnetic parameters of the films in the temperature range 4-300 K. It was found that unusual and unexpected behavior of FMR spectra observed in that films reveal detailed information about the magnetic anisotropies of studied systems. A presence of strong perpendicular anisotropy has been deduced from the fitting of a model to the experimental data. The linear variation of magnetic anisotropy parameters with the temperature has been observed. The results on temperature dependence are discussed in terms of thermal expansion induced magneto-elastic anisotropies.

Key words: ferromagnetic resonance, magnetic anisotropy, ultra-thin magnetic films

1. INTRODUCTION

The interest to ultra-thin magnetic multilayers has been steadily increasing. This is motivated by the fact that magnetic properties of this type of structures are the real technological issues in mass production of data storage devices and magnetic random access memories. A good grasp of the fundamental physics of the magnetization dynamics, in the head as well as the medium, becomes of essential importance to sustain the exponential growth of device performance factors.

The magnetic anisotropy of the thin films is of crucial importance in applications. It is well known that the ferromagnetic resonance (FMR) is one of the most sensitive and accurate techniques to determine magnetic

B. Aktaş et al. (eds.), Nanostructured Magnetic Materials and their Applications, 213–227.
© 2004 *Kluwer Academic Publishers. Printed in the Netherlands.*

anisotropy fields of very this magnetic films [1], [2]. In this paper we study the magnetic anisotropies and relaxation processes in single ferromagnetic layer system, Au/Fe/GaAs(001), using FMR. We observed unconventional FMR spectra in the experiments. They are explained and interpreted based on the model proposed in this study. Consistent fitting of all essential features and behaviors of the FMR spectra in the temperature range 4-300 K allowed us to determine accurately the cubic, uniaxial and perpendicular components of magnetic anisotropy, as well as to establish directions of easy and axes for magnetization. The origin and temperature dependence of magnetic anisotropy fields are extensively discussed in terms of the lattice mismatch between materials in a contact and thermal evolution of the stresses induced by difference in thermal expansion coefficients of materials in the study.

2. EXPERIMENTAL RESULT

2.1 Samples and FMR measurements procedure

The ultrathin film structure Au/Fe/GaAs(001) was prepared by Molecular Beam Epitaxy (MBE) on the (4×6) reconstructed GaAs(001) substrates. A brief description of the sample preparation procedure is as follows. GaAs(001) single-crystalline wafers were prepared by annealing and sputtering cycles and monitored by means of reflection high energy electron diffraction (RHEED) until a well-ordered (4×6) reconstruction appeared [3]. Then GaAs substrates were heated to approximately 500 C in order to desorb contaminants. Residual oxides were removed using a low-energy Ar^+ bombardment (0.6 keV) under grazing incidence. Substrates were rotated around their normal during the sputtering.

Fe films were deposited at room-temperature from a resistively heated piece of Fe at the base pressure of $1×10^{-10}$ Torr. The film thickness was monitored by a quartz crystal microbalance and by means of RHEED intensity oscillations. The deposition rate was adjusted at about one mono-layer (ML) per minute. The gold layer was evaporated at room temperature at the deposition rate of about one monolayer per minute. RHEED oscillations were visible for up to 30 atomic layers. Single Fe ultrathin films with the thicknesses $d = 8, 11, 16, 21, 31$ ML were grown directly on GaAs(001). All the films under study were covered by a 20 ML thick Au(001) cap layer for protection in ambient conditions The example of the prepared film and the coordinate system and principal vector directions are shown in Fig. 1. More details of the sample preparation techniques are given in Ref. [3].

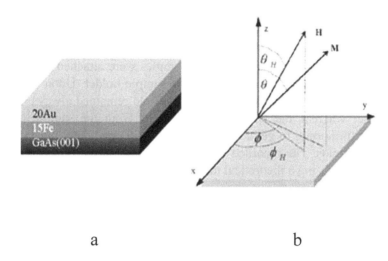

a   b

Figure 1. The sketch of the samples studied in the paper (a) and relative orientation of reference axes system with respect to the thin film (b).

FMR measurements have been carried out using the commercial *Bruker EMX* X-band ESR spectrometer equipped by an electromagnet which provides a DC field up to 16 kG in the horizontal plane. The small amplitude modulation of the field is employed to record field-derivative absorption signal at the temperature range 4-300 K. The microwave and DC components of the external field remain always mutually perpendicular to each other. An *Oxford Instruments* continuous helium-gas flow cryostat was used to cool the sample down to measurement temperatures, and the temperature was controlled by the commercial *Lake Shore 340* temperature-controller. A goniometer was used to rotate the sample around the rod-like sample holder in the cryostat tube, the sample-holder was always perpendicular to the DC magnetic field and parallel to the microwave magnetic field. The samples were placed on the sample-holder in two different geometries. For the in-plane angular studies the films were attached horizontally on the bottom edge of the sample holder. During rotation, the normal to the film plane remained parallel to the microwave field, but the external DC magnetic field held different directions in the sample plane. This geometry is not conventional and gives quite asymmetric absorption curves still at the same resonance field as in the conventional geometry for in-plane measurements (both DC and microwave magnetic fields always lie in the film plane). Thus, we could be able to study at least the angular dependence of the resonance field and magnetic anisotropies using the FMR data taken from the unconventional in-plane geometry. We have also recorded some FMR data in the conventional, in-plane geometry for some specific crystallographic direction to test the correctness of the data obtained in the unconventional geometry.

For the out-of-plane measurements the samples were attached to the flat platform precisionally cut at the cheek of the sample holder. Upon rotation of the sample the microwave component of the field remained always in the sample plane, whereas the DC field was rotated from the sample plane towards the film normal to get additional data for accurate determination of the anisotropy fields. The recorded FMR spectra were stored in a computer and then analyzed using a theoretical model described below.

2.2   In-plane FMR measurements

First we measured the in-plane FMR in the 20Au/15Fe/GaAs(001) sample. Fig. 2a illustrates the temperature evolution of the in-plane FMR spectra taken in the direction of the DC magnetic field $H\|a$ (a is the [100] crystallographic axis of the substrate, see Fig. 1). The single, relatively narrow and intensive signal is observed at very low magnetic fields in the entire temperature range. There is significant temperature dependence of the FMR resonance field. As the temperature is decreased starting from the room temperature the resonance field of FMR signal is monotonically shifted from ~320 Oe for 300 K down to about 130 Oe for 5 K. The line width of the resonance increases with decreasing the temperature as well.

Contrary to the measurement along the a axis, the in-plane FMR spectrum in the b direction of the substrate consist of three signals. It is a unique observation of three FMR signal from a single epitaxial ferromagnetic film, which has not been reported in a previous literature. Usually, a single resonance peak (mode) is expected from such a very thin (15 monolayers) ferromagnetic layer, since the higher order spin-wave modes are expected to shift to negative field due to very high excitation energies of short-wave-length standing spin waves across the thickness of the ultrathin film. This means that the observed multi-component spectra do not belong to the spin-wave modes. The temperature dependence of FMR spectrum for $H\|b$ axis is shown in Fig. 2b. All three peaks are observable in the entire temperature range. The high-field signal has the largest intensity at

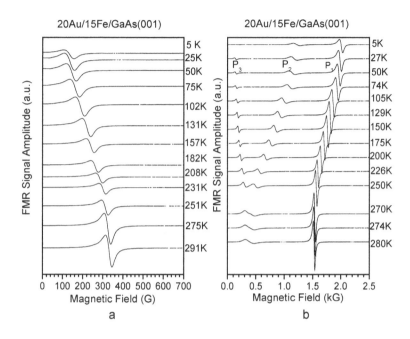

Figure 2. Temperature dependence of in-plane FMR spectra taken for **H**∥**a** (a) and for **H**∥**b** (b).

all temperatures. At room temperature the two low-field peaks overlap and almost merge into the common signal of a distorted shape. As the temperature decreases, the low-field signal separates into two signals, which shift one from the other in the opposite directions. The intense high-field signal shifts monotonically to higher fields upon lowering the temperature. At lowest measurement temperature (5 K) the overall splitting of the FMR spectra reaches ~1700 Oe. It should be noted that the direction of the shift of this main mode for **H**∥**b** is opposite to that for **H**∥**a** axis. This implies that the easy direction of the magnetization in sample plane is the *a* axis (and the *b* axis is the hard direction. The precise determination of the easy axis will be done after theoretical analysis of the experimental data below.

In order to study the anisotropic behavior of the system in the plane of the film, we rotated the DC magnetic field in the plane of the film (*a-b* plane). The room temperature FMR spectra at some selected orientation of the external field with respect to the *a* axis are given in Fig. 3. As can be seen from this figure, the FMR spectra exhibit strong anisotropy. There are multi-component and asymmetric (not pure Lorentzian or Gaussian) peaks for most of the orientation of the external DC field (it should be reminded here that the microwave component of the external magnetic field is perpendicular to the sample plane, in this geometry one does not expect symmetric resonance lines).

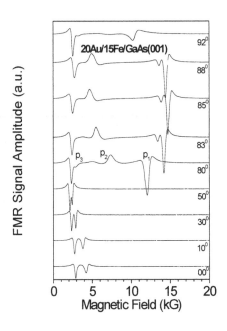

Figure 3. In-plane (**ab** plane) angular dependence of the FMR spectra.

As it can be seen from Fig. 3, the number of the absorption peaks is varied with the in-plane rotation angle. The relative intensity of the signals is also angular-dependent. In fact the FMR spectra show an overall periodicity of 180 degrees. This implies that the system must have at least uniaxial symmetry in the film plane. The unusual splitting of the spectra on three components allows us to suppose that a considerable cubic anisotropy component is superimposed. We will see later from our simulations of the FMR spectra that it is actually the case.

In order to increase the reliability of the FMR spectra interpretation we have also studied the frequency dependence of the in-plane FMR spectra between 9-36 GHz at room temperature. The Figure 4 shows the variation of the resonance fields with microwave frequency. The parabolic dependence of experimental points on the frequency is obviously seen in the figure. This

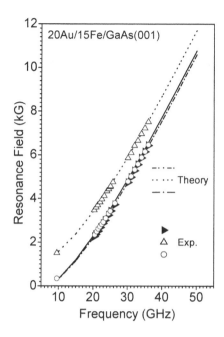

Figure 4. Frequency dependence of the in-plane FMR spectra for **H∥b.**

dependence is a general frequency behavior of a resonance field in the systems with strong anisotropy (the anisotropy energy is comparable with the Zeeman energy at low-frequency region). It is expected that, as the frequency is increased, the Zeeman energy dominates, and the frequency dependence becomes more and more linear.

We have not given frequency dependence of the resonance field for the out-of-plane geometry of measurements, since the resonance field goes far beyond the experimentally attainable field due to the demagnetization.

2.3 Out-of-plane FMR measurements

We have also measured the out-of-plane angular dependence of the resonance fields when the DC magnetic field is rotated from the easy, *a*, or the hard, *b*, axes in the film plane towards the normal direction to the film plane. The Figure 5 shows the evolution of the FMR spectra upon changing the polar angle in the *b-c* plane. Multi-peak FMR spectra are observed again for this geometry. The seperation of the modes from each other steadily increase with increasing the angle towards 90 degrees (where the field becomes parallel to the film normal). In the range of our field sweep we could observe the angular dependence up to 80 degrees. As it will be clear later, this lower field peaks do not belong to the higher order spin wave modes while the first mode at highest field is the first order spin wave mode. Therefore, the seperation of modes can not be used to extract an information about the exchange coupling. However, their angular dependence can be still used to clarify the nature of the magnetic anisotropy and obtain the anisotropy parameters from computer fitting.

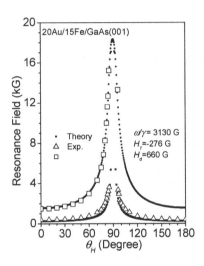

Figure 5. Out of-plane (in **bc** plane) angular dependence of resonance field of FMR spectra.

3. MODEL AND COMPUTER SIMULATIONS OF THE FMR SPECTRA

3.1 Formulation of the model

The FMR data are analyzed using the model free energy expansion given as

$$E_T = -\mathbf{M} \cdot \mathbf{H} + \left(2\pi M_0^2 - K_p\right)\alpha_3^2 + K_1\left(\alpha_1^2\alpha_2^2 + \alpha_2^2\alpha_3^2 + \alpha_1^2\alpha_3^2\right) + K_u\alpha_1^2. \quad (1)$$

Here, the first term is the Zeeman energy in the external DC magnetic field, the second term is the demagnetization energy term including the effective perpendicular anisotropy as well, the third term is the cubic anisotropy energy characterized by the parameter K_1, and the last term is the uniaxial anisotropy energy. In this equation α_I's represent directional cosines of the magnetization vector M with respect to the crystallographic axes (see Fig. 2), M_0 is the saturation magnetization at the temperature of measurement. It should be remembered here that one of the crystallographic axes is always perpendicular to the sample plane, and the remaining two lie in the sample plane. That is why we could combine demagnetizing and perpendicular anisotropy terms in a single effective term (second term) using only the α_3 directional cosine. The relative orientation of the reference axes, sample geometry and various vectors relevant in the problem are given in Fig. 1b.

The fields for resonance are obtained using the well known equation [4]:

$$\left(\frac{\omega_0}{\gamma}\right)^2 = \left(\frac{1}{M_0}\frac{\partial^2 E_T}{\partial\theta^2}\right)\left(\frac{1}{M_0\sin^2\theta}\frac{\partial^2 E_T}{\partial\phi^2}\right) - \left(\frac{1}{M_0\sin\theta}\frac{\partial^2 E_T}{\partial\theta\partial\phi}\right)^2 + \frac{1}{\gamma^2 T_2^2}. \quad (2)$$

Here $\omega_0 = 2\pi\nu$ is the circular frequency of the ESR spectrometer, γ is the gyro-magnetic ratio for the material of the magnetic film, θ and ϕ are the usual polar and azimuthal angles of the magnetization vector M with respect to the reference system. Finally, T_2 represents the effective homogeneous relaxation time of the magnetization that contribute to the line width of the FMR signal. We do not consider standing spin-wave excitations in the film because the film thickness is too small (20-40 Å). In this condition we expect that the spin waves to be visible far beyond our DC field range from zero to 16 kG.

The imaginary component of the dynamic magnetic susceptibility, that corresponds to the absorbed microwave energy by the sample, is given by [9]:

$$\chi_2 = 4\pi\frac{m_\phi}{h_\phi} = 4\pi M_0\left(\frac{\partial^2 E_T}{\partial\theta^2}\right)\frac{2\omega}{\gamma^2 T_2}\left\{\left[\left(\frac{\omega_0}{\gamma}\right)^2 - \left(\frac{\omega}{\gamma}\right)^2\right]^2 + \frac{4\omega^2}{\gamma^4 T_2^2}\right\}^{-1} \quad (3)$$

Here $\omega_0/\gamma = g\mu_B H$. We deduce the model parameters as a result of the fitting of the experimental data using the above two equations for computer simulations.

3.2   Discussion of the in-plane FMR measurements

For the in-plane geometry of measurements the both polar angles in Eqs. (2) and (3): of magnetization, θ, and of DC external magnetic field, θ_H, are fixed at $\theta = \pi/2$ and the azimuthal angle of magnetization ϕ is obtained from static equilibrium condition for the directions ϕ_H of the external field varied in the range from zero to π. The angular variation of the experimentally obtained resonance fields for various modes in the in-plane FMR spectra are plotted in Fig. 6.

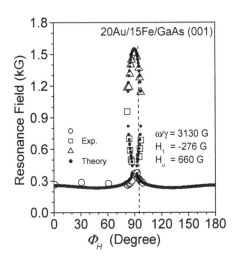

Figure 6. In-plane (**ab** plane) angular dependence of resonance field of FMR spectra.

The calculated points are obtained using the theoretical model, Eq. (2). The best-fit parameters are given in the figure capture. The developed calculation procedure has also been used to simulate the experimental FMR spectra at different temperatures. The simulated spectra are plotted together with the experimental ones on Fig. 7. There is quite good agreement between the calculated and the experimental angular dependence of the resonance field in Fig. 6 and the temperature evolution of the FMR spectrum in Fig. 7.

Computer simulations revealed that the angular dependence of the field for resonance shows unusual behavior. In the main domain of the angles there is only single resonance line. However, when the direction of the DC magnetic field approaches the (hard) axis **b** (the angle ϕ varies $\pm 10°$ around 90°) triple or double resonance absorption is observed (see Figs. 6 and 7).

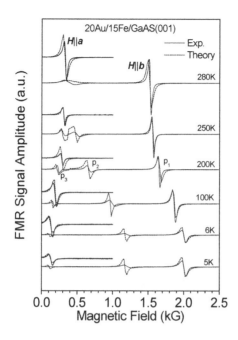

Figure 7. Simulated FMR spectra taken as some selected temperatures for **H∥b**.

This point can be made more clear if we follow the dash vertical line in the Figure 6. Moving along the line from zero magnetic field in the upward direction corresponds to the direction of the DC magnetic field sweep during actual record of an FMR spectrum. Upon the field sweep the dash line intersects the figure of the angular dependence twice in the close vicinity of 90°. If we shift our dash line on approximately ± 10° apart of 90°, it intersects the figure three times, as it is actually shown in Fig. 6. At last, it is clear that there is only one intersection in the remaining domain of angles. Every intersection of the dash line with the figure of angular dependence gives the FMR signal. For the triple FMR line these intersections are marked as *p1*, *p2* and *p3* in Fig. 7.

In the case of double or triple FMR lines, the lower one(s) corresponds to the *non-aligned* situation, when the condition for resonance, Eq. (2), is satisfied at the direction of magnetization, which is not parallel to the external magnetic field. The high-field signal may be called as an *aligned* signal, when the magnetization and the external DC field are practically parallel. To our best knowledge, no observations of *two* non-aligned FMR resonances in a single ferromagnetic film had been reported in the literature. The double-peak FMR spectra in a single iron film have been reported previously (see, for example, Refs. [1], [5]-[7]). We believe that the very specific relation between the cubic and the uniaxial anisotropies is realized in the studied system, which gives rise to the observed unusual behavior of the FMR spectra.

3.3   Discussion of the out-of-plane measurements

For the out-of-plane geometry of measurements the azimuthal angle ϕ_H is fixed either at $\phi_H = 0$ (easy axis), or $\phi_H = \pi/2$ (hard axis) directions, while the polar angle θ_H is varied from $\pi/2$ to zero. The polar and azimuthal angles of magnetization for each direction of the external field have been obtained from static equilibrium condition corresponding to the minimum free energy of the system. Then, the resonance field and the spectrum were found using Eqs. (2) and (3). The Figure 7 shows both the experimental and the calculated angular dependencies of the resonance fields. Multi-peak FMR spectra appeared also for the out-of-plane geometry when measured from the hard in-plane direction. The intesity of the low-field mode is considerably pronounced. The separation of the modes steadily increases with increasing the angle towards the film normal. Again, as in the case of the in-plane geometry of measurements, we calculated not only the resonance fields using the expression (2) and condition of equilibrium for the magnetization direction. We also simulated the FMR spectra using Eq. (3) and compared them with the actual records from the spectrometer. The simulated spectra are shown in Fig. 7. They show fairly good agreement at all angles, temperatures and frequencies of the measurements. One has to emphasize here, that in spite of the separate discussion of the in-plane and out-of-plane geometries of measurements, the magnetic parameters that we quote here are the result of simultaneous fitting of FMR spectra taken from both geometries of measurements.

3.4   Temperature dependence of the magnetic parameters

The cumulative data on the temperature variations of the magnetic parameters are given in Fig. 8. The effective magnetization is referred to the left axis, while the magnetic anisotropy parameters - to the right axis. All the parameters are given in magnetic filed units (Gauss). It should be recalled that the effective magnetization, M_{eff}, includes perpendicular anisotropy (see Eq. (1)). That is why the values of M_{eff} are essentially reduced compared with the bulk magnetization (~1.7 kG). This means that very big, perpendicular to the film, surface anisotropy field (about 5 kG) is induced in the epitaxial ultra-thin film at room temperature. It is also a general feature that both the uniaxial (along the **b** axis of the GaAs substrate) and the cubic anisotropy fields are significantly large. That is why the anisotropy energy dominates the Zeeman energy and causes such unusual and surprising triple-peak FMR spectra. In fact this is the first observation for three peaks all of which correspond to first order spin wave resonance mode.

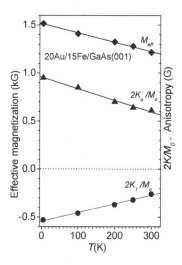

Figure 8. Temperature dependence of the effective magnetization and magnetic anisotropy parameters.

It can be seen from Fig. 8 that all magnetic anisotropy parameters strongly depend on temperature. As the temperature is decreased the effective magnetization is increases. This is partly due to the increase of saturation magnetization of Fe according to the Bloch law, and partly due to the decrease of easy-axis perpendicular anisotropy. The ferromagnetic transition temperature of bulk iron is about 980 K, that is why even at room temperature the magnetic moment is almost fully saturated. Using the literature data on the temperature dependence of iron magnetic moment in the ferromagnetic phase [8] we estimate the increase of the iron magnetization in the range from 300 K to 5 K as much as ~64 Oe. The temperature variation is expected $\propto T^{3/2}$ according to the Bloch law. Obviously, the observed magnitude (~300 Oe) and almost linear temperature dependence of the effective magnetization do not follow the figures described above. That is why we conclude that main contribution to the temperature variation of the effective magnetization comes from the temperature dependence of the perpendicular anisotropy. As the temperature is decreased the perpendicular anisotropy relaxes.

The absolute values of the both, in-plane uniaxial (easy direction along the **b** axis), and the cubic anisotropies increase with decreasing the temperature. The sign of cubic anisotropy parameter is negative, making all of the three crystalline axes easy for magnetization. However the sign of more strong uniaxial anisotropy along the **b** axis is positive, making the **b** axis a hard direction for magnetization.

Table 1. Thermal expansion coefficient and lattice parameters of Au/Fe/GaAs(001) samples

Samples	Thermal Expansion Coefficient				Lattice Parameters	
	298K	523K	1273K	Factor	a=b=c Å	
Iron (Fe)	11.8	15	24	$*10^{-6}K^{-1}$	2.8665	
Gold (Au)	14.2	14.6	16.7	$*10^{-6}K^{-1}$	4.0782/1.41= =2.8923	Strain
Chromium (Cr)	6.2			$*10^{-6}K^{-1}$	2.91	Big strain
Palladium (Pd)	11.8	12.2	13.9	$*10^{-6}K^{-1}$	3.8907/1.41= =2.759	Stress
GaAs	5.73			$*10^{-6}K^{-1}$	5.654/2= =2.827	Stress
Copper (Cu)					3.6149	

As can be noticed from Fig. 8, all of the magnetic anisotropy parameters depend on the temperature almost linearly. This implies that there is a common physical reason behind this unified behavior. Taking into account the big differences between the lattice constants for the bulk Fe and the epitaxially prepared Fe film on the GaAs substrate (see the Table) one may attribute the temperature dependence of the anisotropy parameters to the magneto-elastic effect.

Actually, there is about −1.5% misfit between lattice parameters of Fe, Au and GaAs substrates. In a result, the Fe film is under a compressive strain. When temperature is lowered down to 4K from the room temperatures, this compressive stress decreases by about 40 percent, as can be calculated by using the thermal expansion coefficient of Fe, Au and GaAs crystals, given in the Table. As can be seen, the anisotropies also change approximately the same value in this temperature range. The lattice parameters vary linearly with the temperature, and in the first approximation we expect the anisotropy parameters vary linearly with temperature as a result of different thermal expansion coefficients of materials in the contact (see the Table) and the magneto-elastic coupling.

4. CONCLUSION

We studied the magnetic anisotropies of epitaxial, ultra-thin iron films grown on GaAs substrate (Au/Fe/GaAs(001)). The ferromagnetic resonance (FMR) technique has been explored to determine magnetic parameters of the films in the temperature range 4-300 K. The triple-peak behavior of FMR spectra was observed for the first time, which brings detailed information about the magnetic anisotropies of studied systems. Presence of strong perpendicular anisotropy has been deduced from the fitting of the model to the experimental data. The linear variation of magnetic anisotropy parameters with the temperature has been observed. The results on temperature dependence have been discussed in terms of thermal-expansion-induced magneto-elastic contribution to the magnetic anisotropies.

ACKNOWLEDGMENT

L.R.T. acknowledges BRHE grant REC-007 for partial support.

REFERENCES:

[1] B. Heinrich and J. F. Cochran, Adv. Phys. 42, 523 (1993).

[2] M. Farle, Rep. Prog. Phys. 61, 755 (1998).

[3] X1-T. L. Monchesky, B. Heinrich, R. Urban, K. Myrtle, M. Klaua, and J. Kirshner, Phys. Rev. B 60, 10242 (1999).

[4] H. Suhl, Phys. Rev. 97, 555 (1955).

[5] J.J. Krebs, F.J. Rachford, P. Lubitz, and G.A. Prinz, J. Appl. Phys. 53, 8058 (1982)

[6] Yu. V. Goryunov, N. N. Garifyanov, G. G. Khaliullin, I. A. Garifullin, L. R. Tagirov, F. Schreiber, Th. Mühge, and H. Zabel, Phys. Rev. B 52, 13450 (1995).

[7] Th. Mühge, I. Zoller, K. Westerholt, H. Zabel, N. N. Garifyanov, Yu. V. Goryunov, I. A. Garifullin, G. G. Khaliullin, and L. R. Tagirov, J. Appl. Phys. 81, 4755 (1997).

[8] *Numerical Data and Functional Relationships in Science and Technology*, Landolt-Börnstein, New Series, vol. III/19A (Springer, Heidelberg, 1986).

[9] B. Aktaş and M. Özdemir, Physica B, 193, 125 (1994).

ESR STUDIES ON Sm-Co/Fe EXCHANGE-SPRING MAGNETS

F. Yıldız [a], O. Yalçın [a,b], B. Aktaş [a], M. Özdemir [c], J.S. Jiang [d]

[a]*Department of Physics, Gebze Institute of Technology,41400 Gebze, Kocaeli Turkey*
[b]*Department of Physics,Gaziosmanpaşa University, 60110, Tokat Turkey*
[c]*Faculty of Art and Science, Physics Department, Marmara University, Istanbul Turkey*
[d]*Materials Science Division, Argonne National Laboratory, Argonne, Illinois USA*

Abstract: The magnetic properties of Cr/Sm-Co/Fe/Cr multilayers epitaxially grown on MgO (110) substrate system have been studied by electron spin resonance (ESR) technique. ESR measurements have been carried out for various direction of DC magnetic field with respect to the crystalline axes. Both overall magnetization and ESR spectra exhibit very strong angular dependence. The ESR spectra have been recorded after saturating the hard *Sm-Co* layer in a field of 16 kOe applied parallel to the easy axis lying in the hard magnetic Sm-Co layer. When the field is swept up to the switching value of *Sm-Co* layer along the direction of saturation magnetization of *Sm-Co* no absorption was observed while there are two broad peaks in the spectrum for opposite direction of the measurement field. As the external field direction is rotated from in plane toward film normal the curves steadily broaden and completely disappeared when the field becomes parallel to the film normal. Analysis of the results shows that the easy axis of hard Sm-Co layer is not perfectly parallel to the film plane; rather it is slightly tilted (15 degrees) toward film normal. The deduced value for exchange coupling parameter between hard Sm-Co and soft Fe layers is about 1 kOe.

Key words: Exchange spring magnets, magnetic anisotropies, ferromagnetic resonance, magnetic thin film.

1.   INTORDUCTION

Exchange coupling between an antiferromagnetic and a ferromagnetic films or between two ferromagnetic films has been extensively studied since the phenomenon was discovered by Meiklejohn and Bean [1]. The

B. Aktaş et al. (eds.), Nanostructured Magnetic Materials and their Applications, 229–237.
© 2004 *Kluwer Academic Publishers. Printed in the Netherlands.*

"exchange-spring" expression was used for the first time by Kneller and Hawing [2]. Spring magnet films consist of hard and soft layers that are coupled at the interfaces due to strong exchange coupling between relatively soft and hard layers. The soft magnet provides a high magnetic saturation, whereas the magnetically hard material provides a high coercive fild. So far some structures like, soft ferromagnetic NiFe alloy and hard antiferromagnetic MnFe alloy structures [3,4], hard $CoFe_2O_4$ ferromagnetic and soft $(Mn,Zn)Fe_2O_4$ bilayer system [5], $DyFe_2/YFe_2$ system [6] etc. have been investigated. The recent investigation of Sm-Co/Fe structures has become popular due to their higher exchange energy between hard Sm-Co and soft Fe layers [7]. Exchange-spring coupled magnets are promising systems for applications in perpendicular magnetic data recording-storage devices and permanent magnets [8-10]. Skomski [11] and Coey [12] explored the theory of exchanged coupled films and predicted that a huge energy about three times of commercially available permanent magnets (120MOe) can be induced.

After saturating hard layers, if a reverse magnetic field higher than exchange field, (but still smaller than coercivity of hard layer to prevent its switching), is applied, the magnetic reversal proceeds via a twisting of the magnetization only in the soft layer. The spins which are sufficiently close to the interface are pinned by the hard layer, while those in a deep region of soft layer (far from the hard layer) rotate up to some extent to follow the applied field. To be more specific, the angle of the rotation depends on the distance to the hard layer. That is the angle of rotating in a spiral spin structure similar to that of a Bloch domain wall. If the applied field is removed, the soft spins rotate back into alignment with the hard layer

In this work, we report the results of ESR investigations for CoSm/Fe exchange magnetic films.

2.   THEORETICAL MODEL

For theoretical analysis, the exchange-spring magnet Sm-Co/Fe is divided into subatomic multi-layers (d=2 A), and the spins in each layer are characterized by the average magnetization M_i, and the uniaxial anisotropy constant K_i (Fig. 1a). Sublayers are coupled by an exchange constant $A_{i,i+1}$ [13]. The total free energy of the sublayer system is given

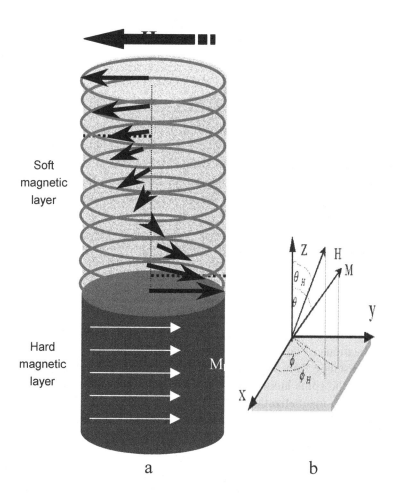

Soft magnetic layer

Hard magnetic layer

a

b

Fig.1.a) Representation of magnetizations of individual sublayers in soft magntic layer after saturating hard layer and applied reverse static magnetic field. b) Relative orientations of the coordinat system used in both theoretical calculations and experimental studies.

$$E_t = -H \sum_{i=1}^{N} M_i (\sin \theta_i \sin \theta_H \cos(\phi_i - \phi_H) + \cos \theta_i + \cos \theta_H)$$

$$+2\pi \sum_{i=1}^{N} M_i^2 \cos^2 \theta_i - \sum_{i=1}^{N} K_i \sin^2 \theta_i \cos^2 \phi_i$$

$$+\sum_{i=1}^{N-1} \frac{A_{i,i+1}}{d^2} (\sin \theta_i \sin \theta_{i+1} \cos(\phi_i - \phi_{i+1}) + \cos \theta_i \cos \theta_{i+1})$$

The spin direction characterized by ϕ_i in i^{th} layer can be obtained from the static equilibrium condition $dE/d\phi_i = 0$. For in-plane measurement the field is applied perpendicular to the film normal. If the field is applied parallell to the saturation magnetization of hard layer, than ($\theta_i = \theta_H = \pi/2$) the equlibrium asimutal angle for the spins in i^{th} layer become

$$\tan \phi_i' = \left[\frac{A_{i,i-1} \sin \phi_{i-1} + A_{i,i+1} \sin \phi_{i+1} + d^2 H M_i \sin \phi_H}{A_{i,i-1} \cos \phi_{i-1} + A_{i,i+1} \cos \phi_{i+1} + 2d^2 K_i \cos \phi_i + d^2 H M_i \cos \phi_H} \right].$$

Computer simulations of this expression are given for different applied magnetic field and different angle of the applied field in Fig. 2a and b. As can be seen from the figures, in the soft magnetic layer the spins (magnetizations of the hard magnetic individual sublayers) twisted more and more as distance from hard layer increases. While magnetic field increases, rotation of the spins increases for a particular angle of the applied field with respect to the magnetization of the hard layer (Fig. 2a). If the angle of the applied field (respect to the magnetisation of the hard layer) is approach to the magnetization of the hard layer for constant static applied magnetic field, then twisting decreases (Fig. 2b). The hard layer is also appreciably perturbed by the rotation of the soft Fe layers. As applied static magnetic field increases, the interficial Sm-Co spin is also increasingly rotated and a domain wall is slowly shifted into the hard layer.

The ferromagnetic resonance relations for each layer have been derived from the magnetic free energy by using Bloch-Bloembergen damping term

$$\frac{1}{\gamma} \frac{\partial \vec{M}}{\partial t} = -\vec{M} \times \vec{H}_{eff} - \frac{\vec{M}_{\theta,\phi}}{\gamma T_2} - \frac{\vec{M}_z - \vec{M}_0}{\gamma T_1}.$$

Then, the imaginary part of the magnetic susebtility and dispersion relation can be obtained as follows [14,15].

$$\chi_2 = 4\pi \frac{m_\phi}{h_\phi} = \frac{4\pi M_0 \left(\dfrac{\partial^2 E}{\partial \theta^2}\right) \dfrac{2\omega}{\gamma^2 T_2}}{\left[\left(\dfrac{\omega_0}{\gamma}\right)^2 - \left(\dfrac{\omega}{\gamma}\right)^2\right]^2 + \dfrac{4\omega^2}{\gamma^4 T_2^2}}$$

$$\left(\frac{\omega(i)}{\gamma}\right)^2 = \left(H_i^{eff}\cos 2\theta_i + H\cos(\theta_i - \theta_H) + \left(\frac{A_{i,i+1}}{d^2 M_i}\right)\cos(\theta_i - \theta_{i+1})\right) \times$$

$$\left(H_i^{eff}\cos^2\theta_i + H\cos(\theta_i - \theta_H) + \left(\frac{A_{i,i+1}}{d^2 M_i}\right)\cos(\theta_i - \theta_{i+1})\right) + \left(\frac{1}{\gamma T_2}\right)^2$$

where $\gamma = g\mu_B \hbar$ is the gyromagnetic ratio, g is the Landé factor, μ_B is the Bohr magneton, and $H_i^{eff} = 2K_i/M_i - 2\pi M_i$ is the effective magnetic field acting on M_i. In this equation, θ_i and ϕ_i are the equilibrium angles for the magnetization of each layer.

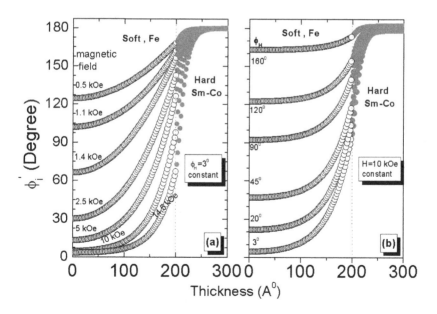

Figure 2. a) Calculated equilibrium spin configuration of a Fe (200Å)/Sm-Co bilayer structure at several reversal fields. b) Spin configuration obtained from the theoretical model calculation for different magnetic field angle is presented

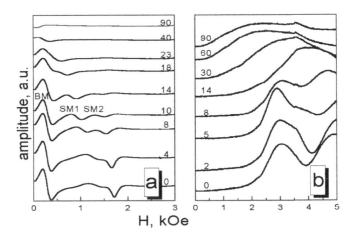

Figure 3. In plane FMR spectra for both samples

3.   EXPERIMENTAL RESULT AND DISCUSSION

Sm-Co (200 Å)/Fe (200 Å and 100 Å) bilayers have been grown on the epitaxial 200 A Cr(211) buffer layer on single crystal MgO(110) substrates by magnetron sputtering technique [16],[6]. To prevent oxidation Sm–Co/Fe film was coated with a 100 A thick Cr layer. ESR spectra of the samples have been registered at room temperature, for both in-plane, IPG (where the applied field is always perpendicular to the film normal) and out-of-plane geometry, OPG (where the field is applied along a general direction with respect to the film normal). The ESR spectra have been taken after saturating the magnetization of the hard Sm-Co layer in DC magnetic field parallel to easy axis of the hard magnetic Sm-Co layer. ESR signals can only be observed when the static magnetic field is applied opposite to the hard layer magnetization vector. If the static field is increased beyond the switching value (~5.5 kOe) for hard layer magnetization, then, the spectra disappear. In this case the sample should be rotated 180 degrees in order to recover the original spectrum. Ferromagnetic resonance (FMR) spectra of 200 Å and 100 Å Fe samples for different angles of the applied magnetic field in the film plane are presented in Fig. 3a and b respectively. There are three peaks: one of them corresponds to the bulk mode and the remaining to the surface modes for 200 Å Fe sample. As the applied field is rotated toward the direction of the Sm-Co magnetization vector, the bulk mode shifts to lower magnetic fields for both samples and then disappears at the angle of about 40 degrees for the 200 Å Fe sample. For 100 Å Fe sample, the FMR line width is broadened as the applied field is rotated from the easy to the hard direction of Sm-Co layer magnetization.

The angular variations of FMR spectra of both samples for OPG case are given in Fig. 4 a and b. Here the angle is measured with respect to the hard direction of the soft layer magnetization. axis in the sample plane and rotated toward the film normal direction. Again three peaks are observed for OPG case of the 200 Å Fe sample and one peak for the 100 Å Fe. Some paramagnetic peaks about 3200-3500G are gotten due to impurities in the MgO substrate.

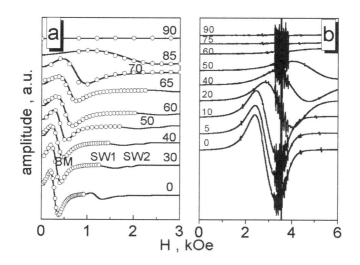

Figure 4. Out of plane FMR spectra for a 200 Å Fe and b) 100 Å Fe samples.

The simulated ESR signal and the overall (average) magnetization have been obtained by a computer modeling (Fig. 4a). The following values for magnetic parameters (of hard/soft layer) have been deduced: $H_i^{eff} = -3,2$ kOe for the Fe layer, $\omega/\gamma = 3,15$ kOe, T_2 ($1 \pm 0.16.10^{-8}$s), $M_s = 1700$ emu/cm^3, $M_h = 550$ emu/cm^3, $A_s = 2.8 \times 10^{-6}$ erg/cm, $A_h = 1.2 \times 10^{-6}$ erg/cm, $A_{int} = 1.8 \times 10^{-6}$ erg/cm, $K_s = 1.0 \times 10^3$ erg/cm^3 and $K_h = 5.0 \times 10^7$ erg/cm^3. A_{int} is the exchange coupling constant between the hard and soft layers.

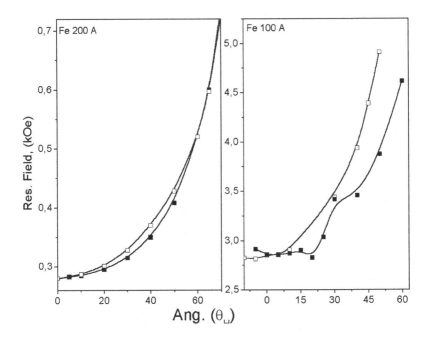

Figure 5. Angular dependance of resonance field for out of plane geometry.

The angular dependence of resonance field for OPG case are given in Fig. 5. It has been found that the easy direction for the magnetization of the hard Sm-Co layer grown on MgO(110) is not completely parallel to the sample plane but declined from the film plane by an angle of about 15 degrees. As a result, angular dependence of the ESR spectra for out-of-plane measurements is not symmetric with respect to the film plane (Fig. 5). A very strong exchange coupling has been observed between the hard and soft layers, as well.

REFERENCES

1. W.H. Meiklejohn and C.P.Bean, Phys. Rev. 102, (1956) 1413.
2. Ekart F. Kneller and Reinhard Hawing, IEEE Trans. Magn. 27 (1991) 3588.
3. C. Tsang, IEEE Trans. Magn. MAG-25, (1989) 3672.
4. W.C. Cain, D.C. Markham, and M.H. Kreyder, IEEE Trans. Magn. MAG-25, (1989) 3695.
5. Y.Suzuki, R.B. van Dover, E.M. Gyorgy, Julia M. Phillips, and R.J. Felder, Phys. Rev. B 53 (1996),14016.
6. J.M.L. Beaujour, G.J. Bowden, S. Gordeev, P.A.J. de Groot, B.D. Rainford, R.C.C. Ward, M.R. Wells, Jurn. Magn. Mag. Mat.226-230 (2001) 1870.

7. Eric E. Fullerton, J.S. Jiang, Christine Rehm, C.H. Sowers, S.D. Bader, J.B. Patel, X.Z. Wu Appl. Phys. Lett.71 (1997) 1579

8. T. Schrefl, H. Kronmüller, and J. Fidler, J. Magn. Magn. Mater. 127 (1993) L275.

9. K. Mibu, T. Nagahama, T. Ono, T. Shinjo, J. Magn. Magn. Mater. 177-181(1997) 1267.

10. J. S. Jiang, H. G. Kaper and G. K. Leaf, Discrete and Continuous Dynamical Systems-Series B 1, No. 2 (2001) 219.

11. R. Skomski Phys. Rev. B 48 (1993) 15812.

12. J.M.D. Coey, Solid State Commun. 102 (1997) 101.

13. Eric E. Fullerton, J.S. Jiang, M. Grimsditch, C.H. Sowers, and S.D.Bader, Phys. Rev.B 58 (1998) 12193.

14. B. Aktas, Thin Solid Films 307 (1997) 250.

15. M.Ozdemir, Y.Oner, B. Aktas, Physica B 252 (1998) 138.

16. S. Wüchner, J. C. Toussaint and J. Voiron, Phys. Rev. B 55 (1997)

5

PART 5: ADVANCED MAGNETIC MATERIALS

PHASE TRANSITION IN MOLECULAR NANOMAGNET Mn$_{12}$

B. Bakar and L. F. Lemmens
Universiteit Antwerpen Departement Natuurkunde Groenenborgerlaan 171 B2020 Antwerpen Belgie

Abstract: In this work we want to report the recently proven equivalence between the structure of Yang-Lee zeros in the grand-canonical description of a system and the bimodality of the probability density of the order parameter in the canonical description. The model we have chosen is a simplified many spin model for molecular magnets. It can be considered as an adaptation of the Curie-Weiss model for finite systems. This model is exactly solvable and there is a one to one mapping of the states of the model and the order parameter i. e. the magnetization.

Key words: Molecular nanomagnet, magnetization.

1.   INTRODUCTION

Phase transitions in finite systems have been a considerable attractive field in theoretical physics. Since the finite systems are far away from the thermodynamic limit there have been several classification schemes in order to interpret phase transitions and their order in the non-extensive systems. These works, started from Yang-Lee who gave a description of phase transitions by analyzing the distributions of zeros of the grand- canonical partition function in the complex temperature plane, are still current research subjects. The influence of the non-extensiveness on phase transition is a field which should be further explored.

Recently, the phase transitions in the finite systems have been tried to define by the topological anomalies of the event distribution in the space of observations [1]. In this suggestion, the first order phase transition is defined

B. Aktaş et al. (eds.), Nanostructured Magnetic Materials and their Applications, 241–249.
© 2004 *Kluwer Academic Publishers. Printed in the Netherlands.*

by the bimodality of the distribution of the order parameter and related to the Yang-Lee theorem [2].

Molecular magnets have attracted considerable attention in recent years as mesoscopic systems where macroscopic quantum phenomena are observed, such as relaxation by tunneling of magnetization through a potential well at the molecular level [3, 4]. Molecular magnets are formed by magnetic molecules usually within a crystal. All the molecules have the same total spin, same anisotropy and the same controlled orientation. These issues make molecular magnets the most promising materials for future quantum computing implementations [5].

Molecular magnets can be characterized theoretically by the following Hamiltonian containing the components of a single large spin:

$$H = D(S_z^2) + E(S_x^2 - S_y^2) + g\mu_B hS \,. \tag{1}$$

Most interpretations of experiments and almost all theoretical approaches use a Hamiltonian of this type to deduce the properties of molecular magnets. There is however an alternative to this approach. Instead of using a single spin with a large z-component, one can introduce many spins with length 1/2 and derive a Hamiltonian with exactly the same energy spectrum.

The paper is organized as follows. In Sec. II we introduce our model for molecular magnets. In Sec. III the model we suggested here is adapted to the Mn_{12}-cluster to calculate the probability density of local magnetization using Metropolis algorithm. In the last section we discuss our results.

2.   TWO MODELS FOR THE MOLECULAR MAGNET

In this section we consider first a cluster of N identical spins in a uniaxial environment characterized by an exchange integral with absolute value $\|D\|$. When we neglect the rhombic terms $E(S_x^2 - S_y^2)$ in the favor of the uniaxial ferromagnetic terms $\| D \| S_z^2$ (D<0) the Hamiltonian of such a cluster can be written as:

$$\frac{H_{MS}}{\| D \|} = -2\sum_{j=1}^{N-1} \sum_{k=j+1}^{N} s_j s_k - h\sum_{j=1}^{N} s_j + \sum_{j=1}^{N} s_j^2 \,. \tag{2}$$

The total z-component of the spins in the cluster is given by

$$S_z = \sum_{j=1}^{N} s_j. \tag{3}$$

For a spin 1/2 cluster, using the completion of the square, one may rewrite the Hamiltonian (2) into the form:

$$\frac{H_{LS}}{\|D\|} = -S_z^2 - hS_z. \tag{4}$$

Incorporating D in the effective temperature, the Hamiltonian (4) is reduced to contain only one parameter proportional to the external field:

$$h = b\frac{g\mu_B}{\|D\|}. \tag{5}$$

Denoting a configuration by the number of spins pointing up $N-m$ and the number of spins pointing down m, the energy spectrum is obtained:

$$\varepsilon_m = hm + m(N-m) - h\frac{N}{2} - \frac{N^2}{4}, \tag{6}$$

and the magnetization is given by:

$$S_z|c\rangle = \frac{1}{2}(N - 2m)|c\rangle, \tag{7}$$

where the configurations

$$|c\rangle = \left|\{s_i\}_{i=1}^{N}\right\rangle \tag{8}$$

form a basis set of states for many-spin model. The magnetization given by (7) ranges from $-N/2$ to $N/2$, or the number of different values, that the magnetization of the cluster can take, is: $N+1$. Thus, the number m is the only number that is needed to derive the magnetization and the energy spectrum of a cluster with a given number of sites. One may label the configuration by that number. But one has still to take into account the number of configurations represented by that label. Here there are 2^N possible configurations. When considering summations over the micro-states of the system this degeneracy of the configurations is important. Indeed, one can consider a model with a Hamiltonian given by (4) describing a single large spin (7). In such a model there are only N+1 configurations. This model will be indicated by the large spin model (LS) in order to distinguish it from the cluster with many spin degrees of freedom (MS).

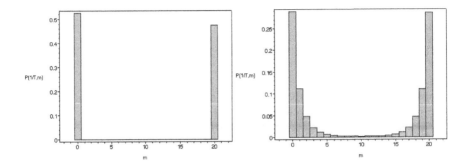

Figure 1. The probability density of local magnetization for the large spin model. Left: T=2K. Right: T=20K.

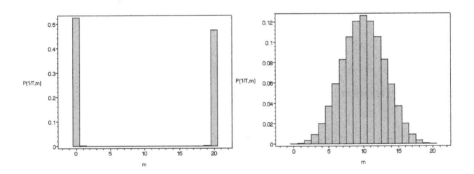

Figure 2. The probability density of local magnetization for the Curie-Weiss model. Left: T=2K. Right: T=20K.

2.1   The density of magnetization

The partition function for the Curie-Weiss model of the spin cluster is easily obtained by a direct calculation:

$$Z_{MS}(\beta) = \sum_{m} \binom{N}{m} \exp(-\beta \varepsilon_m), \tag{9}$$

where the binomial term counts the combinations and β is the inverse temperature in reduced units.

The same calculation is easily performed for the phenomenological large spin model:

$$Z_{LS}(\beta) = \sum_m \exp(-\beta\varepsilon_m), \tag{10}$$

leading for both models to the probability that the system is in the state $|c_m\rangle$ given by:

$$p_{MS}(m) = \binom{N}{m} \frac{\exp(-\beta\varepsilon_m)}{Z_{MS}(\beta)}, \tag{11}$$

$$p_{LS}(m) = \frac{\exp(-\beta\varepsilon_m)}{Z_{LS}(\beta)}, \tag{12}$$

respectively.

The density of magnetization for both models is derived directly from the probabilities.

$$P_{MS}\left[S_z = \frac{1}{2}(N-2m)\right] = p_{MS}(m), \tag{11}$$

$$P_{LS}\left[S_z = \frac{1}{2}(N-2m)\right] = p_{LS}(m), \tag{13}$$

For $\beta=0$ the Curie-Weiss model has a binomial density, the large spin model - a uniform density $1/(N+1)$. For small β the Curie-Weiss model has still a density that is similar to binomial density in the sense that there is only one characterized by m. The large spin model becomes bimodal through the dependence on the energy spectrum: the states with largest probability are 0 or N. However for large β the energy spectrum becomes the dominant factor in the density of the Curie-Weiss model and the density will be bimodal. As explained in the introduction this indicates a phase transition in the magnetic behavior of the cluster.

3.   THE CURIE-WEISS MODEL OF THE Mn$_{12}$-CLUSTER

In this section we replace the Curie-Weiss model by an uni-axial model with the symmetry of the well-known Mn$_{12}$-cluster, abbreviation for Mn$_{12}$(CH$_3$COO)$_{16}$(H$_2$O)$_4$O$_{12}$ 2CH$_3$COOH 4H$_2$O. This model contains an inner ring with 4 Mn^{+4} ions (S=3/2) and an outer ring with 8 Mn^{+3} ions (S=2). The antiferromagnetic interaction between the rings leads to the well-

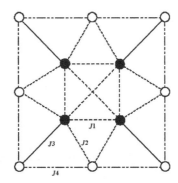

Figure 3. The geometric representation of the magnetic interactions of the Mn$_{12}$-cluster. The values of exchange integrals are: J_1~215K, J_2~J_3~86K, and J_4~30K.

known $m=\pm 10$ ground state. The schematic diagram of the magnetic interactions is given by Fig. 3. The values of exchange integrals are taken from experiment.

3.1   The magnetization density

In order to obtain the density of order parameter, in this case the magnetization, we will simulate the possible configurations of the cluster using Metropolis algorithm. Once the configuration is obtained, we determine the magnetization and count the number of times we obtain that value of the magnetization. Using the frequency of occurrence the probability density of the magnetization can be estimated. All parameters in the Hamiltonian are taken from experiment and we have restricted ourselves to $h=0$ leaving us the inverse temperature β as the parameter to vary.

To estimate the magnetization density of Mn$_{12}$-cluster one has initially to calculate the joint probability distribution of the possible spin configurations. In an Ising magnetic cluster with n sites, and spin at a site i is characterized by s_i, m_i, the configuration is given by $|\{s_i, m_i\}, i=1..n>$. Using the Gibbs assumption $p\sim\exp(-\beta\varepsilon_m)$ in combination with the model Hamiltonian (4) the local energy of a configuration can be obtained. This configuration of the local energy enables us to estimate the local magnetization density as a function of inverse temperature β and spin states. At high temperature it is seen that there is only one mode for the magnetization density indicating that state with $m=0$ is the most probable state. For low temperature two modes are found indicating that the ground states of Mn_{12} with $m=\pm10$ are equally probable. Our results obtained by a Monte Carlo method at high and low temperatures are shown in Fig. 4 and Fig. 5, respectively. The simulated spin

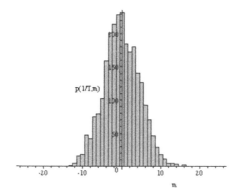

Figure 4. The simulated probability density of local magnetization for Mn_{12}-cluster at $T=25K$.

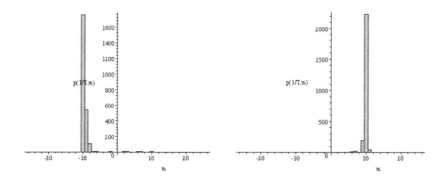

Figure 5. The simulated probability density of local magnetization for Mn_{12}-cluster. Left: $T=0.25K$. Right: $T=0.22K$.

state configuration of local magnetization as a function of inverse temperature is shown in Fig. 6.

4.   DISCUSSION

In our case Mn$_{12}$-cluster, the probability density of the order parameter in a Curie-Weiss model changes from bimodal at low temperature to a density with a single mode at high temperature, indicating a first order phase transition. This change in density is important, because magnetic clusters are only able to store information if the density is bimodal. If the cluster is described phenomenologically by replacing the spins by a single large spin, the density of magnetization remains bimodal over the entire temperature range. Comparing these two models, we conclude that they have the same energy spectrum, but different density of states: the

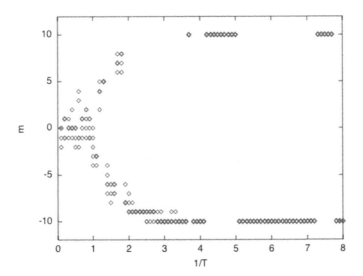

Figure 6. The simulated spin state configuration of Mn$_{12}$-cluster versus inverse temperature.

phenomenological model has a uniform density of states, the Curie-Weiss model attaches combinatorial coefficients.

In order to take the tunneling into account a quantum mechanical description is necessary. This is essential and will be studied in the future.

ACKNOWLEDGMENTS

This work has been performed partly in the framework of a RAFO project/1 LEMMLU/KP99 and of the GOA BOF UA 2000 projects of the Universiteit Antwerpen.

REFERENCES

1. F. Gulminelli, Ph. Chomaz, and V. Duflot, Phys. Rev. E 64, 046114 (2001).
2. C. N. Yang and T. D. Lee, Phys. Rev. 87, 404 (1952).
3. J. R. Friedman, M. P. Sarachik, J. Tejeda, and R. Ziola, Phys. Rev. Lett. 76, 3830 (1996).
4. L. Thomas, F. Lionti, R. Ballou, D. Gatteschi, R. Sessoli, and B. Barbara, Nature 383, 145 (1996).
5. M. N. Leuenberger and D. Loss, Nature 410, 789 (2001).

ONEREFRINGENT BIANISOTROPIC SUPERLATTICES IN THE LONG WAVELENGTH APPROXIMATION

E.G. Starodubtsev, O.D. Asenchik
Gomel State Technical University, October av. 48, Gomel, 246746, Belarus
Fax: + 375 - 232 - 479165; e-mail: starodub@tut.by, asenchik@gstu.gomel.by

Abstract: General conditions of linear birefringence compensation in bianisotropic layered-periodic structures or superlattices at proportional effective tensors of dielectric permittivity and magnetic permeability are investigated. Calculations and numerical analysis are carried out in frame of the long wavelength approximation for electromagnetic field. Rather wide range of structure properties are found when the considered effect can take place.

Key words: bianisotropic layered-periodic structures, superlattices, linear birefringence compensation.

1. INTRODUCTION

The effect of linear birefringence compensation in anisotropic magnetic monocrystals at mutually proportional tensors of dielectric permittivity ε and magnetic permeability μ was predicted by F.I. Fedorov [1] and investigated in a number of works (e.g., in [2-4]). It was shown, that at condition $\mu = k\varepsilon$ (where k is a proportionality coefficient) in case of symmetric ε, μ, and at $\tilde{\mu} = k\varepsilon$ (or $\tilde{\varepsilon} = k\mu$) in case of nonsymmetric ε, μ (tilde denotes transposition of a tensor), refractive index of such media has the only value which is a function of the wave normal. In this case the medium is an onerefringent anisotropic medium. This effect is due to a peculiar compensation of the medium dielectric anisotropy by the magnetic anisotropy. The considered effect can take place only for the definite light frequency due to frequency dispersion of crystal dielectric and magnetic properties. That can be used in devices with frequency selective light

B. Aktaş et al. (eds.), Nanostructured Magnetic Materials and their Applications, 251–262.
© 2004 *Kluwer Academic Publishers. Printed in the Netherlands.*

transmission [5]. Moreover, at the birefringence compensation various effects caused by the medium gyrotropy become apparently more strong [4]. In bianisotropic monocrystals such onerefringence is a rather rare effect, as the relations between components of tensors ε and μ are very rigid and being determined only by the nature of the crystal.

There are more possibilities of onerefringence realization in periodic layered crystal structures or superlattices (SL), in cases when SL properties can be described by effective tensors of dielectric permittivity and magnetic permeability. For such structures dielectric and magnetic characteristics of the effective medium can be controlled varying geometry and properties of layers. A particular case considered in [5], where onerefringent SL was constructed from the layers of uniaxial crystals with coincident optical axis oriented perpendicular to the layers boundaries. However, the general problem of onerefringence conditions search for bianisotropic SL from arbitrary crystallographic symmetry components has not been solved, as far as we know.

The work aims an investigation of general conditions for onerefringence in short-period layered-periodic structures or SL satisfying the effective medium approximation. Calculations and numerical modeling were carried out in the frame of the long wavelength approximation for electromagnetic field, when $D \ll \lambda$, where D is a period of the structure formed by the plane parallel layers, λ is a wavelength.

2. GENERAL EXPRESSIONS

Long wavelength approximation for electromagnetic field is satisfied for SL with periods of tens Angstroms and for the optical wave band. In this case one can assume that the field in the structure does not change at distances of the order of the SL period $D = \sum_i d_i$, where d_i are thicknesses of single crystalline layers within the SL period. Let us choose a cartesian coordinate system with axis Z which is perpendicular to the boundaries of the layers in SL (Fig.1).

From the conditions of additivity of electric and magnetic moments in the SL volume [5, 6] one can gain relations connecting the normal components of the fields strength $E_3^{(e)}, H_3^{(e)}$ and tangential components of the inductions $D_1^{(e)}, D_2^{(e)}, B_1^{(e)}, B_2^{(e)}$ of electromagnetic field in the SL and in the layers:

$$Q^{(e)} = \sum_{i=1}^{n-1} x_i Q^{(i)} + (1 - \sum_{i=1}^{n-1} x_i) Q^{(n)}, \qquad (1)$$

where $Q = E_3, H_3, D_1, D_2, B_1, B_2$, $x_i = d_i / D$ is a relative thickness of i-th layer ($0 \le x_i \le 1$, $i = 1,2,..,n$), the upper indexes in brackets e and $1,2,....,n$ point to quantities characterizing the effective medium and layer i in the SL period, respectively.

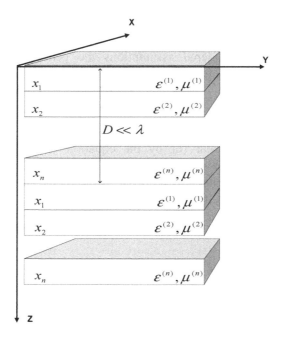

Figure 1. Geometry of the problem

On defining the effective SL characteristics, we shall suppose that the phenomenological material equations [3]

$$\vec{D}^{(s)} = \varepsilon^{(s)} \vec{E}^{(s)} + \alpha^{(s)} \vec{H}^{(s)}, \quad \vec{B}^{(s)} = \mu^{(s)} \vec{H}^{(s)} + \beta^{(s)} \vec{E}^{(s)}, \tag{2}$$

with the complex nonsymmetric tensors $\varepsilon^{(s)}, \alpha^{(s)}, \mu^{(s)}$ and $\beta^{(s)}$ for media with various kinds of anisotropy, gyrotropy and absorption both for the layers forming SL and for the effective medium: $s=1,2,\ldots,n,e$.

According to the methods proposed in [6], gaining from Eqs. (2) quantities E_3, H_3 and taking into account continuity of quantities $E_1, E_2, H_1, H_2, D_3, B_3$ on the layer boundaries and Eq. (1) we have expressions [7]

$$\frac{1}{\varepsilon_{33}}, \frac{\varepsilon_{3j}}{\varepsilon_{33}}, \frac{\varepsilon_{r3}}{\varepsilon_{33}}, \varepsilon_{rj} - \frac{\varepsilon_{r3}\varepsilon_{3j}}{\varepsilon_{33}}, \frac{1}{\mu_{33}}, \frac{\mu_{3j}}{\mu_{33}}, \frac{\mu_{r3}}{\mu_{33}}, \mu_{rj} - \frac{\mu_{r3}\mu_{3j}}{\mu_{33}} \qquad (3)$$

Here and thereafter $r,j=1,2$. Quantities (3) (summation over repeating indexes is not made), when substituted instead of parameter Q in Eq. (1), define all the components of complex nonsymmetric effective tensors $\varepsilon^{(e)}, \mu^{(e)}$ for arbitrary crystallographic symmetry of the layers.

In frame of the considered assumptions the general onerefringence condition for SL is determined by the following tensors equation

$$\tilde{\mu}^{(e)} = k\varepsilon^{(e)} \qquad (4)$$

where k is a complex scalar (or by the equation $\tilde{\varepsilon}^{(e)} = k\mu^{(e)}$ which is equivalent to Eq. (4)). Obviously, for the solution of the problem it is necessary to determine general conditions for components of the tensors $\varepsilon^{(i)}, \mu^{(i)}$, to gain and to solve equation [4] with respect to quantities x_i satisfying Eq. (4).

3.   ONEREFRINGENCE IN TWO COMPONENT SL

Let us consider the case when $D = d_1 + d_2$ (the SL period includes two layers). In this case Eq. (1) has the form $Q^{(e)} = xQ^{(1)} + (1-x)Q^{(2)}$, where $x = x_1$ is a relative thickness of the first layer in the SL period. Substituting expressions (3) to Eq. (1) and taking into account Eq. (4) we have equations

$$x(m_{r3}^{(1)} - m_{r3}^{(2)} - e_{3r}^{(1)} + e_{3r}^{(2)}) + m_{r3}^{(2)} - e_{3r}^{(2)} = 0, \qquad (5)$$

$$x(m_{3j}^{(1)} - m_{3j}^{(2)} - e_{j3}^{(1)} + e_{j3}^{(2)}) + m_{3j}^{(2)} - e_{j3}^{(2)} = 0, \qquad (6)$$

$$x^2 \left(\frac{1}{\mu_{33}^{(1)} \mu_{33}^{(2)}} \left(\Delta\mu_{3r} \Delta\mu_{j3} - \Delta\mu_{jr} \Delta\mu_{33} \right) - \frac{1}{\varepsilon_{33}^{(1)} \varepsilon_{33}^{(2)}} \left(\Delta\varepsilon_{3j} \Delta\varepsilon_{r3} - \Delta\varepsilon_{rj} \Delta\varepsilon_{33} \right) \right) +$$

$$x \left(\begin{array}{l} \dfrac{1}{\mu_{33}^{(1)} \mu_{33}^{(2)}} \left(\mu_{jr}^{(2)} \Delta\mu_{33} - \mu_{33}^{(1)} \Delta\mu_{jr} - \Delta\mu_{3r} \Delta\mu_{j3} \right) - \\[2mm] \dfrac{1}{\varepsilon_{33}^{(1)} \varepsilon_{33}^{(2)}} \left(\varepsilon_{rj}^{(2)} \Delta\varepsilon_{33} - \varepsilon_{33}^{(1)} \Delta\varepsilon_{rj} - \Delta\varepsilon_{3j} \Delta\varepsilon_{r3} \right) \end{array} \right) + m_{jr}^{(2)} - e_{rj}^{(2)} = 0, \tag{7}$$

where the definitions have been used

$$m_{pq}^{(i)} = \frac{\mu_{pq}^{(i)}}{\mu_{33}^{(i)}}, \quad e_{pq}^{(i)} = \frac{\varepsilon_{pq}^{(i)}}{\varepsilon_{33}^{(i)}},$$

$$\Delta\mu_{pq} = \Delta\mu_{pq}^{(21)} = \mu_{pq}^{(2)} - \mu_{pq}^{(1)}, \quad \Delta\varepsilon_{pq} = \Delta\varepsilon_{pq}^{(21)} = \varepsilon_{pq}^{(2)} - \varepsilon_{pq}^{(1)} \tag{8}$$

and i is a layer number in the SL period (in the given case $i=1,2$). Here and thereafter p, $q=1,2,3$.

In general case of nonsymmetric tensors $\varepsilon^{(i)}, \mu^{(i)}$ Eqs. (5)-(7) represent a system of four linear and four quadratic equations with respect to unknown x (at the symmetric tensors Eqs. (5) and Eqs. (6) are equivalent and we have the system of two linear and three quadratic equations). This system is also added by the condition $0 < x < 1$.

Let us consider some features of the system (5)-(7). In general case of unknown tensors $\varepsilon^{(i)}, \mu^{(i)}$ (36 complex components) and parameter x the system has an infinite number of solutions. Defining values of 29 components, for example, components of tensors $\varepsilon^{(1)}, \varepsilon^{(2)}, \mu^{(1)}$ and 2 components of tensor $\mu^{(2)}$ one can find the rest 7 ones and the parameter x. At symmetric tensors $\varepsilon^{(i)}, \mu^{(i)}$, definition of 20 tensor components allows us to find the rest 4 ones and parameter x.

At nonzero coefficients at x in Eqs. (5) or Eqs. (6), the Eqs. (7) can be reduced to linear in x equations or to equations without x.

Components of tensors $\Delta\varepsilon, \Delta\mu$ characterizing difference of dielectric and magnetic properties of the layers can satisfy conditions $\Delta\varepsilon_{pq} \ll \varepsilon_{33}^{(i)}, \Delta\mu_{pq} \ll \mu_{33}^{(i)}$. Then in Eqs. (7) coefficients at x have the first and coefficients at x^2 have the second order on small parameters $\Delta\varepsilon_{pq} / \varepsilon_{33}^{(i)}, \Delta\mu_{pq} / \mu_{33}^{(i)}$. At these conditions in a wide range of parameters and with a rather high accuracy Eqs. (7) can be replaced by linear equations omitting the quadratic in x terms. After corresponding replacements (ξ, η), Eqs. (7) can be resulted in the form $\xi x^2 + \eta x + 1 = 0$. Numerical analysis shows that, in particular, at $\xi/\eta = 0.1$ and $\eta < -3$ one of quadratic equation roots and a root of the linear equation $\eta x + 1 = 0$ lie in the range of $0 < x < 1$ and differ in magnitude about one percent and less (with increasing $|\eta|$).

At conditions $m_{r3}^{(1)} - e_{3r}^{(1)} \neq 0, m_{3j}^{(1)} - e_{j3}^{(1)} \neq 0$ from Eqs. (5), (6) we have one of the necessary SL onerefringence conditions

$$\frac{m_{r3}^{(2)} - e_{3r}^{(2)}}{e_{3r}^{(1)} - m_{r3}^{(1)}} = \frac{m_{3j}^{(2)} - e_{j3}^{(2)}}{e_{j3}^{(1)} - m_{3j}^{(1)}} = \frac{x}{1-x}. \tag{9}$$

As $x/(1-x) > 0$, relations (9) denote that magnetic properties anisotropy is predominant in one layer and dielectric anisotropy is predominant in the other layer. Analogous results can be gained also for the quantities $m_{jr}^{(i)}, e_{rj}^{(i)}$.

The other necessary condition of onerefringence can be derived by eliminating x from Eqs. (5) and Eqs. (7) (or from Eqs. (6) and Eqs. (7)). For example, at $m_{3j}^{(1)} - e_{j3}^{(1)} \neq 0$ from Eqs. (6), (7) we have four equations

$$(m_{jr}^{(2)} - e_{rj}^{(2)})(m_{3j}^{(1)} - e_{j3}^{(1)}) + (m_{3j}^{(2)} - e_{j3}^{(2)})(\frac{\varepsilon_{rj}^{(1)}}{\varepsilon_{33}^{(2)}} + \frac{\varepsilon_{rj}^{(2)}}{\varepsilon_{33}^{(1)}} - \frac{\Delta\varepsilon_{3j}\Delta\varepsilon_{r3}}{\varepsilon_{33}^{(1)}\varepsilon_{33}^{(2)}} -$$
$$\frac{\mu_{jr}^{(1)}}{\mu_{33}^{(2)}} - \frac{\mu_{jr}^{(2)}}{\mu_{33}^{(1)}} + \frac{\Delta\mu_{j3}\Delta\mu_{3r}}{\mu_{33}^{(1)}\mu_{33}^{(2)}} + \frac{m_{jr}^{(1)} - e_{rj}^{(1)}}{m_{3j}^{(1)} - e_{j3}^{(1)}}(m_{3j}^{(2)} - e_{j3}^{(2)})) = 0 \tag{10}$$

Conditions (9), (10) show, in particular, that for the SL onerefrigence components of the tensors $\varepsilon^{(i)}, \mu^{(i)}$ should satisfy definite relations.

Let us consider some particular cases following from solutions of Eqs. (5)-(7) system.

SL originated from layers characterized by scalar tensors ε, μ, for example, made from layer of cubic crystals, can not be onerefringent. For such media the system of Eqs. (5)-(7) has only trivial solutions $x=0;1$.

SL from onerefringent layers can be onerefringent only at conditions

$$\tilde{\mu}^{(1)} = k\varepsilon^{(1)}, \tilde{\mu}^{(2)} = k\varepsilon^{(2)}, \tag{11}$$

when the proportionality coefficient between tensors $\tilde{\mu}$ and ε of different layers is the same.

At $x=0.5$ and conditions

$$\tilde{\mu}^{(1)} = k\varepsilon^{(2)}, \tilde{\mu}^{(2)} = k\varepsilon^{(1)} \tag{12}$$

SL can be onerefringent (in this case the layers are not onerefringent).

When SL is originated from uniaxial components with coincident optical axes, which are parallel to OX axis, the tensors characterizing the layers have a diagonal form: $\varepsilon^{(i)} = diag(\varepsilon_e^{(i)}, \varepsilon_o^{(i)}, \varepsilon_o^{(i)})$, $\mu^{(i)} = diag(\mu_e^{(i)}, \mu_o^{(i)}, \mu_o^{(i)})$, $i=1,2$, where indexes e and o point to the quantities corresponding to extraordinary and ordinary waves. In this case the SL can be onerefringent, in particular, at conditions

$$\frac{\mu_e^{(2)}}{\mu_o^{(1)}} + \frac{\mu_e^{(1)}}{\mu_o^{(2)}} = \frac{\varepsilon_e^{(2)}}{\varepsilon_o^{(1)}} + \frac{\varepsilon_e^{(1)}}{\varepsilon_o^{(2)}}, \quad \frac{\mu_o^{(2)}}{\mu_o^{(1)}} + \frac{\mu_o^{(1)}}{\mu_o^{(2)}} = \frac{\varepsilon_o^{(2)}}{\varepsilon_o^{(1)}} + \frac{\varepsilon_o^{(1)}}{\varepsilon_o^{(2)}},$$

$$x = \left[1 + \sqrt{\left(\frac{\mu_e^{(1)}}{\mu_o^{(1)}} - \frac{\varepsilon_e^{(1)}}{\varepsilon_o^{(1)}} \right) \bigg/ \left(\frac{\varepsilon_e^{(2)}}{\varepsilon_o^{(2)}} - \frac{\mu_e^{(2)}}{\mu_o^{(2)}} \right)} \right]^{-1},$$

(13)

where the radicand must be positive.

4. ONEREFRINGENCE IN SL FROM THREE OR MORE COMPONENTS

In case when $D = d_1 + d_2 + d_3$ Eq. (1) has the form $Q^{(e)} = x_1 Q^{(1)} + x_2 Q^{(2)} + (1 - x_1 - x_2) Q^{(3)}$, where x_1 and x_2 are thicknesses of the first and the second components in the SL period. Then analogous to Eqs.(5)-(7) conditions have the form

$$A_1 x_1 + A_2 x_2 + m_{r3}^{(3)} - e_{3r}^{(3)} = 0,$$

(14)

$$B_1 x_1 + B_2 x_2 + m_{3j}^{(3)} - e_{j3}^{(3)} = 0,$$

(15)

$$C_1 x_1^2 + C_2 x_2^2 + Dx_1 x_2 + E_1 x_1 + E_2 x_2 + m_{jr}^{(3)} - e_{rj}^{(3)} = 0,$$

(16)

where denotations (8) and the following denotations are used

$$A_i = m_{r3}^{(i)} - m_{r3}^{(3)} - e_{3r}^{(i)} + e_{3r}^{(3)}, B_i = m_{3j}^{(i)} - m_{3j}^{(3)} - e_{j3}^{(i)} + e_{j3}^{(3)},$$

$$C_i = \frac{1}{\mu_{33}^{(i)}\mu_{33}^{(3)}}\left(\Delta\mu_{j3}^{(3i)}\Delta\mu_{3r}^{(3i)} - \Delta\mu_{jr}^{(3i)}\Delta\mu_{33}^{(3i)}\right) - \frac{1}{\varepsilon_{33}^{(i)}\varepsilon_{33}^{(3)}}\left(\Delta\varepsilon_{r3}^{(3i)}\Delta\varepsilon_{3j}^{(3i)} - \Delta\varepsilon_{rj}^{(3i)}\Delta\varepsilon_{33}^{(3i)}\right),$$

$$D = \frac{1}{\mu_{33}^{(2)}\mu_{33}^{(3)}}\left(\Delta\mu_{j3}^{(32)}\Delta\mu_{3r}^{(32)} - \Delta\mu_{jr}^{(31)}\Delta\mu_{33}^{(32)}\right) + \frac{1}{\mu_{33}^{(1)}\mu_{33}^{(3)}}\left(\Delta\mu_{j3}^{(31)}\Delta\mu_{3r}^{(31)} - \Delta\mu_{jr}^{(32)}\Delta\mu_{33}^{(31)}\right) -$$

$$\frac{\Delta\mu_{j3}^{(21)}\Delta\mu_{3r}^{(21)}}{\mu_{33}^{(1)}\mu_{33}^{(2)}} - \frac{1}{\varepsilon_{33}^{(2)}\varepsilon_{33}^{(3)}}\left(\Delta\varepsilon_{r3}^{(32)}\Delta\varepsilon_{3j}^{(32)} - \Delta\varepsilon_{rj}^{(31)}\Delta_{33}^{(32)}\right) -$$

$$\frac{1}{\varepsilon_{33}^{(1)}\varepsilon_{33}^{(3)}}\left(\Delta\varepsilon_{r3}^{(31)}\Delta\varepsilon_{3j}^{(31)} - \Delta\varepsilon_{rj}^{(32)}\Delta\varepsilon_{33}^{(31)}\right) + \frac{\Delta\varepsilon_{r3}^{(21)}\Delta\varepsilon_{3j}^{(21)}}{\varepsilon_{33}^{(1)}\varepsilon_{33}^{(2)}},$$

$$E_i = \frac{1}{\mu_{33}^{(i)}\mu_{33}^{(3)}}\left(\mu_{jr}^{(3)}\Delta\mu_{33}^{(3i)} - \mu_{33}^{(i)}\Delta\mu_{jr}^{(3i)} - \Delta\mu_{j3}^{(3i)}\Delta\mu_{3r}^{(3i)}\right) -$$

$$\frac{1}{\varepsilon_{33}^{(i)}\varepsilon_{33}^{(3)}}\left(\varepsilon_{rj}^{(3)}\Delta\varepsilon_{33}^{(3i)} - \varepsilon_{33}^{(i)}\Delta\varepsilon_{rj}^{(3i)} - \Delta\varepsilon_{r3}^{(3i)}\Delta\varepsilon_{3j}^{(3i)}\right)$$

Characteristic properties of Eqs. (14)-(16) system correspond to the ones considered for Eqs. (5)-(7) system. In particular, coefficients C_1, C_2, D in Eqs. (16) have also the second order and coefficients E_1, E_2 have the first order on small parameters $\Delta\varepsilon_{pq}/\varepsilon_{33}^{(i)}, \Delta\mu_{pq}/\mu_{33}^{(i)}$, that can be used for transformation of Eqs. (16) to linear equations in x_1, x_2.

One can show that at any number of layers in the SL period the main system of equations determining the onerefringence conditions is analogous to the system of equations (14)-(16). It will include the following equations with respect to quantities $x_1, x_2, ..., x_n$: 4 linear equations with coefficients being functions of components $\mu_{p3}, \varepsilon_{3q}, \mu_{3p}, \varepsilon_{q3}$ of the tensors characterizing the layers, 4 equations of the second order with coefficients depending on the components $\mu_{pq}, \varepsilon_{pq}$.

5. NUMERICAL ANALYSIS

Numerical analysis was carried out for the two-component SL made from uniaxial non-absorbing components (each of tensors ε, μ characterizing the layers is real and has two principal values). Note, that for all cases considered below there are solutions of Eqs. (5)-(7) system also for absorbing layers (at complex ε, μ). Solutions for non-onerefringent layers were sought.

At first let us consider the case when optical axes of SL components are parallel to the layers boundaries. It is known that at $\vec{n} \perp \vec{c}$, where $\vec{n}$ is a wave normal, $\vec{c}$ is a parallel to the optical axis vector, optical anisotropy of uniaxial crystals is exhibited most strongly. In this case difference between refraction indexes of ordinary and extraordinary waves in crystals is maximum. So, the onerefringence in this case will be a significant effect leading to "disappearance" of the greatest possible anisotropy effect on the radiation properties. Fig. 2 illustrates solutions of Eqs. (5)-(7) system for the principal values of tensors $\varepsilon_e^{(i)}, \varepsilon_o^{(i)}, \mu_e^{(i)}, \mu_o^{(i)}$ at various x for the case when both SL components optical axes are parallel to OX axis. In this case, at definite x values, for example, at x=0.5, conditions (13) are satisfied.

When the first SL component optical axis is parallel to OX axis and the second one is parallel to OY axis, only approximate solutions of Eqs. (5)-(7) system are calculated. In this case effective tensors $\varepsilon^{(e)}, \mu^{(e)}$ are diagonal and values of expressions $\varepsilon_{pp}^{(e)} / \mu_{pp}^{(e)}$ are equal to each other within accuracy of 5-10 percents at various x.

Possibility of conditions (5)-(7) realization at SL components optical axes which are parallel to OZ axis is characterized by Fig.3. For the data in Figs. 2, 3 at the majority of x values inequalities $\varepsilon_e^{(1)} < \varepsilon_o^{(1)}$, $\mu_e^{(1)} > \mu_o^{(1)}$, $\varepsilon_e^{(2)} > \varepsilon_o^{(2)}$, $\mu_e^{(2)} < \mu_o^{(2)}$ are satisfied. These conditions mean, that the structure is formed by layers of optically positive and optically negative uniaxial crystals, and signs of dielectric and magnetic anisotropy for each layer are opposite.

The question about possibility of creating onerefringent SL from the same material non-onerefringent layers is interesting. Conditions (5)-(7) are not satisfied at numerical analysis for uniaxial layers with the non-parallel optical axes for two cases: i) optical axes are perpendicular to OZ axis; ii) optical axes directions are near to OZ axis.

We note, that the solutions illustrated by Figs. 2, 3 are not unique, as they depend essentially on a choice of the initial conditions for calculation (initial principal values of tensors $\varepsilon^{(i)}, \mu^{(i)}$). On the basis of the numerical analysis it is possible to assume that realization for the onerefringence conditions uniaxial SL components should have parallel (or near to parallel) optical axes.

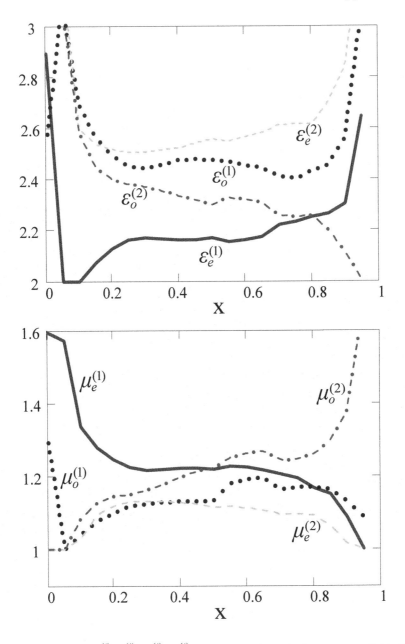

Figure 2. Values of $\varepsilon_e^{(i)}, \varepsilon_o^{(i)}, \mu_e^{(i)}, \mu_o^{(i)}$. SL components optical axes are parallel to OX axis.

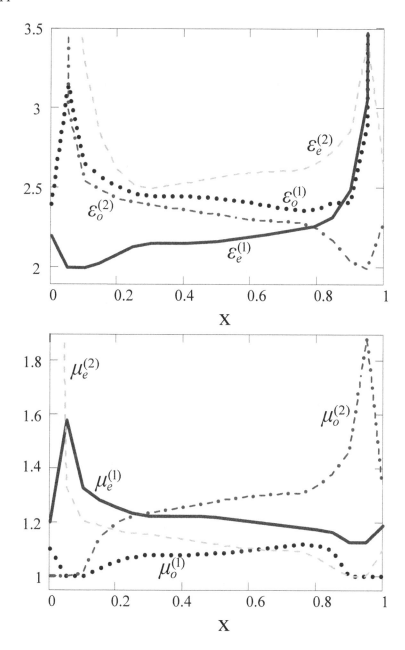

Figure 3. Values of $\varepsilon_e^{(i)}, \varepsilon_o^{(i)}, \mu_e^{(i)}, \mu_o^{(i)}$. SL components optical axes are parallel to OZ axis.

6.   CONCLUSION

General conditions when bianisotropic layered-periodic structures or SL satisfying the long wavelength approximation can be onerefringent were found and investigated. It is shown, that there are rather wide ranges of dielectric and magnetic properties of the layers, when onerefringence conditions are satisfied for non-onerefringent layers. The layers can be not only crystalline, it is important that their properties could be described by phenomenological tensors ε, μ. Derived relations are true not only for optical radiation, but for more longer electromagnetic waves (and thicker layers). The usage of our results is possible also in cases of changing effective tensors $\varepsilon^{(e)}, \mu^{(e)}$ by external effects, for example, by magnetic [7] or electric [8] fields. In these cases regimes of dynamic "switching on - switching off" onerefringence in the considered structures are possible.

REFERENCES

1. F.I. Fedorov. Optics of anisotropic media. Minsk, 1958.
2. A.M. Goncharenko. Izvestiya Akademii Nauk BSSR (Series of Physics and Mathematics), 1970, N 2, p. 132.
3. V.V. Gvozdev, A.N. Serdyukov. Doklady Akademii Nauk BSSR, 1979, N 4, pp. 319-321.
4. F.I. Fedorov. Theory of gyrotropy. Minsk, 1976.
5. V.E. Gaishun, I.V. Semchenko, A.N. Serdyukov. Kristallografiya, 1993, Vol. 38, N 3, pp. 24-27.
6. Djafari Rouhani B., Sapriel J. Phys. Rev. B. 1986. Vol. 34, N 10, pp. 7114-7117.
7. E.G. Starodubtsev, I.V. Semchenko, G.S. Mityurich. Advances in Complex electromagnetic materials, NATO ASI Series 3. High Technology, 1997, Vol. 28, pp. 169-176.
8. E.G. Starodubtsev. Proc. of Intern. Conf. "Bianisotropics'97", Glasgow, Great Britain, 1997, pp. 297-300.

ESR STUDY OF DEFECTS IN SPIN-GAP MAGNETS

A. Smirnov
P. L. Kapitza Institute for Physical Problems RAS, 117334 Moscow, Russia

Abstract: Impurities in spin-gap systems provide a local destruction of the singlet state, resulting in formation of spin clusters with the local antiferromagnetic order. The ESR measurements enable one to estimate the size of spin clusters and to determine the effective spin of the clusters. The special attention is payed to the impurity induced magnetic ordering on the singlet and disordered background of a spin-gap magnet. A special kind of the microscopic phase separation is observed at this impurity induced ordering.

Key words: Spin-gap magnets, spin clusters, phase separation

1.   INTRODUCTION

Spin-gap magnetic systems, exhibiting disordered ground states on a regular crystal lattice, provide a unique possibility to create nanoscale areas of antiferromagnetically correlated spins just by inserting impurities, which substitute magnetic ions. Several examples of the spin-gap systems are among the dielectric crystals containing chains of magnetic ions coupled by the antiferromagnetic exchange interaction. The ground state of the regular one-dimensional spin chain is magnetically disordered: the average value of a spin projection at a site of the crystal lattice is zero. The $S = 1/2$ chains have a critical ground state - the spin-spin correlation length is infinite and correlations decay in a power law [1]. This type of spin-liquid systems may be switched to the ordered state by weak interactions like inter-chain exchange or anisotropy. The $S = 1$ chains have the disordered ground state, which differs through the final correlation length. This state is stable against perturbations because of the energy gap (so called Haldane gap) [2]. The spin-gap opens also in the spin S=1/2 chain when the chain is dimerized, i.e. when the exchange integral becomes alternated, taking in turn two values:

B. Aktaş et al. (eds.), Nanostructured Magnetic Materials and their Applications, 263–273.
© 2004 *Kluwer Academic Publishers. Printed in the Netherlands.*

$J \pm \delta J$ [3]. The dimerization may take place due to the structural reasons or arises spontaneously, by the so-called spin-Peierls transition [4,5]. Other spin gap systems are described in [6,7].

Due to the spin-gap Δ magnetic excitations are frozen out at low temperatures, and these magnetic systems appear to be nonmagnetic in moderate fields. In the magnetic fields exceeding a threshold value a magnetization arises. The described disordered states represent an alternative to well known ordered ground states, e.g. ferromagnets or antiferromagnets. Besides, the problem of defects in these crystals is of fundamental importance, providing a possibility to create new multi-spin nanoscale objects.

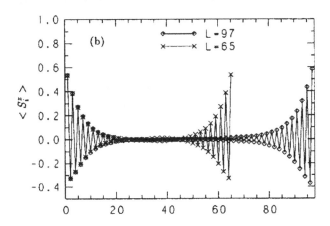

Figure 1. The spin configuration of the fragment of a Haldane chain for the odd number of spins and the total spin projection $S^z_{tot} = 1$. Figure from Ref. [10].

Impurities, breaking the spin chains, destroy locally the nonmagnetic singlet state and restore nonzero spin projections in an area around impurity. These spin projections are correlated antiferromagnetically and compose areas of local antiferromagnetic order, or "staggered" magnetization. Thus, the substitution of one ion provides a multi-spin cluster formed by the spins of the main matrix. The longitudinal size (along the spin chain) of these correlated areas is of about the correlation length ξ. The correlation length ξ is related to the spin-gap value by the natural relation $\xi \sim \hbar v / \Delta$, here v is the spin wave velocity. Different numerical and analytical methods (see, e.g. [8,9,10]) predict for the Haldane magnet the value $\Delta = 0.4 J$ and the correlation length $\xi = 7a$, a is the interspin distance and J is the exchange integral corresponding to the Heisenberg Hamiltonian:

$$H = \sum_{i} J S_i S_{i+1} . \tag{1}$$

The formation of the multi-spin antiferromagnetically correlated clusters near the chain ends was studied by theoretical modeling for the spin-Peierls system [11] and for the Haldane system [10]. The spin configuration obtained by the Monte-Carlo simulation for fragments of the Haldane chain is shown in Fig.1

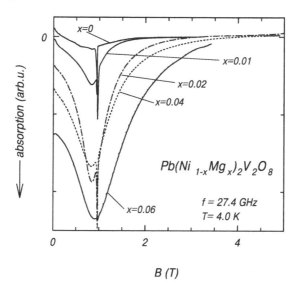

Figure 2. ESR absorption lines of the ceramic samples of $Pb(Ni_{1-x}Mg_x)_2V_2O_8$. Curves are normalized to unit weight. The narrow resonance line is the DPPH-label corresponding to g-factor 2.00.

The staggered magnetization has a maximum value near the spin chain end and decays nearly exponentially with a characteristic distance equal to ξ at going apart from the chain end. The chain ends in a Haldane magnet attracted much interest also in connection with the problem of the effective spin of the magnetic degree of freedom at a chain-end. The consideration of the valence-bond-solid model for the Haldane chain resulted in the surprising suggestion, that the end-chain spin configuration has an effective spin $S_{eff} = 1/2$ [12]. The first experiments on this problem were done with the Haldane compound $[Ni(C_2H_8N_2)_2(NO_2)]ClO_4$ (abbreviated as NENP). A small portion of $S = 1$ Ni^{2+}-ions was substituted by the $S = 1/2$ Cu^{2+}-ions. The ESR signals detected three coupled $S = 1/2$ degrees of freedom: one from the Cu-ion and two from chain ends. The numerical simulation [10] (see Fig. 1) confirms the concept of the effective spin

$S = 1/2$: the summation of the spin projections near the left or near the right chain-end gives the total projection of the chain-end cluster $S_{end} = 1/2$.

The next important problem related to broken chains in a spin-gap magnet is the effect of the impurity-induced magnetic ordering. The effect was predicted by Shender and Kivelson [13]. They considered the areas of local antiferromagnetic order near the cahin ends as "magnetic molecules". The staggered magnetization within these magnetic molecules is correlated by intrachain and weak interchain exchange, providing the creation of three-dimensional long range ordering (see also [11,14]). The effect of the impurity-induced ordering was then found in spin-Peierls magnet $CuGeO_3$ (see, e.g. [15]), Haldane magnet $PbNiV_2O_8$[16] and other spin-gap compounds [17].

The detailed review of previous magnetic resonance investigations of the spin-gap and related objects is presented in Ref. [18]. In the present paper we shall describe recent experiments [19-21] aimed on the detection of the magnetic degrees of freedom of the nanoscale spin clusters formed near impurities in two spin-gap magnets: in the spin-Peierls magnet $CuGeO_3$ and in the Haldane magnet $PbNi_2V_2O_8$. The size of the area of the local antiferromagnetic order will be evaluated, the hypothesis on the effective spin of the chain-end in the Haldane magnet will be checked experimentally. Than we shall describe experiments on the impurity-induced ordered phases – we deduce the value of the order parameter and reveal the spatial modulation of the order parameter in this nonuniform phase.

2.   EFFECTIVE SPINS OF THE ENDS OF BROKEN HALDANE CHAINS

The discovery of the inorganic Haldane magnets like Y_2BaNiO_5 or $PbNi_2V_2O_8$ enables one to substitute magnetic ions by impurity ions in a wide concentration range. Inserting of nonmagnetic impurities instead of magnetic ions produces a given number of breaks in spin chains. The problem of the effective spin S_{eff} at the chain end may be effectively studied

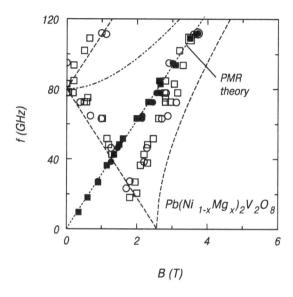

Figure 3. Frequency-field dependence of the spin resonance absorption for the ceramic samples of $Pb(Ni_{1-x}Mg_x)_2V_2O_8$. Filled symbols represent the resonance frequencies at T=4.0 K, i.e above the maximum value of the Néel temperature, open symbols represent the data taken at T=1.3 K, i.e. in the ordered phase: circles - $x = 0.04$, squares - $x = 0.06$

by the electron spin resonance (ESR) technique, because the spin sub-levels of a spin $S = 1/2$ are not split by the crystal field, in contrast to magnetic objects with the effective spin $S > 1/2$ (see, e.g. [22].) The Haldane magnet $PbNi_2V_2O_8$ is characterized by the intrachain exchange $J = 9$ meV. The single-ion anisotropy constant, corresponding to the Hamiltonian term $D(S_i^z)^2$, was reported as $D = -0.23$ meV in [16] or $D = -0.45$ meV [23]. The discrepancy between these two values of D is due to different models of the magnetic interactions used for the analysis of excitation spectrum. Even the smaller value should result in a zero-field splitting of the spin $S = 1$ resonance mode of about 50 GHz.

The ESR lines of the powder samples of the doped Haldane magnet $Pb(Ni_{1-x}Mg_x)_2V_2O_8$ are shown in Fig. 2. These absorption curves are taken in the paramagnetic phase, at the temperature 4.0 K, which is above the highest temperature of the impurity-stimulated ordering.

The frequency-field dependence for this kind of magnetic resonance is given in Fig. 3 by filled symbols. We see the perfect linear dependence demonstrating zero frequency in zero field. The upper limit of the zero-field frequency is 0.5 GHz. The increase of the ESR intensity with the concentration of nonmagnetic ions proves the creation of the magnetic degrees of freedom by breaking spin chains. The linear frequency-field dependence and the absence of the zero-field ESR splitting confirms the effective spin $S_{eff} = 1/2$ for impurity-generated magnetic degrees of freedom.

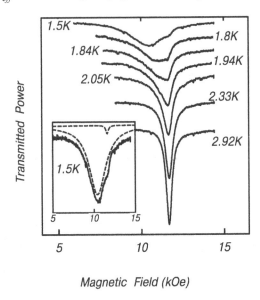

Figure 4. The evolution of the ESR line of the sample of $Cu_{0.983}Mg_{0.017}GeO_3$ at cooling through the Néel temperature. Magnetic field is oriented parallel to b-axis. Microwave frequency is 36 GHz. The inset demonstrates the fitting by two Lorentzians.

Note, that for the effective spin of the chain end $S_{eff} = 1$ we should expect a wide field-band of the microwave absorption at the microwave frequency in the range 10-50 GHz, because even the smaller of two D-values, given above should result in a zero-field splitting of about 50 GHz.

The linewidth ΔH appears to be concentration-dependent: the ΔH at $x = 0.02$ is for 0.5 T larger, than at $x = 0.01$. This large increase could not be ascribed to long-distance dipole-dipole interaction (this effective field is of the order of 0.005 T). Therefore we ascribe this increase in the linewidth to contacts between areas of local antiferromagnetic order and we can estimate the length of the chain-end cluster L_{cl}. At the length of the chain fragment $L_{frag} < 2L_{cl}$ two chain-end clusters should overlap. For small concentrations the portion of spin-chain segments which are shorter than $2L_{cl}$ equals approximately $2xL_{cl}/a$. Thus, at the reasonable value of $L_{cl} \sim 10a$ a significant portion of chain segments (about 0.4 at $x = 0.02$) should contain strongly interacting clusters, touching each other. This interaction may cause the broadening of the ESR line because the exchange interaction along with the single-ion anisotropy results in an effective anisotropic exchange giving rise to the line broadening. The estimation obtained for L_{cl} corresponds well to the theoretical value of $\xi \approx 7a$.

3. IMPURITY-STIMULATED ANTIFERROMAGNETIC ORDERING.

At cooling the samples of $Pb(Ni_{l-x}Mg_x)_2V_2O_8$ we can follow the impurity-stimulated antiferromagnetic transition by observation of the transformation of the ESR spectrum. The temperature evolution of the spectrum at $x > 0.03$ corresponds well to the expected modification of the ESR line in a powder sample of a conventional antiferromagnet (see, e.g. [24]): the paramagnetic resonance line at g=2.2 spreads into the band of absorption corresponding to the antiferromagnetic resonance in differently oriented single-crystal particles. The boundaries of the absorption band taken at different frequencies are plotted on the frequency-field diagram in Fig.3 by open symbols.

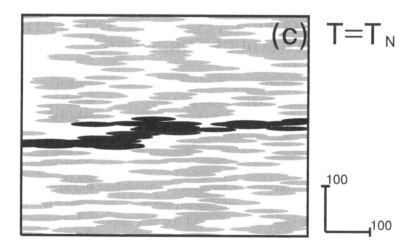

Figure 5. Two-dimensional modeling of the microscopic phase separation at the impurity induced antiferromagnetic ordering in a spin-gap magnet. Areas of local antiferromagnetic order are marked by grey color, the disordered space is shown as white, black area is the largest region of coherent order parameter, percolating through the sample. The scale is given in interspin distances.

The boundaries of the absorption band should correspond to the antiferromagnetic resonance field at the orientations of the magnetic field along the principal crystal axes. These data were fitted by the known frequency-field dependence of the easy-axis antiferromagnet [25], as shown in Fig. 3 by the dashed lines. The agreement may be considered as satisfactory taking into account the large linewidth which smoothes the edges of the absorption band. From this frequency-field diagram we can derive the gap of the spin wave spectrum, it's maximum value is $\omega_0/2\pi$ =80 GHz at $x = 0.04$. Using the molecular field approximation [25] and the above values of J and $D = -0.23$ meV we can estimate the average value of the ordered spin component in this impurity-induced antiferromagnetic phase. The molecular field approximation yields for the gap of the antiferromagnetic resonance spectrum:

$$\hbar\omega_0 = 4\sqrt{|D||J|\overline{|\langle S_i^z \rangle|}} \tag{2}$$

Thus we obtain the order parameter $\overline{|\langle S_i^z \rangle|} \simeq 0.06$. Here we mean the averaging over the time and crystal sites. This value represents the maximum of the order parameter reached at $x = 0.04$.

4.   MICROSCOPIC PHASE SEPARATION AT THE IMPURITY-INDUCED ORDERING.

The impurity-induced ordering is of special interest in case of rather low concentrations, when the areas of local AFM order do not overlap by regions where the local order parameter is close to the maximum value, i.e. at $a/x >> \xi$. In this case the antiferromagnetic order parameter should be strongly modulated, because the spin projection decays at going apart from the defect.

To study this effect in details we took the advantage of the spin-Peierls samples $Cu_{1-y}Mg_yGeO_3$ which can be prepared in form of high quality crystals. We used the samples studied at the thorough investigation of the $y - T$ phase diagram [26].

Previous investigation showed that at low doping concentration ($y < 0.02$) the transition to the Néel state takes place on the dimerised background, formed at the spin-Peierls transition temperature 14.5 K. At $T \leq 7$ K the value of the spin-gap is close to it's maximum value $\Delta(0)$ =25 K. Our observations at low y revealed that at the Néel point the ESR line transforms from one Lorentzian line into two separate lines, as shown in Fig. 4. We studied the transformation of the ESR spectrum at this transition and followed the transformation of the resonance line from the paramagnetic to the antiferromagnetic one. One can see the coexistence of two separate

Lorentzian resonance lines in a temperature interval below T_N. This coexistence indicate clearly the presence of two magnetic phases in the sample. The presence of two phases can't be attributed to the trivial nonuniform distribution of impurities, because the concentration is uniform within 0.1% [26]. This fact does not allow the diffusion of the Néel temperature in the interval 1.5-2.25 K, where the coexistence of two resonance lines is observable. Besides, the width of the Néel transition, determined from the susceptibility measurements or from the resonance field shift does not exceed 0.1 K.

The effect of the coexistence of the paramagnetic and antiferromagnetic phases is temperature- and concentration dependent. The fitting of the observed ESR lines with two Lorentzians enables one to derive the paramagnetic fraction of the sample as the ratio of the paramagnetic component intensity to the intensity obtained at $T > T_N$. The paramagnetic fraction vanishes at low temperatures when the whole sample shows only antiferromagnetic ESR signal. There is no paramagnetic fraction also at higher concentrations, at $y > 0.03$. The observed coexistence of two resonance lines and hence, of two magnetic phases in the sample indicates a nonuniform phase and is unusual for conventional antiferromagnetic phase transition.

Usually the transformation from the paramagnetic to antifferromagnetic state at the Néel point proceeds as a shift of the ESR line from the paramagnetic resonance field to the antiferromagnetic resonance field. The spin-resonance modes of a conventional antiferromagnet are due to the coherent order parameter and do not allow paramagnetic resonance component. We naturally explain the observed ESR signals and the presence of two magnetic phases in the samples of $Cu_{1-y}Mg_yGeO_3$ starting from the main concept of nanoscale antiferromagnetically correlated areas which are formed in the vicinities of the impurity atoms. At $T = 0$ the local order parameter decays exponentially with the distance from the impurity, according to the correlation length value. Thus, it is naturally to propose, that for nonzero temperatures the correlation of static spin projections will be destroyed by thermal fluctuations at a distance from the impurity of about L, which may be estimated from the relation:

$$k_B T = JS^2 \exp\{-2L/\xi\}. \tag{5}$$

Therefore, at high temperatures (far above T_N), there are areas of the local antiferromagnetic order of a short size, they do not touch each other and form isolated spin clusters. Each cluster has the own magnetic moment equal to μ_B and hence contributes to the paramagnetic ESR signal and magnetic susceptibility. At lowering temperature the size of clusters grows and they begin to touch each other and coalescence, the areas of the coherent antiferromagnetic order parameter cover several impurity centers. There are many isolated clusters along with the conglomerates. Each cluster and conglomerate has the net magnetic moment and if this moment is larger, than the magnetic moment due to the antiferromagnetic susceptibility, the cluster contributes to the paramagnetic signal. Finally, at a temperature near T_N, large areas of the coherent antiferromagnetic order appear, percolating through the sample (Fig. 5). At this moment we should observe the antiferromagnetic resonance signals because the antiferromagnetic susceptibility of the large area results in a magnetic moment prevailing μ_B. The percolation of the area of the coherent antiferromagnetic order in this case is fully analogous to the percolation of mutually interpenetrating spheres in a corresponding problem of percolation theory [27]. As one can see on Fig. 5, isolated clusters are still present at the moment of percolation. There are three kinds of magnetic areas: large antiferromagnetic conglomerates, isolated clusters and the residual of a slightly perturbed spin-gap matrix with negligible order parameter (shown by white color in Fig. 5). The former give the antiferromagnetic resonance signals, the isolated clusters provide the paramagnetic line and the residual of the spin-gap matrix is magnetically silent. This percolation scenario of the phase transition is in the agreement with the Monte-Carlo simulation of the ground state performed in [28]. The two-dimensional modeling of the impurity-induced modeling in spin-Peierls and Haldane systems resulted in the distribution of the order parameter in the form of sharp peaks separated by the areas with much smaller order parameter. Starting from this calculated ground state we arrive to our percolation model by the assumption that at the finite temperature the coherence of the order parameter will be lost in the areas with $|\langle S_i^z \rangle| \ll 1$.

5.   CONCLUSIONS

1. The impurities imbedded in the spin-gap matrix of the spin-Peierls or Haldane magnet result in the formation of the nanoscale areas of the longitudional size of about 10-30 interspin distances.

2. The effective spin of the nanoscale magnetic clusters formed near the broken spin-chain ends in the Haldane magnet is $S_{eff} = 1/2$.

3. The effect of the inducing of the long range antiferromagnetic ordering by the impurities is due to the formation of the nanoscale spin clusters near impurity atoms. The order parameter is much smaller than in conventional antiferromagnet: $<S_i^z>_{max} \simeq 0.06$

4. The nanoscale spin clusters are nucleus of the antiferromagnetic long range ordering. Their chaotic distribution results in a microscale phase separation when the macroscopic antiferromagnetic areas coexist with the single nanoscale clusters surrounded by the disordered rests of the spin-gap matrix. At low concentration of the impurities, which stimulate the long range ordering, the modulation of the order parameter equals 100%.

REFERENCES

[1] H.Bethe Z.Physik **71** 205 (1931).

[2] F. D.M. Haldane, Phys. Rev. Lett. **50**, 1153 (1983).

[3] L.N.Bulaevski, Fiz.Tverd.Tela **11**, 1132. (1969) [Sov.Phys. Solid State].

[4] E.Pytte, Phys. Rev. B , 4637 (1974).

[5] M.Hase, I.Terasaki and K.Uchinokura, Phys. Rev. Lett. , 3651 (1993).

[6] H.Kageyama, K.Yoshimura, R.Stern, et al., Phys. Rev. Lett 82, 3168 (1999).

[7] E.Dagotto and T.M.Rice, Science 271, 618 (1996).

[8] S.V.Meshkov, Phys. Rev. B **48**, 6167 (1993).

[9] T.Kennedy, J. Phys. Condens. Matter. **2**, 5737 (1990).

[10] S.Miyashita and S.Yamamoto, Phys. Rev. B **48**, 913 (1993).

[11] H. Fukuyama, T. Tanimoto and M. Saito, J. Phys. Soc. Jpn. **65**, 1182 (1996).

[12] M.Hagiwara, K.Katsumata, I.Affleck et al., Phys. Rev. Lett. **65**, 3181 (1990).

[13] E.F.Shender and S.A.Kivelson, Phys. Rev. Lett. **66**, 2384 (1991).

[14] M.Mostovoy, D.Khomskii, and J.Knoester, Phys. Rev. B , 8190 (1998).

[15] L.P.Regnault, J.P.Renard, G.Dhalenne and A.Revcolevschi, Europhys. Lett. **32**, 579 (1995).

[16] Y.Uchiyama, Y.Sasago, I.Tsukada et al., Phys. Rev. Lett . , 632 (1999).

[17] A.Oosawa, T.Ono and H.Tanaka Phys. Rev. B , 020405 (2002).

[18] K.Katsumata, J. Phys.: Condens. Matter , R589 (2000).

[19] A.I.Smirnov, V.N. Glazkov, H.-A. Krug von Nidda A. Ya. Shapiro, L. N. Demianets, Phys. Rev. B , 174422 (2002).

[20] V. N. Glazkov, A. I. Smirnov, K. Uchinokura and T. Masuda, Phys. Rev. B , 144427 (2002).

[21] V.N.Glazkov, A.I.Smirnov, G.Dhalenne and A.Revcolevschi, Zh. Eksp. Teor. Fiz , 164 (2001) [JETP 143 (2001)].

[22] S.A.Altshuler, B.M.Kosyrev, Electron Paramagnetic Resonance in Compounds of Transition Elements (Nauka, Moscow 1972; Halsted, New York 1975).

[23] A.Zheludev, T.Masuda, I.Tsukada et al., Phys. Rev. B , 8921 (2000).

[24] A.N.Vasil'ev, L.A.Ponomarenko, A.I.Smirnov, E.V.Antipov, Yu.A.Velikodny, M.Isobe and Y.Ueda, Phys. Rev. B , 3021 (1999).

[25] T.Nagamiya, K.Yosida and R.Kubo, Adv. Phys. , 1 (1955); S.Foner in: Magnetism ed.G.T.Rado and H.Suhl, NY Academic Press 1963 v.1.

[26] T.Masuda, I.Tsukada, K.Uchinokura, Y.J.Wang, V.Kiryukhin and R.J.Birgeneau, Phys. Rev. B , 4103 (2000).

[27] V.K.S.Shante and S.Kirkpatrick, Adv. Phys. , 325 (1971).

[28] C.Yasuda, S.Todo, M.Matsumoto and H.Takayama, Phys. Rev. B , 092405 (2001).

MAGNETIC PROPERTIES OF CrO_2 THIN FILMS STUDIED BY FMR TECHNIQUE

B.Z. Rameev [1,2], L.R. Tagirov [1,2,3], A. Gupta [4], F. Yıldız [1], R. Yilgin [1], M. Özdemir [1], and B. Aktaş [1]

[1] Gebze Institute of Technology, 41400 Gebze-Kocaeli, Turkey
[2] Kazan Physical-Technical Institute, 420029 Kazan, Russia
[3] Kazan State University, 420008 Kazan, Russia
[4] IBM T.J. Watson R.C., Yorktown Heights, New York 10598, USA

Abstract: Epitaxial CrO_2 thin films on TiO_2 (100) single-crystal substrates were fabricated by CVD process with use of solid (CrO_3) and liquid (CrO_2Cl_2) precursors. Magnetic properties of the films were studied by ferromagnetic resonance technique. Strong dependence of magnetic anisotropies on the thickness of CrO_2 films was observed in the series prepared using the CrO_3 solid precursor. It was found that it is due to magnetoelastic anisotropy in the plane of the CrO_2 film caused by lattice mismatch with the TiO_2 substrate. On the contrary, the anisotropy parameters of CrO_2Cl_2-based films demonstrate very weak variations with the thickness. Atomic force microscopy studies showed that the different behaviour apparently relates to various morphologies of the films prepared with solid and liquid precursors. In the CrO_2Cl_2 liquid-precursor series the individual CrO_2 grains, growing epitaxially on the TiO_2 substrate, form fairly regular rectangular-shaped blocks with the lateral size of about 150 nm. Breaks in the nearest to the substrate layer of the epitaxial CrO_2 film result in quick relaxation of the tensile stress, arising due to the lattice mismatch, and decrease greatly the effect of magnetoelastic anisotropy in the CrO_2Cl_2 liquid-precursor films.

Key words: thin films; half-metallic ferromagnet; magnetic anisotropy; ferromagnetic resonance

B. Aktaş et al. (eds.), Nanostructured Magnetic Materials and their Applications, 275–284.

1.   INTRODUCTION

A great interest to the studies of half-metallic ferromagnets is based on the fact that they have highly spin-polarized conduction band. This makes them prospective candidates for applications in magnetic tunnel junctions and other spintronic devices. Among various half-metallic compounds the CrO_2 dioxide is distinguished by the experimentally proved highest polarization of conduction band of almost 100 % [1-5]. The epitaxial CrO_2 films have been synthesized recently by chemical vapour deposition (CVD) technique [6] and there are few papers on electric and magnetic properties of this material [3-10] Very recently, it has been shown that the CVD process with use of a liquid precursor [11] provides the films of better quality compared with a solid precursor used before.

In this work the magnetic properties (magnetization, magnetocrystalline and magnetoelastic anisotropies) of the epitaxial CrO_2 films are studied by the ferromagnetic resonance (FMR) technique as a function of the film thickness. The results received for the films grown using two different CVD process, the CrO_3 solid precursor and CrO_2Cl_2 liquid precursor, are compared.

2.   EXPERIMENTAL PROCEDURES

Epitaxial CrO_2 (100) thin films were fabricated by CVD technique onto TiO_2 (100) single-crystal substrates with use of CrO_3 solid (SP) and CrO_2Cl_2 liquid precursors (LP), as described before in [7,11]. The film thicknesses were in the range 200 Å -1500 Å and 350 Å -1350 Å, respectively. The studied films were about 5×5 mm^2 in size, and they were cut on small rectangular pieces before FMR measurements.

The X-ray diffractometer *Rigaku RINT* 2000 series was used to determine directions of crystallographic axes of TiO_2 substrate. *Digital Instruments NanoScope IV* Scanning Probe Microscope was used to study the surface morphology of the films.

The FMR absorption spectra of the films were obtained on *Bruker EMX* Electron Spin Resonance (ESR) spectrometer at X-band frequency of 9.5 GHz. The spectra were registered at room temperature and various orientations of the crystalline axes with respect to the applied DC magnetic field (*H*). The field derivative of microwave power absorption (*dP/dH*) was recorded as a function of the DC magnetic field.

3. RESULTS AND DISCUSSION

Standard procedure of the FMR measurements of thin films was applied to the CrO_2 films. The procedure consists of measurements in two standard geometries: 1) the direction of DC magnetic field is changed in the plane of the film ("***in-plane***" geometry), 2) the magnetic field is rotated from easy axis direction in the film plane towards the normal to the film plane ("***out-of-plane***" geometry). As it is expected for thin films, a dominant contribution of the shape anisotropy is observed in out-of-plane FMR measurements of all CrO_2 samples. As a result, the easy direction of thin film magnetization lies in plane of the CrO_2 films. On the other hand, the effect of magnetocrystalline anisotropy is clearly observed in the in-plane measurements, where the easy axis of the magnetization corresponds to the minimal in-plane FMR resonance field, and the hard axis corresponds to the maximal in-plane resonance field. An example of the in-plane and out-of-plane FMR spectra is presented in Fig. 1 for the SP sample with the thickness 1500 Å. The in-plane and out-of-plane angular dependencies of FMR resonance fields are shown in Fig. 2 for the thick and thin samples of the SP series. The FMR spectra and angular dependencies of resonance fields in the LP series are presented in Fig. 3 and Fig. 4, respectively.

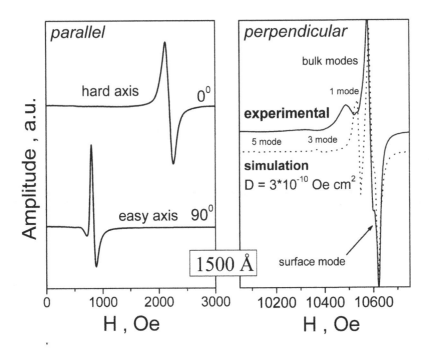

Figure 1. In-plane and out-of-plane FMR spectra for the SP sample with the thickness 1500 Å

Splitting of the FMR signal into the surface and bulk spin-wave modes (right panel in Fig. 1) indicates that there is a surface pinning of the magnetization at the interfaces of CrO_2 films. The surface modes correspond to solutions, which decay into interior of a film. The bulk modes correspond to excitations of standing magnetization waves in the bulk of a film. Computer simulation (Fig. 1) of FMR spectrum for the 1500 Å sample allows us to estimate the exchange stiffness constant, $D \sim 3 \times 10^{-10}$ Oe cm^2, which corresponds well enough to the values reported in the literature [6,12].

In all measured samples the minimal FMR resonance field is achieved at the direction of DC magnetic field parallel to the c axis. Therefore, the easy axis switching from the c direction to the b direction expected from literature [6] was not observed in the both series of the samples.

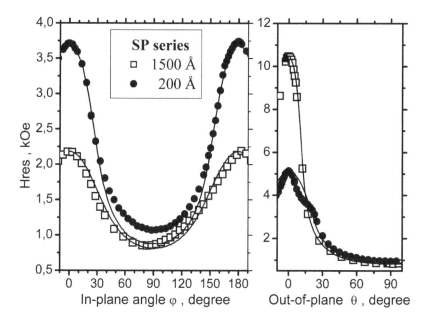

Figure 2. The in-plane and out-of-plane angular dependencies of FMR resonance fields for the samples of the SP series with the thickness 200 Å and 1500 Å. Results of computer modeling are presented by the solid lines.

It is remarkable that strong dependence on the thickness is observed the CrO_3-based films, while very small variation with thickness of in-plane anisotropy is observed in the series of CrO_2Cl_2-grown films. To discuss possible reasons for the different thickness dependencies one have to

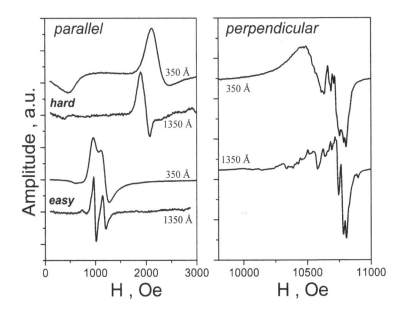

Figure 3. In-plane and out-of-plane FMR spectra for the sample of LP series with the thickness 350 Å and 1350 Å.

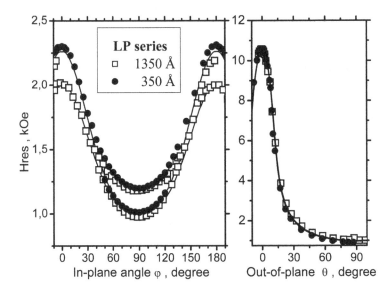

Figure 4. The in-plane and out-of-plane angular dependencies of FMR resonance fields for the samples of the SP series with the thickness 350 Å and 1350 Å. Results of computer modeling are shown by the solid lines.

concern the structural properties of the CrO_2 thin films as well as TiO_2 substrates.

Structural properties of CrO_2 films made by CVD technique have been investigated in Refs. [6,7,9]. Both CrO_2 epitaxial film and TiO_2 single-crystalline (100) substrate have rutile structure with an almost tetragonal unit cell and lattice parameters close to the bulk values: $a = b = 4.421$ Å, and $c = 2.916$ Å. Lattice mismatch with the (100) TiO_2 substrate is -3.79% along the [010] (*b*) direction and is -1.48% along the [001] (*c*) direction. It results in an anisotropic tensile strain in the plane of the CrO_2 film [6]. The unstrained crystal structure of the film results in the uniaxial magnetocrystalline anisotropy with an easy axis parallel to the *c* crystalline axis, while the additional perturbation due to the lattice mismatch produces magnetoelastic anisotropy energy. That is the anisotropy energy takes the form:

$$E_{ani} = K_1 \sin^2 \theta + K_2 \sin^4 \theta + K_3 \sin^4 \theta \cos^2 2\varphi + K_\sigma \sin^2 (\theta - \delta),$$

(1)

where θ is the angle between the magnetization M and the c axis of the CrO_2 crystal, K_i are the anisotropy constants of the crystalline anisotropy, K_σ is the magnetoelastic anisotropy constant, and the stress is applied at the angle δ with respect to the easy (*c*) axis [6,9,10]. The analysis of the in-plane and out-of-plane resonance field angular dependencies was performed making use of the simplified model:

$$E_{ani} = K_{1eff} \sin^2 \theta + K_2 \sin^4 \theta + K_3 \sin^4 \theta \cos^2 2\varphi,$$

(2)

where the single parameter K_{1eff} absorbs the second order magnetocrystalline anisotropy term and the magnetoelastic term in Eq. (1). This is acceptable until the magnetoelastic anisotropy term is smaller than the magnetocrystalline one. Fitting of the out-of-plane angular dependence simultaneously with the in-plane resonance field dependence allows to determine values of the anisotropy constants and the effective magnetization of these films ("effective" means that magnetization can include also another anisotropy fields, which have not been taken into account in our model, Eq. 1). The calculated angular dependencies of the resonance fields are presented in Figs. 2 and 4. A good agreement between the theoretical and experimental values is observed. The obtained values of the anisotropy parameters and the effective magnetization for the both series of the CrO_2 films are collected in the Table 1.

Table 1. Anisotropy constants of CrO_2 thin films.

Thickness, Å	M_{eff}, G	K_{1eff}/M_s, G	K_2/M_s, G	K_2/M_s, G	$K_\perp/2\pi M_s =$ $M_{eff}-M_s$, G
		Solid precursor series of CrO_2 films			
200	90	1070	100	-200	461
600	320	750	-150	250	231
800	470	430	–	–	81
1500	513	372	–	–	38
		Liquid precursor series of CrO_2 films			
350	512	345	–	–	–
1350	530	280	–	–	–

The unfilled cells (–) of the table mean that the parameter is equal or nearly equal to zero.

An increase of the parameter K_{1eff} value for the thinner film of SP series reflects enhancement of the magnetoelastic term in the total anisotropy energy, which is related to the tensile strain in the film plane [6,13]. It is obvious that the strain is expected to be more effective in a thinner film. As it was mentioned above, the easy axis is the crystalline axis (*c*) which is the shortest axis in the rutile structure. Therefore, we may expect that any compression (stretch) will produce anisotropy, which corresponds to easy (hard) axis. Due to mismatch the CrO_2 film is stretched mainly in the *b* direction (−3.8%, in the *c* axis it is only −1.5%). This makes the *b* direction more "hard" with respect to the unstretched film, while the *c* axis becomes more "easy" for the magnetization. This means that the crystalline anisotropy parameter, K_{1eff}, is expected to increase. Furthermore, stretching in both axes in the film plane causes compression in the direction perpendicular to the film (*a* axis). Consequently, an additional perpendicular anisotropy with the easy axis along the normal should be induced. In our model this may be revealed as a decrease of the effective magnetization for the thinner films, that is $M_{eff} = M_s - K_\perp/2\pi M_s$, where M_s is the saturation magnetization at room temperature, and $K_\perp$ is the perpendicular anisotropy constant. For instance, fixing the value of the room temperature saturation magnetization, M_s, at 550 G [6] will result in the values of additional perpendicular anisotropy, $K_\perp/2\pi M_s$, listed in the last row of the Table 1. Thus, the thickness dependence of the both magnetic parameters in the SP series of the films is consistent with our expectations.

Contrary to that, the FMR spectra of LP series show very small variation with thickness (Fig. 3). This indicates that the magnetic anisotropy parameters in the CrO_2Cl_2-based film series depend on the thickness very slightly. We attribute the different behaviour of the magnetic anisotropies to various morphologies of the films prepared using solid and liquid precursors. To study the morphology of the films we performed AFM measurements of the both series. The surface images of the LP series of the CrO_2 film with the thickness of 350 Å received in the contact mode are shown in Fig. 5. The individual CrO_2 grains, growing epitaxially on the TiO_2 substrate, form fairly regular rectangular-shaped blocks with the lateral size of about 150 nm. The AFM scans of the thinnest film of SP series show interconnected grains of spherical shape rather than regular shaped blocks as for the LP series of the CrO_2 films. The roughness of the films decreases drastically for the thicker films of both series, and the grain structure is not clearly observable in AFM scans of the thicker films.

Figure 5. Height-mode (left) and deflection-mode (right) AFM scans of the CrO_2 film with the thickness of 350 Å.

Obviously, the stress (and the magnetoelastic anisotropy) due to the lattice mismatch with the substrate is greatly reduced for the LP film series as a result of the plaquet growth observed in Fig. 5. Besides, the multiple magnetostatic modes observed in FMR (Fig. 1) indicate exchange decoupling at plaquet boundaries as well. Therefore, the breaks in the nearest to the substrate layer of the epitaxial CrO_2 film are evidently relevant for the relaxation of the tensile stress due to the lattice mismatch, decreasing greatly the magnetoelastic anisotropy in the CrO_2Cl_2 liquid-precursor films.

CONCLUSIONS

Room temperature X-band FMR measurements of epitaxial CrO_2 thin films grown by CVD technique with use of the solid (CrO_3) and liquid (CrO_2Cl_2) precursors were performed. The values of the room temperature effective magnetization and the anisotropy field parameters were extracted.

Directions of easy and hard axes of magnetization were determined, and it was shown that at room temperature the easy axis of magnetization is parallel to the c axis of the CrO_2 rutile structure. An essential effect of the magnetoelastic anisotropy was observed upon decrease of the film thickness in the solid precursor series. Contrary to this, the magnetoelastic anisotropy is greatly reduced in the series prepared using the liquid precursor. The different behaviour was attributed to various morphologies of the films prepared using the solid and liquid precursors. The breaks in the nearest to the substrate layer of the films for the liquid precursor series result in quick relaxation of the tensile stress caused by the lattice mismatch. Therefore, the LP CVD technique looks more prospective to grow CrO_2 films with magnetic parameters favourable for potential spintronic applications.

AKCNOWLEDGMENT

The research is supported by Gebze Institute of Technology grant No. 2003-A15. L.R.T. acknowledges partial support by URFI Programme on Basic Research and NIOKR/ANT.

REFERENCES

1. K.T. Kämper, W. Schmitt, G. Güntherodt, R.J. Gambino, and R.Ruf, Phys. Rev. Lett. **59** (1987) 2788.
2. R.J. Soulen, J.M. Byers, M.S. Osofsky, B. Nadgorny, T. Ambrose, S.F. Cheng, P.R. Broussard, C.T. Tanaka, J. Nowak, J.S. Moodera, A. Barry, J.M.D. Coey, Science **282** (1998) 85.
3. R.J. Soulen. M.S. Osofsky, B. Nadgorny et al., Journ. Appl. Phys. **85** (1999) 4589.
4. Y. Ji, G.J. Strijkers, F.Y. Yang, C.L. Chien, J.M. Byers, A. Anguelouch, Gang Xiao, A. Gupta, Phys. Rev. Lett. **86** (2001) 5585.
5. A. Anguelouch, A. Gupta, Gang Xiao, D.W. Abraham, Y. Ji, S. Ingvarsson, C.L. Chien, Phys. Rev. B **64** (2000) 180408(R).
6. X.W. Li, A. Gupta, G. Xiao, Appl. Phys. Lett. **75** (1999) 713.
7. X.W. Li, A. Gupta, T.R. McGuire, P.R. Duncombe, G. Xiao, Journ. Appl. Phys. **85** (1999) 5585.
8. A. Gupta, J.Z. Sun, Journ. Magn. Magn. Mater. **200** (1999) 24.
9. F.Y. Yang, C.L. Chien, E.F. Ferrari, X.W. Li, G. Xiao, A. Gupta, Appl. Phys. Lett. **77** (2000) 286.
10. L. Spinu, H. Srikanth, A. Gupta, X.W. Li, G. Xiao, Phys. Rev. B **62** (2000) 8931.
11. A. Anguelouch, A. Gupta, G. Xiao, D. W. Abraham, Y. Ji, S. Ingvarsson, C. L. Chien, Phys. Rev. B **64** (2001) 180408(R).
12. A. Barry, J.M.D. Coey, L. Ranno, K. Ounadjela, Journ. Appl. Phys. **83** (1998) 7166.
13. B.Z. Rameev, R.Yilgin, B. Aktaş, A. Gupta, L.R. Tagirov, Microelectronic Engineering, **69** (2003) 336–340.

TEMPERATURE DEPENDENCE OF MAGNETIC PROPERTIES OF SINGLE CRYSTALLINE Fe$_3$O$_4$ THIN FILMS EPITAXIALLY GROWN ON MgO

S. Budak [a], F. Yıldız [b], M. Özdemir [c] and B. Aktaş [b]

[a] Fatih University, Faculty of Art and Science, Physics Department, 34900, Istanbul,Turkey
[b] Department of Physics, Gebze Institute of Technology, 41410 Cayirova-Gebze Kocaeli, Turkey
[c] Marmara University, Faculty of Art and Science, Physics Department, Göztepe Istanbul, Turkey

Abstract: Single crystalline Fe$_3$O$_4$ thin film of 110 angstrom in thickness grown on MgO (100) substrate have been investigated by Electron Spin Resonance technique as a function of temperature between 3 and 300K. Experimental spectra taken at different temperatures were analysed by classical Spin Wave Resonance (SWR) technique for different orientations of sample for both in-plane and out-of plane geometry. SWR spectra exhibit unidirectional character upon cooling the sample in the presence of external field, Hc. A strong unidirectional (Exchange) anisotropy field as evidenced from a shift of the resonance field for r-FC case relative to that for n-FC case at low temperature. Using a field-induced unidirectional exchange and bulk anisotropy, with the usual magneto-crystalline cubic anisotropy energy, we have simulated the experimental spectra and angular dependence of resonance field and deduced anisotropy and other magnetic parameters as a function of temperature. Verwey transition for thin film case was found to shift down to around 90 K from 120 K for bulk sample of the same compound.

1.   INTRODUCTION

The existence of a transition in magnetite at about 120°K was evident in 1929 from the work by Weiss and Forrer and Millar, who made measurements of magnetization and of specific heat, respectively. A mechanism was proposed in 1939 by Verwey, who suggested that below the transition the Fe^{++} and Fe^{+++} ions form an ordered lattice, while at higher

B. Aktaş et al. (eds.), Nanostructured Magnetic Materials and their Applications, 285–299.

temperatures disorder is produced by the movement of electrons between the iron ions [1].

The surface effects on magnetic properties of a material are becoming the subject of intense current researches, as the trend to higher magnetic recording densities creates a need for smaller magnetic particles having larger surface-to-volume ratios. If the surface magnetic properties differ from those of the bulk, they can dominate the overall magnetic behavior of small particles or very thin films. In fact, Vassiliou et al. showed that the magnetic anisotropy for the particles in diameter of about 30 Å is two orders of magnitude larger than that of the bulk ferrite (γ-Fe$_3$O$_4$) crystals [2].

Ferrites are used in many applications ranging from permanent-magnet materials to high-frequency cores due to the possibility to tune, e.g. the magnetic anisotropy and spesific resistivity by chemical substitutions. Recent improvements in the deposition technology of the oxide materials enable the fabrication of the artificial oxidic layered structures with precise control over composition and thickness [3].

The members of Fe-O system are FeO, and Fe$_2$O$_3$ and Fe$_3$O$_4$, which are the most important, and most abundant ferromagnetic transition metal oxides. They are extensively used for magnetic storage information. Furthermore, epitaxial films grown on MgO substrates show magneto-resistance (MR) even at room temperature, whereas single crystals do not [4].

The crystal of magnetite Fe$_3$O$_4$ has the cubic inverse spinel structure with a lattice constant of 8.367 Å. In this study, we have investigated single crystalline Fe$_3$O$_4$ thin film of 110 angstrom in thickness grown on MgO (100) substrate by Electron Spin Resonance technique as a function of temperature between 3 and 300K and we have analysed the experimental spectra taken at different temperatures by classical Spin Wave Resonance (SWR) technique for different orientations of sample for both in-plane and out-of-plane geometry.

2.   RESULTS AND DISCUSSIONS

Fe$_3$O$_4$ thin film was prepared by deposition on MgO(100) substrates using reactive DC sputtering technique. Experimental resonance field values were fitted by using the below equation:

$$E = -M \cdot H + K_1(\alpha_1^2 \alpha_2^2 + \alpha_1^2 \alpha_3^2 + \alpha_2^2 \alpha_3^2) +$$
$$+ K_a Cos^2(\varphi) + K_{eff}^2 \alpha_3^2 - K_{1,2}^S \alpha_3^2 + E_{ex}$$

(1)

where E is the magnetic energy density and is valid for a cubic single crystal film in external field quite over the Verwey transition temperature. Here, the first term represents Zeeman energy density, second term is responsible for the cubic magneto-crystal energy density with the anisotropy parameter, K_1, and the direction of cosines of M, α_i 's. The third term was added to account for any possible induced in-plane axial anisotropy due to any possible geometrical asymmetry in sample preparation apparatus. The fourth term $K_{eff}(= K_u - 2\pi M^2)$ represents the effective uniaxial anisotropy parameter including both the shape anisotropy $(2\pi M^2)$ and any perpendicular anisotropy K_u arising from magneto-elastic coupling due to lattice mismatch [5]. The term $K_{1,2}^S \alpha_3^2$ corresponds to easy axis surface anisotropy, and the last term represents the exchange energy.

For a general direction of the magnetic field, H, the basic dispersion relation for FMR is given as below [6,7]

$$\left(\frac{\omega}{\gamma}\right)^2 = \left(\frac{1}{M_s Sin^2\theta} \frac{\partial^2 E}{\partial \varphi^2} + Dk_n^2\right) \times \left(\frac{1}{M_s} \frac{\partial^2 E}{\partial \theta^2} + Dk_n^2\right) -$$
$$- \left(\frac{1}{M_s Sin\theta} \frac{\partial^2 E}{\partial \varphi \partial \theta}\right)^2$$

(2)

here γ is the gyro-magnetic ratio, ω is the microwave frequency, θ and φ are the spherical polar angles for M, D is the exchange stiffness parameter and k_n is the spin wave vector for the nth mode. For the detailed information, one can look at the Ref. [2].

The angular dependence of the spectra taken from 110 Å Fe_3O_4 thin film at 3K is shown in Fig. 1. As seen from the Fig.1a, the intensities of the spectra do not show significant change starting from 5° up to 90°. From 5° to 45°, linewidths show increase, over 45° linewidths of the spectra increases but the intensities do not show much increase. The paramagnetic signal between 45° and 20° is coming from Mg^{2+} and is background. The resonance fields of the spectra, taken from 110 Å Fe_3O_4 thin film at 3K, are shown in Fig.1b. In the figure, by solid circles, shown the experimental values, the continuous curve shows the theoretical best fitt curve by using our model. As seen from the figure, our fit follows experimental values well. At the output of this fitting, the following parameters were deduced: unidirectional anisotropy, H_A=625 G; resonance field of the spectrometer, H_R=3305 G; magnetization, K/M-$2\pi M$=1450 G. As can be seen clearly from the figure, when direction of the applied magnetic field starts to change from the perpendicular geometry to the parallel geometry, resonance field values decrease and go the single minimum value. This is an expected behavior for the ferromagnetic sample. As known very well from the ferromagnetic materials, resonance fields of spectra in the parallel geometry are smaller than the resonance fields of spectra in the perpendicular geometry. The similar behavior of the angular dependence of the resonance field was also seen by Öner *et al.* in $Ni_{76}Mn_{26}$ and $Ni_{74}Mn_{24}Pt_2$ films [8], by Yalçin *et al.* in the alloys of NiFe films at different temperatures [9].

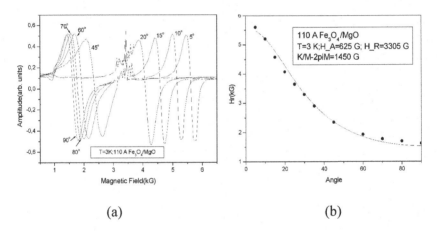

(a)   (b)

Figure 1. (a) Angular behavior of spectra taken from 110A Fe_3O_4 thin film; (b) Fitting of resonance fields by using a model.

To characterize the sample in more detail, the spectra taken at different cooling fields for both in parallel and in perpendicular geometry were searched. When the magnetic field was applied in both the normal in-plane direction and the reverse directions, it was looked at the hysteresis behavior of the sample. For this reason the following steps had been done: Fig. 2a shows the behavior of the spectrum of 110 Å Fe_3O_4 thin film taken under the cooling field of 2 kG in the perpendicular geometry at 3 K. A hysteresis behavior is seen during the application of the magnetic field in the normal and in the reverse directions. The value of the hysteresis between the normal and the reverse cases is 300 G. This hysteresis is the due to the ferromagnetic behavior of the materials. Fig. 2b shows the behavior of the spectrum of 110 Å Fe_3O_4 thin film taken under the cooling field of 4 kG in the perpendicular geometry at 3 K. A hysteresis behavior is seen again during sweeping of the magnetic field in the direction of both the normal and reverse cases. To get more information about the behavior of the sample in the normal and in the reverse directions, the sample was swept again in the normal and reverse directions and we got the second normal and the second reverse spectra. At this condition, hysteresis behavior was also seen again. The magnetic field difference for the second normal and reverse direction cases is smaller with respect to the magnetic field for the first normal and reverse direction cases. This may be due to the saturation of the magnetic moment of the sample after the first sweeping. After the first sweeping, there could be possibility for the magnetic moments of the sample not to go to the starting point. The similar behavior was also seen by Aktaş for the different thickness of the Fe_3O_4 thin film [2]. Fig. 2c shows the behavior of the spectrum of 110 Å Fe_3O_4 thin film taken under the cooling field of 4 kG in the parallel geometry at 3 K. At this case, a hysteresis behavior was also observed as in the case of perpendicular cooling case.

The similar behavior of 110 Å Fe_3O_4 was also seen at 10 K as in the case of the same sample at 3 K. But the behaviors are not the same throughtly like at 3 K. Fig.3a shows the behavior of the intensities with respect to the angle taken in the out-of-plane geometry. Firstly, sweeping was done starting from the parallel geometry ($\theta_H = 0^o$) to the perpendicular geometry($\theta_H = 90^o$) and then the second sweeping was done from the perpendicular geometry to the parallel geometry($\theta_H = 180^o$). As seen from the figure, there is a magnetic field difference between $\theta_H = 0^o$ and $\theta_H = 180^o$. This is the hysteresis behavior of the ferromagnetic materials.

If we look at the intensities of the spectra starting from the parallel geometry ($\theta_H = 0^o$), we see that up to 65°, intensities decrease while shifting to right side and line widths increase; by 75°, the intensities start to show increase and line widths show decrease; After reaching the perpendicular geometry($\theta_H = 90^o$), swipping direction changed and swipping starts towards the parallel direction. At 100°, line width decreased and intensity increased. While going towards the parallel geometry, line widths continued to increase and intensities continued to decrease. After 120°, line widths show decrease and intensities show increase. When $\theta_H = 180^o$, it is seen that this spectrum is not coincide with the spectrum of $\theta_H = 0^o$. This is the hysteresis behavior of the ferromagnetic sample. The magnetic field difference between these two spectra is seen in Fig. 3. The resonance field values of the spectra with respect to the angle taken at T=10K is shown in Fig. 3b. In this figure while the empty circles show the experimental values, the continuous curve shows the best fitting theoretical curve of the experimental values by using the model. As seen from this figure, the theoretical fit is quite acceptable. Some discrepancies can be seen for the θ values between the 0-20° and 160-180° ranges. The antisymmetric behavior of the curve in this region shows that

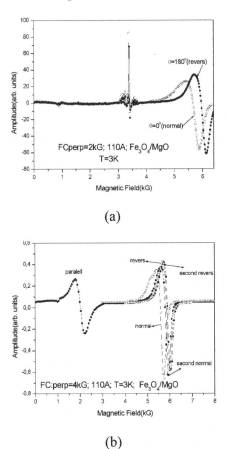

(a)

(b)

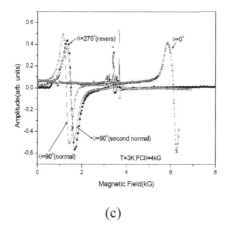

(c)

Figure 2. FMR Spectra of Fe_3O_4 in the paralell and perpendicular cooling conditions at 3 K.

this sample has an anisotropic property. The obtained parameters can be summarized as follows: Unidirectional anisotropy H_A=625 G, induced magnetization K/M-$2\pi M$=1675 G, the exchange stiffness constant D=1.5x10^{-11} Gm^2, the resonance field of the spectrometer H_R=3305 G. As seen from the figure, the resonance field values of the sample in out-of-plane geometry start to increase from 1500 G to 6200 G and then continue to decrease up to 1900 G at 180° at 10 K. As seen from this result, there is approximately 400 G difference between the initial point ant the final point. This is the anisotropy field. The same sample has the similar behavior at room temperature. But in the room temperature, starting point is 1750 G and the maximum value for the resonance field is 8000 G. For the last point, the resonance field value reached to the initial one, and the anisotropy was not seen. But in the low temperature measurements, this anisotropy field can be seen easily. So it can be said that the low temperature measurements give some significant information about the magnetic materials. For the room temperature measurements, one can look at Ref. [4].

Similar behaviors of the resonance fields with respect to the angle have been seen also by van der Heijden *et al.* for 115 nm Fe_3O_4 thin films at 50 K[3].

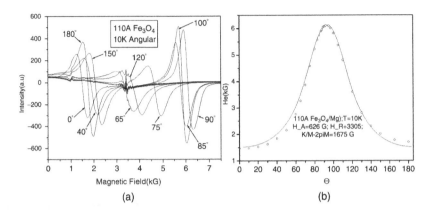

Figure 3. (a) Angular behavior of spectra taken from 110A Fe_3O_4 thin film; (b) Fitting of resonance fields by using a model.

Both the experimental and the theoretical spectra have been shown for the parallel and perpendicular geometries for the 110 Å Fe_3O_4 thin film depending on the temperature in Fig. 4. In the figure, as the empty circles show the experimental spectra, the continuous curves show the best fitting theoretical spectra from the model. As seen from the figure, the theoretical fitting is quite acceptable. While doing the theoretical fitting, we have got magnetic parameters which characterize the sample. As seen from the spectra, the paramagnetic peaks are located at about 3300 G. These are the background coming from the Mg^{2+} ions in the substrate. If the temperature starts to increase starating from T=10 K, the parallel and perpendicular spectra approach to each other. This shows that the induced magnetization decreases when the temperature increases. The magnetic parameters can also give information about the crystal transition temperature, the Vervey temperature, because our sample is ferrite. While the temperature continues to decrease, the crystal structure changes from the cubic structure to the orthorhombic structure at the certain temperature. We have got also magnetic parameters about the Vervey temperature.

The temperature dependence of the resonance fields of the spectra taken from 110 Å Fe_3O_4 thin film for both the parallel and perpendicular geometries is shown in Fig. 5a. As seen from the figure, behaviors of the resonance fields of the parallel and perpendicular geometries are similar to each other. The temperature dependence of the line widths of the spectra taken from 110 Å Fe_3O_4 thin film for the parallel and perpendicular geometries is shown in Fig. 5b. Generally speaking, the behaviors are similar to each other for the parallel and perpendicular geometries. As seen from the figure, the line widths values decrease as the temperature increases. This is characteristic behavior of the ferromagnetic samples. Similar behavior for the line widths was also seen by Öner *et al.*, in $Ni_{76}Mn_{24}$ and $Ni_{74}Mn_{24}Pt_2$ compounds [8]. The sudden increase in line width for the perpendicular geometry at 77 K might be due to proximity to the Vervey temperature or the growing conditions of the film.

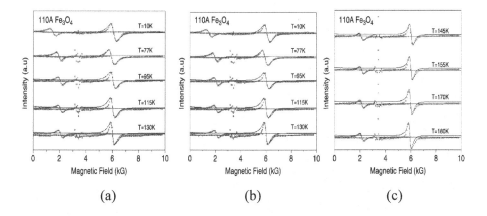

(a)   (b)   (c)

Figure 4. FMR spectra of 110A Fe_3O_4 thin film. Open circles show experimental values, continuous curves show theoretical fitting curves.

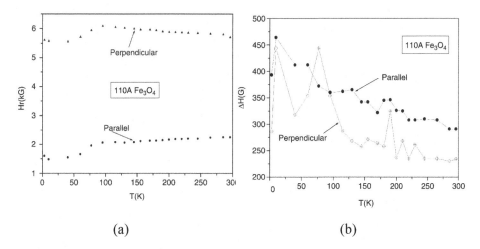

(a)   (b)

Figure 5. Resonance Fields and linewidths variation with respect to the temperature for the perpendicular and the parallel geometries.

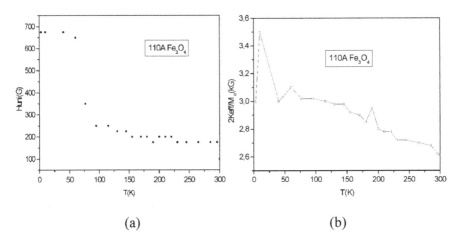

Figure 6. Unidirectioanl anisotropy and magnetization changes with respect to the temperature of 110A Fe_3O_4 thin film.

The temperature behavior of the unidirectional anisotropy can be seen in Fig.6a, and the temperature behavior of the magnetization is seen in Fig. 6b. As seen from the Fig.6a, the unidirectional anisotropy approximately does not change between the 3-60 K. Between 60-95 K, the sudden decrease can be seen.

This behaviour may be due to Verwey transition temperature. The Verwey temperature for the bulk form of the same material is 120 K. But in the thin film case this value decrease to 90 K. The similar behavior for the unidirectional anisotropy is also seen by Aktaş at the different thickness of the same Fe_3O_4 film [2]. But in that sample, there was no plato. This may be due to the difference between the thicknesses of the films. If we look at the Fig. 6b, the magnetization behavior can be seen as the temperature increases. The sudden increase can be seen at T=10 K between 3000 G and 3500 G. But beyond these values, it starts to decrease as the temperature increases. This behavior is also expected from the ferromagnetic materials. Similar behavior of the magnetization was also seen by Budak *et al.*, in LaCaMnO [10] and LaSrMnO systems[11] by Özdemir *et al.*, in $Au_{77}Fe_{23}$ films [12] and by Öner *et al.*, in NiMn films [8].

Fig. 7 shows the spectra taken from 110 Å Fe_3O_4 thin film for the parallel geometry at 13 K when the sample is rotated by 180° for both the increasing and decreasing of the magnetic field. In this figure, as the empty circles show the experimental values, the continuous curve shows the best fit using the model. As seen from the figure, the theoretical fitting follows the experimental values quite good. The derived magnetic parameters are also seen on the figure. If we look at the figure, the hysteresis effect is again seen here. The magnetic difference between these two curves is about 100 G. This hysteresis effect is an expected behavior from the ferromagnetic samples. The fitting magnetic parameters for this curve are as follows: The unidirectional anisotropy is 620 G, the exchange stiffness constant $D=1.5 \times 10^{-11}$ Gm^2 . The resonance field of the spectrometer was 3305 G when the spectra were taken. For these curves, only one parameter was seen as different. This value is the induced magnetization.

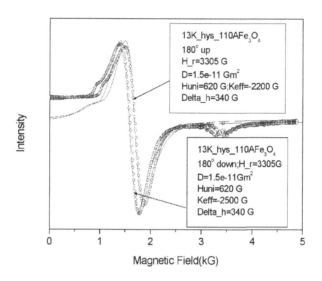

Figure 7. Fitting of hysteresis curve of 110 Å Fe_3O_4 thin film at 13 K for the 180 down and up orientations.

Figs. 8-11 show the temperature dependence of the resonance fields of the spectra taken from 110 Å Fe_3O_4 thin film for both parallel and perpendicular geometries. Fig.8a shows the temperature dependence of the resonance fields for the parallel geometry when magnetic field is swept for the increasing and decreasing directions. Here the empty circles show the increasing (normal) direction of the magnetic field and the filled circles show the decreasing (reverse) direction of the magnetic field. As seen from the figure, the temperature behaviors are similar to each other for the normal and reverse cases.

When temperature increases, the resonance field values increase. But the difference between the increasing and the decreasing cases can be seen easily by considering the figure. This difference of the resonance fields give a signal about the hysteresis

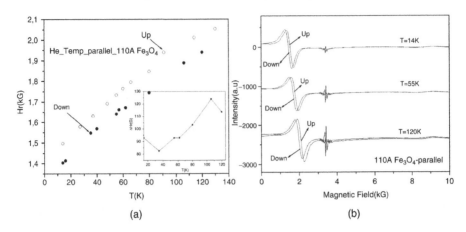

(a)   (b)

Figure 8. Resonance Field variatons of 110 A Fe_3O_4 thin films with respect to the temperature for the paralell case. Inset shows the difference between the up and down resonance fields differences. Also some selected spectra are shown here.

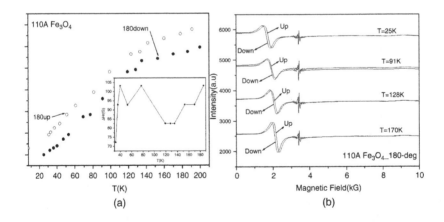

(a)   (b)

Figure 9. Resonance Field variatons of 110 A Fe_3O_4 thin films with respect to the temperature for the paralell case when 180 down and up orientation satisfied. Inset shows the difference between the up and down resonance fields differences. Also some selected spectra are shown here.

behavior of the sample. If there was no hysteresis effect, the resonance field values have to overlap. The inset in the Fig. 8a shows the differences between the reonance fields for the normal and reverse cases. For considering the hysteresis effect very well, the some selected spectra were shown in Fig. 8b. The hysteresis effect can be seen easily by using these spectra. This hysteresis effect says that this behavior is coming from the ferromagnetic properties of the sample.

Fig. 9a shows the temperature dependence of the resonace fields for the rotation of the sample by 180^0 when the sample is in the parallel condition while the sweeping magnetic field is increasing (normal) and decreasing (reverse). Here, the empty circles show the increasing of the sweeping magnetic field and the filled circles show the decreasing of the applied magnetic field. As can be seen from the figure, the temperature dependence of the resonance fields are similar to each other for both up and down positions.

The inset in the Fig. 9a shows the temperature dependence of the difference of the resonance field values of up and down positions. As seen from this figure, there is a plato between 120-140K. This behavior may be related with the Vervey temperature of the bulk form of the same sample, because bulk form of the Fe_3O_4 sample has the Verwey temperature of 120 K. After this temperature, the crystal structure of the sample changes. Some of the spectra taken from the sample are shown in Fig. 9b. As seen from the figure, there is a hysteresis behavior starting from T=25 K to T=170 K.

Fig. 10a shows the temperature dependence of the resonace fields of the sample when the sample is in the perpendicular geometry while the sweeping magnetic field is increasing (normal) and decreasing (reverse). Here, the empty circles show increasing of the sweeping magnetic field and the filled circles show decreasing of the applied magnetic field. The hysteresis effect can also be seen here. As seen from the figure, the temperature dependence of the resonance fields are similar to each other for both up and down positions. The resonance fields increase up to the temperature of about 120 K. After 120 K, the values of the resonance fields start to decrease for both conditions. At about 120 K, one can see one maximum peak. This behavior may be related to the Verwey transition of the bulk form of the same sample. As it was mentioned before, the Verwey temperature of the bulk form of the same sample is 120 K. But this does not mean that every property has to belong to the thin films. One can see that some properties of the materials don't change in the bulk and thin forms.

The inset in the Fig. 10a shows the temperature dependence of the difference of the reonance field values of up and down positions. The hysteresis effect can be seen more clearly in Fig. 10b.

Fig. 11a shows the temperature dependence of the resonace fields for the rotation of the sample by 180^0 when the sample is in the perpendicular condition while the sweeping magnetic field is increasing (normal) and decreasing (reverse). Up to at about 120 K, the resonance fields for two conditions increase, after 120 K these values start to decrease. Here, the empty circles show the increasing of the sweeping magnetic field and the filled circles show the decreasing of the applied magnetic field. As seen from the figure, there is an information about the Verwey temperature of the bulk form of the same sample. The hysteresis effect can be seen more clearly in Fig. 11b. As seen from this figure, the hysteresis effect is seen in all the selected spectra. Due to the space limitation, we have chosen only a few of them.

The measurements taken after the rotataion of the sample when the sample is in both the parallel and perpendicular geometry do not show significant changes. This gives some explanation about the growth

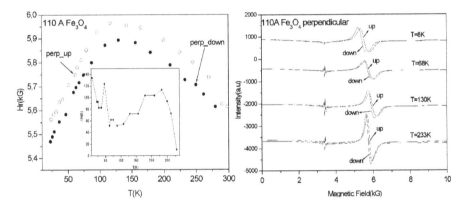

Figure 10. Resonance Field variatons of 110 A Fe_3O_4 thin films with respect to the temperature for the perpendicular case . Inset shows the difference between the up and down resonance fields differences. Also some selected spectra are shown here.

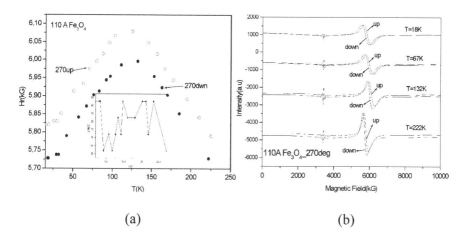

(a)   (b)

Figure 11. Resonance Field variatons of 110 A Fe_3O_4 thin films with respect to the temperature for the perpendicular case when n 270 down and up orientation satisfied. Inset shows the difference between the up and down resonance fields differences. Also some selected spectra are shown here.

conditions of the film. As seen from the measurement, only small changes can be seen according to the sweeping directions of the film.

REFERENCES

[1] Williams H.J., Bozorth R.M., Goertz M., Phys. Rev. B 91, 1107 (1953)
[2] Aktaş B., Thin Solid Films 307, 250 (1997)
[3] Heijden P.A.A., Opstal M.G., Swüste C.H.W., Bloemen P.H.J., Gaines J.M., Jonge W.J.M., J.Magn.Magn.Mater. 182, 71 (1998)
[4] Budak S., Yıldız F., Özdemir M., Aktaş B., J.Magn.Magn.Mater. 258-259, 423 (2003)
[5] Margulies D.T., Parker F.T., Spada F.E., Goldman R.S., Li J., Sinclair R., Berkowitz A.E., Phys. Rev. B 53, 9175 (1996)
[6] Maksymowicz L.J., Sendorek D., J. Magn. Magn. Mater. 37, 177 (1983)
[7] Speriosu V.S., Chen M.M., Suzuki T., IEEE 25, 3875 (1969)
[8] Öner Y., Özdemir M., Aktaş B., Topaçlı C., Harris E.A., Senoussi S., J. Magn. Magn. Mater. 170, 129 (1997)
[9] Yalçın N., Aktaş B., Özdemir M., Durusoy H.Z., Physica B 233, 251 (1997)
[10] Budak S., Özdemir M., Aktaş B., Physica B (submitted revised version)
[11] Budak S., Özdemir M., Aktaş B., Physica B (in print)
[12] Özdemir M., Aktas B., Öner Y., Sato T., and Ando T., J. Phys. Cond.Mat. 9, 6433 (1997)

6

PART 6:MAGNETIC NANOPARTICLES AND NANOWIRES

EFFECT OF MPEG COATING ON MAGNETIC PROPERTIES OF IRON OXIDE NANOPARTICLES: AN ESR STUDY

Y.Köseoğlu [a], F.Yıldız [b], D.K. Kim [c], M. Muhammed [c] and B. Aktaş [b]

[a] Fatih University, Department of Physics, 34900 Istanbul, Turkey
[b] Gebze Institute of Technology, Kocaeli, Turkey
[c] Royal Institute of Technology, Materials Chemistry Division, Stockholm, Sweden

Abstract: Metoxy polyethylene glycol (MPEG) coated and uncoated Fe_3O_4 (SPION) nanoparticles have been investigated by Electron Spin Resonance (ESR) technique at the temperature range between 10-300K. ESR measurements on powdered samples have been carried out by using X-band ESR Bruker EMX spectrometer. A strong and broad single ESR signal has been observed at all temperatures. Temperature dependence of ESR linewidth and resonance field is studied. The linewidth increases and the resonance field decreases by decreasing temperature while the ESR signal intensity remeains almost independent on the temperature. The shift in the resonance field value is a clear indication of the induced field (exchange anisotropy field), causing the frustration (disorder) of any magnetic system. A strong effect of coating on magnetic properties of SPION nanoparticles has been observed as well.

Key words: SPION, MPEG coating, Ferromagnetism, ESR

1. INTRODUCTION

Fine magnetically ordered particles have long been the focus of intensive study. Growing interest on this subject has been due to extraordinary manifestations of the magnetic properties of these particles as well as to the necessity to solve numerous related applied problems [1-4]. Magnetic nanoparticles have been used as contrast agents in magnetic resonance (MR) for localization and diagnosis of brain and cardiac infarcts, liver lesions or tumors. It was suggested that the magnetic particles could be conjugated to

B. Aktaş et al. (eds.), Nanostructured Magnetic Materials and their Applications, 303–312.

various monoclonal antibodies, peptides or proteins to achieve target-directed magnetic-resonance imaging (MRI) [5,6]. Based on their unique mesoscopic physical, tribiological, thermal, and mechanical properties, superparamagnetic nanoparticles offer a high potential for several applications in different areas such as ferrofluids, color imaging, magnetic refregiration, detoxification of biological fluids, magnetically controlled transport of anti-cancer drugs, magnetic resonance imaging contrast enhencement and magnetic cell separation [7-9]. The intent of this article is to discuss the temperature and frequency dependence of magnetic properties of MPEG coated superparamagnetic iron oxide nanoparticles (SPION) by using Electron Spin Resonance technique.

2.   EXPERIMENTAL

MPEG modified SPION. The SPION suspension was washed several times with analytical grade methanol (99.5%) with a help of a strong external magnet. An aliquot of 5 mL with a concentration of 10 mg/mL SPION was dispersed in toluene/methanol (1:1 v/v) mixture and heated at 95°C under N_2 until 50 vol % of the solution was evaporated. After evaporation, methanol was added in equal volume and the mixture was re-evaporated to one half. This procedure was repeated three times, until residual water was thoroughly removed. A solution of 3-aminopropyltrimethoxy silane (APTMS) was added to the suspension. APTMS acts as a coupling agent, where silanization takes place on the particle surfaces bearing hydroxyl groups in the organic solvent. This results in the formation of a three-dimensional polysiloxane networks. The ferrofluid suspension was stirred and heated under N_2 atmosphere at 110°C for 12 hrs. The silanization, taking place during the reflux of APTMS results in the formation of APTMS coating with a thickness of two or three molecular layers, tightly cross-linked with a large surface density of amines. Afterward, the silanized SPION were subsequently sonicated for 2 min in a mixture of toluene/methanol (1:1 v/v). After cooling the solution temperature until 80°C, 50 mg MPEG 5000 was added to coat the surface of the SPION.

For AFM imaging of MPEG coated SPION, silicon wafer was used as a substrate. Silicon wafer was cut into pieces of 1x1 cm. These wafers were washed with acetone first and N_2 was flown for drying. Piranha solution was used to de-grease and de-oxygenate the substrate surface. The dried wafers were then silanized in the following way: The silicon pieces were placed on a hot plate in a vacuum chamber. On the plate was also a small 3-neck flask

loaded with 200 µL of APTMS in 20 mL toluene. In low vacuum, the hot
plate was heated to 60 °C for 10 min to saturate the silicon substrate with the
solution. The temperature was then increased to 150 °C for 1 hr to form the
silane molecules on the silicon surface. The silanized silicon was immersed
in MPEG coated SPION suspension for 10 min, rinsed in deionized water
and blown dry by N_2. The samples were characterized by an atomic force
microscope (AFM) and the sample size were determined to be 200 nm.
Sample preparation technique has been described in more details elsewhere
[6].

Then we have studied the temperature and frequency dependence of
magnetic properties of metoxy polyethylene glycol (MPEG) immobilized
super-paramagnetic iron oxide nanoparticles (SPION). The polycrystalline
powders placed in nonmagnetic paraffin just above its melting temperature
have been tried to orient in the presence of a strong magnetic field (15 kG).
The samples were subsequently cooled down below the melting temperature
of paraffin in this field to have magnetic orientation. Then a sample with
dimensions 1.5x2x2.5 mm was cut from this ingot for ESR measurements.
The ESR measurements were performed on a conventional X-band ($f \cong 9.7$
GHz) Bruker EMX model spectrometer employing an ac magnetic field
modulation technique. The static magnetic field was scanned from 0 to
10000 G. The modulation field used had amplitude of 10 G. An Oxford
Instruments continuous helium gas flow cryostat has been used, allowing the
X-band microwave cavity to remain at ambient temperature during ESR
measurements at low temperatures. The temperature was stabilized by a
conventional Lakeshore 340 temperature controller within an accuracy of 1
degree between 10-300 K. A goniometer was used to rotate the sample with
respect to the external magnetic field in order to observe angular variations
of the ESR spectra. K-and Q-Band ESR spectra taken at room temperature
by our home-made high frequency spectrometers that are capable to measure
only absorption spectra.

3.   RESULTS AND DISCUSSION

Some of the ESR spectra taken from MPEG coated and uncoated Fe_3O_4
samples are shown in Figure 1.

As one can see, there are significant temperature dependence of the ESR
spectra. Furthermore the spectra are obviously affected by MPEG coating.
As the temperature decreases, the resonance field of the ESR signal shifts to
lower fields for both coated and bare samples. The broad signals are
approximately symmetric around the resonance field H_r for uncoated

samples at all temperatures. Such a symmetric signal may, therefore, be attributed to a single type of paramagnetic (or superparamagnetic) center for both samples. However, there seems to be significant distortion in the ESR spectra of MPEG coated samples and additional absorptions occure at low field side of the main peaks especially around 100 K. The lines also become broader at lower temperatures. The relatively very weak and narrow signal at appeared at 3380 Oe belongs to the DPPH standard used for more accurate g-factor calculations.

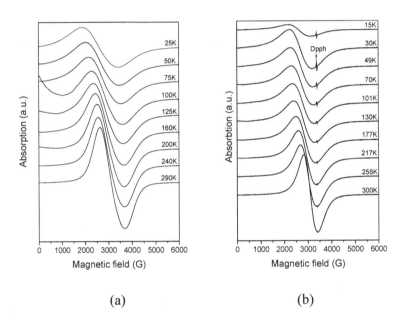

Figure 1. Temperature dependence of X-band ESR spectra of
(a) MPEG coated and (b) uncoated SPION.

The shift of the resonance field can be seen more clearly in Figure 2. While the resonance field value is about 3100 G for the room temperature; it decreases gradually as the temperature is decreased. For coated sample some small and relatively smooth changes at about 30 K, 90 K and 140 K are appeared as well. But the changes for uncoated sample are observed only at around 25 K and 50 K. These small changes can also be seen in figure 4 which shows the temperature variation of the efffective g-value. The amount of the shift in the resonance field is 500 G at low temperatures with respect to that at 300K for both samples. This result is correlated with the line width behavior. It should be noted that, in any ESR measurement, gyro magnetic (Larmour) precession frequency is observed in an effective field. Therefore, this shift in the resonance field value can be taken as a clear indication of the temperature-induced field (exchange anisotropy field), which is the main cause of the frustration (disorder) of any magnetic system. That is the increase in microscopic fields at low temperatures reveals itself as the line broadening, line shift and a decrease in the signal intensity at low temperatures.

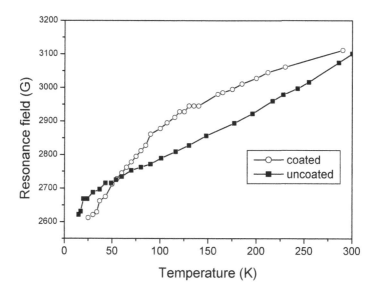

Figure 2. Temperature dependence of resonance field for MPEG coated SPION.

Temperature dependence of the peak-to-peak line width of the ESR spectra is shown in Figure 3. As seen from the this figure, some remarkable changes in the linewidth values are also seen at the temperatures where also line intensity and resonance field values significantly vary. But there is one more change at around 180 K. The ESR line gets broadened as the temperature decreases. This line broadening can be attributed to spin disorder (frustrations) possibly coming from mainly antiferromagnetic interactions between the neighboring spins in magnetic layers of both samples. This frustration might be partially connected to antiferromagnetic interactions between the magnetic clusters. This can also be related to the behavior of the whole system as a solid respect to spin dynamics and this disordered. The frozen spin profile manifest themselves especially in line broadening of ESR spectrum.

The line broadening effects might arise also from the dipolar interactions between the superparamagnetic nanoparticles. Normally, the internal magnetic field originating from magnetic entities is expected to be more uniform as a result of highly ordered magnetic moments at low temperatures, giving narrower ESR line in contrary in our case.

However, at low temperatures, one may expect the order of the system and decrease in the inhomogenity of the dipolar fields. Experimental results show the reverse of this (means at low temperatures the disorder of the dipolar fields is increasing because magnetization is not increasing with the same ratio). It can be said that some of the antiferromagnetic interactions becomes more effective and this effect enhances level of magnetic (spin) disorder of the system.

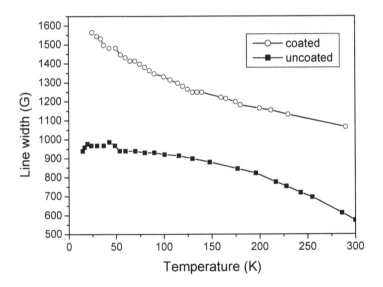

Figure 3. Temperature dependence of ESR line width of coated and uncoated samples.

The line width of the coated sample is grater than that of uncoated sample. When the particles are chemically coated, the line width is getting larger with decreasing the temperature. Without coating of surfactant on the particles, due to the increase in the large ratio of surface area to volume, the attractive force between the nanoparticles will increase, and agglomeration of the particles will take place. These agglomerated nanoparticles act as a cluster, resulting in an increase of the blocking temperature and a decrease in the line width of the ESR signal. In contrast, the surfactant-coated nanoparticles are more freely aligned with the external field than the uncoated nanoparticles resulting a wider line width. The repulsive force between hydrophobic surfactant molecules coated on single particles can prevent them from agglomeration. The total effective magnetic moment of such coated particles is found to decrease, which is most likely due to a non-collinear spin structure originated from the pinning of the surface spins and coated surfactant at the interface of nanoparticles. The measured magnetic moment is also found to decrease due to the contribution of the volume of the diamagnetic coating mass to the total sample volume [9].

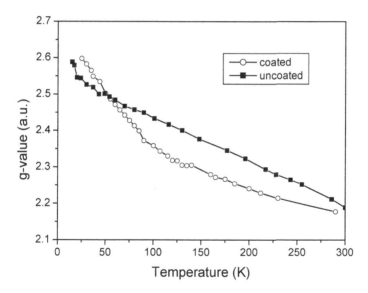

Figure 4. Temperature variation of effective g-value for coated and uncoated samples.

Effective g-values for the samples (can be derived from the measurement of the magnetic field at the centre of the ESR spectra) are given in figure 4. The effective g-values of the samples are increasing slowly by decreasing temperature. There is a step like change at around 140K for the coated sample. The increase of the g-value (the spectrum is shifted to the lower fields) shows the increase in the internal fields. This result correlates with the line broadening at this temperature range. Therefore, the increase in the effective microscopic fields at low temperatures is obvious.

Frequency dependence of the resonance field for the coated sample is shown in Figure 5. An almost linear dependence of resonance field to frequency has been observed for K- and Q-band spectra. As the frequency increases, the resonance field also increases. The linear frequency dependence of the resonance field at high frequencies suggests that the susceptibilities of the samples are sufficiently low at high temperatures. Since we do not have low temperature facility for high frequency apparatus, we could not measure the frequency dependence of the resonance field at low temperatures. As the H_r-ν curve is extrapolated to lower frequencies, it cuts the vertical axis at a positive value. When the experimental values are fitted to the theoretical resonance equation $h\nu = g\beta(H_r + H_i)$, it is seen that there are internal magnetic fields in the samples even in zero frequency.

By the analysis of the frequency dependent resonace field of EPR signal, the effects of the internal field is refined and thus the spectroscopic g-factor and internal field were determined to be $g = 1.9773$ and $H_{int} = 104G$ respectively. However, a very small non-linearity in H_r versus frequency curve was atributed to the external field dependency of the internal field through magnetic susceptibility.

In conclusion, MPEG coated and uncoated Fe_3O_4 nanoparticles show super-paramagnetic behaviour and there are dipolar interactions between the particles. According to positions of the particles around the dipolar interactions, they encourage ferromagnetic or antiferromagnetic order. Because of these opposite interactions, it can be said that the system is forced to show disordered behaviour (actually disorder-spin glass behaviour). However, correlation of spin-glass behaviour to dipolar interaction to explain this effect is uncertain. Especially, line width for MPEG coated SPION is increasing up too high values as 1500 G. If this value is correlated to inhomogeneous distribution, dipolar fields must have as high as this value. In our experiments we have tried to dilute the powders in paraffin (wax), but we are not sure how much we have achieved.

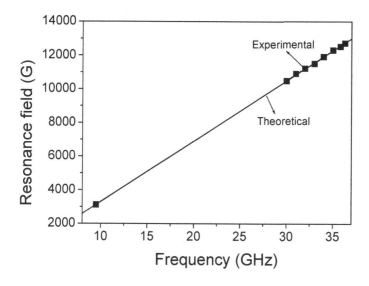

Figure 5. Frequency dependency of Resonance field of MPEG coated SPION.

On the other hand, it is known that at low temperatures, the structure of ferrites is changed from cubic to orthorhombic structures and magnetization is decreased a little. In fact, small deviations from stochiometry cause difficulties to reach the maximum magnetization, and it is shown that the hysteric curve becomes wide and shallow.

It is well known that all the interactions like these are antiferromagnetic but because of the difference of the spin values of the ions contributing to magnetization with various magnetic moments, there is a ferrimagnetic order and as a result of this, it can be reached to huge magnetization values. As the exchange interactions are antiferromagnetic and dipolar interactions are either ferromagnetic or antiferromagnetic, it is not surprising to get a disordered spin profile known as spin-glass. The increase in the line width of the ESR spectra can be correlated to distribution of exchange interactions.

Hence, MPEG coating decreases agglomeration and the inter-particle interactions. The surfactant-coated nanoparticles are more freely aligned with the external field than the uncoated nanoparticles resulting a wider line width. The repulsive force between hydrophobic surfactant molecules coated on single particles can prevent them from agglomeration. The total effective magnetic moment of such coated particles is found to decrease, which is most likely due to a non-collinear spin structure originated from the pinning of the surface spins and coated surfactant at the interface of nanoparticles. The measured magnetic moment is also found to decrease due to the contribution of the volume of the diamagnetic coating mass to the total sample volume

REFERENCES

[1]   V.I. Nikolaev, T.A. Bushina, K.E. Chan, J. Magn. Magn. Mater. 213 (2000) 213.

[2]   Juh-Tzeng Lue, J. Phys. Chem. Solid. 62 (2001) 1599.

[3]   R.H. Kodama, J. Magn. Magn. Mater. 200 (1999) 359.

[4]   M.F. Hansen, S. Morup, J. Magn. Magn. Mater. 184 (1998) 262.

[5]   D.K.Kim, Y.Zhang, J. Kehr, T. Klason, B. Bjelke, M. Muhammed, J. Magn. Magn. Mater. 225 (2001) 256.

[6]   D.K. Kim, Ph.D. Thesis, MATCHEM, MET, KTH, 2002.

[7]   J. Popplewell, L. Sakhnini, J. Magn. Magn. Mater. 149 (1995) 72.

[8]   K. Raj, B. Moskowitz, R. Casciari, J. Magn. Magn. Mater. 149 (1995) 174.

[9]   D,K, Kim,Y. Zhang, W. Voit, K.V. Rao, M. Muhammed, J. Magn. Magn. Mater. 225 (2001) 30.

THERMAL MAGNETIC DECAY IN MAGNETIC NANOPARTICLES

Y. Öztürk and I. Avgin

Electrical and Electronics Engineering Department, Ege University, Bornova, Izmir 35100 Turkey

Abstract: Thermal stability of magnetic storage systems has a well known physical constraint: the superparamagnetic effect. Thermally activated magnetization reversal over the energy barrier causes a slow decay in magnetization. We investigate here the thermal relaxation of the magnetic films with perpendicular anisotropy. A relaxation model with interactions treated on a mean field level without disorder is used. This model assumes that the interactions between grains drag the system from an initial magnetized state towards the equilibrium magnetic state in a reverse applied field. In this work, the reversal mechanism involves a coherent rotation (Stoner-Wohlfarth model) and the anisotropy of the grains is parallel to the field. Time dependence of remnant magnetization and magnetic viscosity are calculated.

Key words: Magnetic nanoparticles, thermal decay, coherent rotation.

1.   INTRODUCTION

Enhancement of the magnetic storage density is usually achieved by reducing key dimensions. As the written bits become denser, the individual grains on which the bits are recorded become smaller. The technological progress of the magnetic media towards higher information densities with lower noise and faster access times demands an unending advancement in magnetic systems with particulate or film media. The most indispensable properties that any new magnetic material must display are larger coercivity and smaller grain size. Apart from the technical difficulty of producing such materials, the grains may exhibit a so-called super-paramagnetic effect which is the main threat to the integrity of the recorded data. The thermal

B. Aktaş et al. (eds.), Nanostructured Magnetic Materials and their Applications, 313–325.

energy supplied by the environment causes magnetic reversals consequently spoiling the recorded signal. From a technological point of view, the high density recording material must be designed in such a way as to control the thermal relaxation. Thermal decay is also a concern in basic physics since diverse physical systems such as spin glasses, trapped flux in superconductors and charge carriers in amorphous semiconductors [1] all exhibit similar decay. Therefore it is vital to understand the physical mechanisms behind the thermal fluctuations. In this work we examined the effect of a mean demagnetizing field [1] on the slow decay. For simplicity, we assume all grains are identical.

The Debye relaxation or Arrehenius-Néel model predicts an exponential time decay of the magnetization for a fixed energy barrier; as such, the lifetime of the decay depends on the energy barrier. However quasi logarithmic magnetization decay is observed instead of an exponential decay. Usually two approaches are taken to explain the slow decay: one assumes that the barriers have some distribution producing a range of random time constants. The magnetization will then be a superposition of the exponential decays generated by these random time constants which gives rise to a quasi logarithmic decay [1]. The sources of the distribution can vary depending on material, design and the kind of interactions. For example, the distribution in the grain volume or in the anisotropy energy may result in a distribution of the energy barriers. But the nature of the distribution is still debated; for instance, for logarithmic decay a wide distribution of disorder is necessary. However some materials such as magneto-optical recording systems [2] show very narrow distributions. This model assumes that the grains do not interact magnetically yet the grains in thin-film media are coupled by exchange and magnetostatic forces. A logarithmic decay can be obtained also by an alternative model (without disorder) including the interactions, in particular the demagnetizing effects with a grain volume as an adjustable parameter [1,3]. In this work the mean field approach is taken when treating the magnetostatic interactions. This method [1,3] as applied to the CoCr films confirmed the experimental results for the number of practical time decades related to the long term storage. Recently Wood [4] *et al.* used this model to investigate the thermal effects in films with a over layer both perpendicular and longitudinal media in the presence of the demagnetizing effects (due to a recorded pattern). Rizzo [5] *et al.* also applied the same method for patterned longitudinal thin films assuming a very sharp transition and calculated the magnetic viscosity. Concerning the longitudinal films one has a very small demagnetizing effect, and clearly the mechanism suggested in refs. [1,3] is absent so that the magnetization would be predicted to show exponential decay, whereas the observed magnetization also decays logarithmically . This suggests that there are other key ingredients leading to logarithmic decay, and we will be working in that direction in the future.

2. RELAXATION MODEL WITH INTERACTIONS

The growing need for high density recording has required a higher signal-to-noise ratio (SNR) of the thin film media used in either perpendicular or longitudinal recording. Any magnetization reversal is thermally driven, and hence the remnant magnetization and consequently the recorded signal decays with time. In fact, the previous generations of media have negligibly small thermal decay which caused little concern for the recorded data. The magnetization decay in high density recording media with good SNR is very troublesome so it is required that the recorded information must be thermally stable [6]. The experimental setup for the magnetization decay is as follows: first a large external field is applied to saturate the system, then this field is turned off and another smaller field antiparallel to the magnetization direction is turned on. The magnetization reversal may take place in a number of mechanisms depending on material conditions. In this work we only deal with coherent reversals [7], known as the Stoner-Wohlfarth model. For the coherent rotations the anisotropy axis may be parallel to the applied field, but in the longitudinal case the anisotropy axis may have random orientations in the film plane. This is a quite simplified but powerful model that can be successfully applied to a number of cases, and it can serve as an insightful step for probing the effect of the interactions. In this model magnetic grains have a magnetic moment (single domain) with a uniaxial easy axis and the coupling of the grains is through long-range magnetostatic interactions. The ground state of the system has no magnetization in zero external fields. If the system is saturated in a direction say $+z$, the mean field arguments result in a demagnetizing field in $-z$ for the perpendicular anisotropy. This reverse internal field drives the system to demagnetized state. However, as the magnetization of the system decays, the demagnetizing field also gets smaller as time progresses leading to a quasi logarithmic magnetization decay.

Let's consider uniaxial identical particles with magnetization in the $+z$ direction. A particle will have anisotropy energy $-K_u v \sin^2 \vartheta$ where K_u, v, ϑ are the anisotropy energy, the volume of the grain and the angle between the magnetization direction and the easy axis. This particle will have two minima at angles at 0 and π, and one maximum at $\pi/2$. The energy is degenerate at these minima unless a reverse field is applied. The total energy is then $E = -K_u v \sin^2 \vartheta + MvH \cos \vartheta$ where M is the magnetization of the grain and the H is the magnetic field.

This system is indeed a two level system with shallow and deep well. The collection of identical magnetic grains will be distributed in each of the wells at a given time. There are always particles changing from one direction to the other even in equilibrium due to the thermal fluctuations. Since there are two magnetic states available, thermally induced magnetization switches can occur from state 1 to state 2 and vice-versa where there are two switching rates involved. The effect of interactions, the magnetostatic and the exchange, can be seen from the simulations of the magnetic hysteresis loops; for instance see Tagawa [8] *et al.* We can obtain irreversible magnetization by solving the nonlinear relaxation differential equations numerically. First we present an approximate analytical solution [4] including back reversal. Taking the notation used in [4], we set the instantaneous magnetization to $M = M_+ - M_-$, and the saturation magnetization $M_s = M_+ + M_-$. It is convenient to define the normalized magnetization $x = M / M_s$ so that the relaxation differential equation is given by

$$\frac{dx}{dt} = -f_0[(1+x)e^{-\beta K_u v\left(1-\frac{H+DM}{H_K}\right)^2} - (1-x)e^{-\beta K_u v\left(1+\frac{H+DM}{H_K}\right)^2}]$$ (1)

where f_0 (10^9 sec^{-1}) is called attempt frequency, D is the demagnetizing coefficient equal to 4π for a perpendicular film and close to zero for longitudinal films with uniform magnetization. H is the external field and $H_K = 2K_u / M_s$ is the anisotropy field. The attempt frequency is related to the precession frequency of the grains. In general, for a free spin, the precession frequency depends on the applied field; $10^9 s^{-1}$ is approximately the spin-flip scattering rate for a moment in the metal so that would make it a good number to use for a single spin. Surprisingly, it is observed that its value does not alter the solution of the differential equation.

3.   THE LONG TIME RESULTS AND MAGNETIC VISCOSITY

In the absence of the applied field one can solve this differential equation for small normalized-magnetization. The approximate solution for perpendicular [4] and longitudinal [5] cases are carried out in the presence of the demagnetizing field. It is convenient to write Eq. (1) in the following form

$$\frac{dx}{d\tau} = -2e^{-\beta K_u v\left(h^2 + (M_s Dx/H_k)^2 + 2M_s Dx/H_K\right)}(x\cosh qx + \sinh qx) \tag{2}$$

For small x (the long time behavior) in zero field the differential equation can be arranged to read

$$\frac{dx}{d\tau} = -2(x\cosh qx + \sinh qx),$$

where $q = 2DM_s\beta K_u v/H_K = DM_s^2 v/k_B T$ (at room temperature this number will be between 10 and 100) and the normalized time $\tau = t f_0 e^{-\beta K_u v}$. If we scale x with q the first term in the left side of Eq. (2) will be 10 to 100 times larger than the second term so one can omit the first term without adding considerable error. Then the approximate solution of Eq. (2) can be obtained as

$$\ln(\tanh(qx/2)) = -2q(\tau + \tau_0), \tag{3}$$

where τ_0 is the constant from the solution of the differential equation that will be determined from the initial condition: $\tau_0 = -\ln\tanh(q/2)/2q$ if at initial time x is set to 1 which is very small quantity (note that initial time is not negative since tanh(large number) < 1, but very close to 1). The normalized magnetization as a function of time is written as

$$x = -\frac{1}{q}\ln(\tanh(\tau + \tau_0)) \approx -\frac{1}{q}\ln q(\tau + \tau_0), \tag{4}$$

where the constant $\tau_0 \propto 10^{-5}$ since the time is multiplied by a number on the order of 10^{-13}. We recall that this solution is valid only for the zero field case, and the nonzero field case is considered in the next section. Magnetic viscosity can be calculated from

$$S = -\frac{dM}{d\ln(\tau + \tau_0)} = 1/q,$$

and it depends on basic material parameters. If we identify the time dependent interaction field, we can deduce the fluctuating field from the arguments presented in ref. [4] where thermally induced random fluctuations can be represented by $H(t) = H_f(Q + \ln t)$ with Q constant. Here H_f is the so-called fluctuating field. The time dependent magnetization can be obtained from the irreversible susceptibility as $M(t) = \chi H(t) = M_0 - S \ln t$. Adopting to the present case with zero applied field,

$$H_f = -\frac{dH(t)}{d \ln(\tau + \tau_0)} = \frac{k_B T}{M_s v}. \tag{5}$$

So a magnetized film composed of uniaxial grains will have a time dependent magnetic field decaying logarithmically with a rate given by the fluctuation field. For more complex cases, where the magnetization is not parallel to the anisotropy axis, the decay is faster [4,9] than that given here. The numerical solution to Eq. (1) is given in Fig.1. for two different perpendicular media, and their related parameters are given in the Table.1. For both films, the remnant magnetization decay is calculated and checked against the data (see Fig. 1). We note here for these films there is a demagnetizing field and an exchange interaction whose role is to weaken the demagnetizing field. Weakened demagnetizing effects were observed in numerical simulations [8] and can be attributed to the exchange effect. The slope of M-H loop can provide an estimate of the decreased demagnetizing coefficient that is given by $D_{eff}^{-1} = dM / dH$. The estimated demagnetizing coefficients are also given in Table 1.

Table 1. [Perpendicular magnetic material parameters]

Samples	δ (nm)	V (estimated)' $10^{-18} cm^3$	H_c (kOe)	M_s (emu/cc)	H_k (kOe)	K_u(erg/ cm^3)	D_{eff} / 4_i	SQ
Multilayer CoPd	24	1.1	1.6	540	6.85	1840000	0.43	0.66
Multilayer CoCr/Pd	30.4	2.6	1.1	370	3.75	694000	0.2626	0.82

From Fig. 1 it is clear that both Co/Pd and CoCr/Pd films decay logarithmically, and the model reproduces the data for fitted volumes given in Table.1. Doubtless, the interaction causes logarithmic decay. The inset shows the magnetic viscosity S vs. temperature. A surprising feature of the magnetic viscosity at higher temperatures [1] is that the viscosity is expected to increase for any thermally activated process yet the viscosity goes to zero after reaching a maximum. At high temperatures the thermal energy available is so great that the system relaxes before the measurement can be made and thus the decay cannot be observed. Whereas at low temperatures, the available thermal energy is so small that it cannot cause any detectable decay.

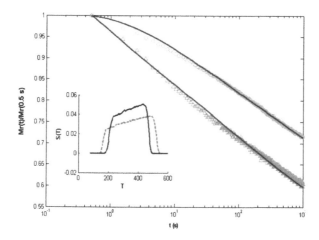

Figure 1. Magnetization decay for materials given in the Table 1. The top curve belongs to Co/Pd and the bottom curve belongs CoCr/Pd. The inset shows the magnetic viscosity (decay slopes) as a function of temperature. The larger slope belongs to CoCr/Pd.

Figure 2 shows the magnetic viscosity as a function of field for two sets of temperatures. The slope is obtained at $t = 100$ sec by taking the derivative of magnetization with respect to $\ln(t)$. For low saturation magnetization and anisotropy field the decay slope is higher and the width narrower. Clearly, the decay slope reaches zero when applied field is less than H_k. This is compatible with the effect of demagnetization since it reduces the energy barrier further compared to the barrier with applied field alone. After a certain field (less than H_k) the barrier reduces to zero, and thus the moments instantly flips to the other direction. The viscosities shown in the figures have an almost rectangular shape which is not always observed. This behavior is contrary to some experimental results where a maximum decay slope is observed at the coercive field. This disagreement may be due to the dispersion in the grain volume or the axes may have random orientations. So this simple model results in almost the same decay slope at fields between $\pm H_k$. The viscosity results shift toward the lower reduced field direction as the temperature is increased and clearly temperature does not affect the shape of the viscosity.

In Figure 3, the viscosity vs. $\ln t$ behavior is exhibited. The magnetization decay is obviously logarithmic only for couple of decades, and the viscosity has different time dependences for high and low fields. For positive fields, the magnetization does not start to decay immediately, showing nearly zero viscosity for some time; however, for negative fields, the magnetization rapidly falls to a certain value and then logarithmic decay starts.

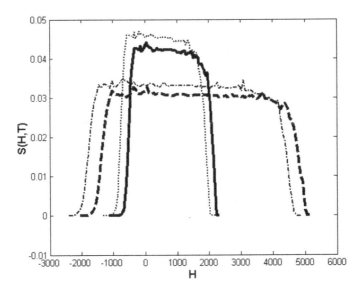

Figure 2. The magnetic viscosity curves. The field is measured in Oe. The narrow curve belongs to CoCr/Pd. Results marked dark are at room temperature the other at 350 Kelvin.

In Figure 3, the viscosity vs. $\ln t$ behavior is exhibited. Obviously, the magnetization decay is logarithmic only for couple of decades and the viscosity has different time dependences for high and low fields. For positive fields, the magnetization does not start to decay immediately, showing nearly zero viscosity for some time; however, for negative fields, the magnetization rapidly falls to a certain value and then logarithmic decay starts.

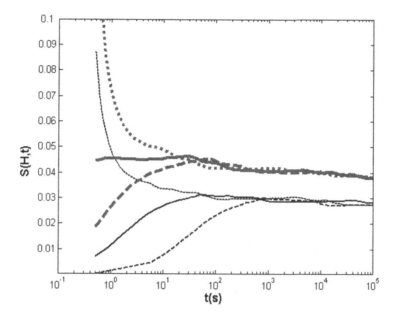

Figure 3. Magnetic viscosity as function of time and field. The lower plots belong CoPd for H/H_K= -0.0365 (dotted), 0 (solid) and 0.0365 (dashed).and the Upper CoCr/Pd H/H_K= -.0133 (dotted), 0 (solid) and 0.0133 (dashed).

For CoCr/Pd there is logarithmic decay starting at 10 seconds with almost the same slope for all fields given in the plot, but for Co/Pd it starts at close to 800 sec again with almost the same slopes. The slopes approach zero at long times.

4.   THE FLUCTUATING FIELD

The model reproduces the overall behavior but not the right shape for the magnetic viscosity for all fields, and it seems that other ingredients such as a distribution of volumes or anisotropies must be included. From the time-dependent irreversible magnetization curves one can measure the magnetic viscosity. Independently, magnetic field change also induces a change in the irreversible magnetization, from which one may obtain the irreversible susceptibility. It was argued [9] that the thermal fluctuations may be coveniently associated with a fictitious field, the so-called fluctuating field. This field can be obtained by independently measuring the magnetic viscosity and the irreversible susceptibility from which one obtains

$H_f = S / \chi_{irr}$. The fluctuating field along with other measurable quantities can be inferred by assuming a constitutive relation of the form $f(M,H,t) = 0$ and using thermodynamic relations [10]. The interesting feature of the fluctuating field is that it does not depend on the number of reversed magnetic grains, but rather is related to the activation mechanisms. Wohlfarth therefore argued for its fundamental importance, and it has attracted much attention [7,9,10] ever since. The small normalized magnetization as a function of field and time can be obtained with another approach [5]. If we use a similar approximation to that given in Ref. [5], assuming negligible back reversal, we obtain from Eq. (1)

$$x + 1 \approx (t_n / t)^{1/\alpha}, \tag{6}$$

where $t_n = e^{\beta E(H)} / f_0 \alpha$, $\alpha = 2 D M_s \beta K_u v (1 - H / H_K) / H_k = b \sqrt{E(H)}$. The energy barrier as function of applied field is given by $E(H) = K_u v (1 - H / H_k)^2$. $t_n(h)$ $(h = H / H_k)$ is a very large quantity then $t_n / t \gg 1$. The normalized viscosity and irreversible susceptibility can be obtained from Eq. (6). The viscosity is given by [10]

$$S = -\frac{dx}{d \ln t}\bigg|_h = \frac{M_s}{\alpha}(x + 1) , \tag{7}$$

and the irreversible susceptibility can be written

$$\chi = \frac{dx}{H_k dh}\bigg|_t = \frac{M_s (x+1)}{H_K} \left\{ -\frac{1}{\alpha^2} \frac{\partial \alpha}{\partial h} \ln \frac{t_n}{t} + \frac{1}{\alpha} \frac{\partial}{\partial h} \ln t_n \right\} \tag{8}$$

The fluctuating field is defined as [9,10]

$$H_f = \frac{S}{\chi} = \frac{(1 - h)kT}{E + kT \ln t + kT \ln b\sqrt{E} - kT} . \tag{9}$$

If time dependence is replaced with the critical energy barrier as argued in [2] (which is taken to be 25 kT), the fluctuation field reduces to

$$h_f = \frac{H_f}{H_k} = \frac{kT(1-h)}{E + E_c + kT \ln b\sqrt{E} - kT} , \tag{10}$$

where the critical energy barrier is given by [2, 9]

$$E_c = kT \ln(t f_0).$$

Eq. (10) can be compared with the Eq. (13) of ref. [2]

$$h_f = \frac{(1-h)}{2E_c}$$

for the same S-W model with a distribution of the barriers. Eq. (11) can also be determined from another more general thermodynamic relation [2,10]

$$H_f = -\frac{dH}{d \ln t}\bigg|_M .$$

The behavior of the fluctuating field obtained from both studies is displayed in Fig.4.

It is clear that for h close to 1, both results approach the same limit. The noninteracting result is linear with the reduced field and tends toward zero at $h = 1$; whereas interacting result starts with a value almost half the value of the noninteracting case and gradually decreases eventually approaching zero.

Another way of analysing magnetic decay data is connected to the activation volume associated with the magnetic reversal but is distinct from the Barkhausen volume or domain volume [9, 10]. This volume was shown to be related to the volume swept out by rotation during a reversal as it changed between minimum and maximum energy values which, through dimensional analyses, is formulated as

$$v_{act} = \frac{kT}{M_s H_f}$$

This concept [9, 10] was also generalised to other reversal mechanisms. For larger size particles the activation volume is less than the particle volume; whereas with nanoparticles both volumes are on the same order [9, 10].

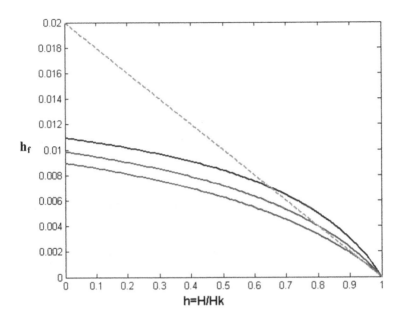

Figure 4. Fluctuating fields normalized to anisotropy field. The dashed line is from ref. [2]. The solid lines are the predictions of interacting model with different critical barriers top to bottom: 25, 35, 55 (in units of kT).

5.   SUMMARY AND DISCUSSION

Here we have investigated the mean-field interaction effects, which are described by a demagnetizing field. We have seen that quasi-logarithmic time decay is observed. We have also applied this model to perpendicular films with zero applied field and obtained relatively good results. When the system is in zero applied field the results agreed with experiment; however, the shape of the magnetic viscosity is not an expected result. There may be several causes for this disagreement, especially disorder in the systems resulting from inhomogeneities. This will inevitably lead to a distribution in time constants. Here we have assumed uniaxial anisotropy with magnetization and field perfectly aligned. In reality, particularly with longitudinal films, the axis may have random deviations from the field directions. This will cause considerable complexity in the problem. The grains may also reverse with other modes as well through incoherent rotation or with a collective mode. The thermal relaxation remains an interesting problem and further study is needed.

ACKNOWLEDGEMENTS

One of us (I.A) would like to thank to Dr. K. Ouchi, director of (AIT) *Akita Research Institute of Advanced Technology* for his kindness and invitation to the center. I would like to thank also Dr. N. Honda for introducing me in this interesting subject and helpful discussions. Throughout my stay in Akita the help from S. Takahashi and discussions with T. Suziki and L. Wu are greatly appreciated. The financial support by the JISTEC (Japan International Science & Technology Exchange Center) is gratefully acknowledged. We also thank Prof. D. L. Huber for critically reading the manuscript and helpful discussions.

REFERENCES

1. E. Dan Dahlberg, D. K. Lottis, R. M. White, M. Matson and E. Engle, 'Ubiquitous nonexponential decay: the effect of long-range coupling?, J. Appl. Phys. 76 (1994) 6396. (See references given here.)
2. R. W. Chantrell, J. D. Hannay, C. N. Coverdale, G. V. Roberts, and A. Lyberatos, 'The physics of the fluctuation field and activation volume', J. Magn. Soc. Japan, 23 (1999) 2058. (See references given here.)
3. D. K. Lottis, R. M. White and E. Dan Dahlberg, 'The magnetic aftereffect in CoCr films: a model', J. Appl. Phys. 67 (1990) 5187.
4. R. Wood, J. Manson, and T. Coughlin, 'On the thermal decay of magnetization in the presence of demagnetizing fields and a soft magnetic layer', JMMM, 193 (1999) 207.
5. N. D. Rizzo and T. J. Silva, 'Thermal relaxation in the strong-demagnetizing limit', IEEE Trans. Magn., 34 (1998) 1857.
6. N. Honda, K. Ouchi, and Shun-ichi Iwasaki, 'Design Consideration of Ultrahigh-Density Perpendicular Magnetic Recording Media', IEEE Trans. Magn., 38 (2002) 1615.
7. E. C. Stoner and E. P. Wohlfarth, 'A mechanism of magnetic hysteresis in heterogeneous alloys', Trans. Roy. Soc. (London) A240 (1948) 599.
8. I. Tagawa and Y Nakamura, 'Magnetic recording simulation considering mean field interaction', IEEE Trans. Magn. 29 (1993) 3981.
9. D C Crew P G McCormick and R Street 'The interpretation of magnetic viscosity', J. Phys. D: Appl. Phys. 29 (1996) 2313.
10. A. Lyberatos and R. W. Chanrell, 'The fluctuation field of ferromagnetic materials', J. Phys.: Cond. Matter 9 (1997) 2623.

MAGNETIZATION REVERSAL OF STRIPE ARRAYS

Katharina Theis-Bröhl[a], Till Schmitte[a], Andreas Westphalen[a], Vincent Leiner[a,b], Hartmut Zabel[a]

[a] *Institut für Experimentalphysik/Festkörperphysik, Ruhr-Universitaet, D 44780 Bochum, Germany*
[b] Institut Laue Langevin, Grenoble, France

Abstract: Magnetic stripes with different aspect ratios provide control over the remanent domain state, the coercivity, and over different types of reversal mechanisms. We discuss two methods for evaluating the magnetization vector during reversal: vector magneto-optical Kerr effect and polarized neutron reflectivity. Both methods are compared and critically discussed.

Key words: Magnetic stripe arrays, magnetization reversal, magneto-optical Kerr effect, polarized neutron reflectivity

INTRODUCTION

Advances in lithographic techniques for structuring magnetic materials and an increasing number of potential device applications have spurred the interest in artificially structured magnetic films. It is therefore of utmost importance to properly characterize the remanent state and the magnetization reversal mechanisms of these new systems using different experimental techniques. A number of powerful imaging tools are available for analyzing the magnetic domain state in real space, such as magneto-optical Kerr effect microscopy (MOKE-microscopy) [1], Lorentz-microscopy [2], scanning electron microscopy with polarization analysis (SEMPA) [3], polarized electron emission microscopy (PEEM) [4], magnetic force microscopy (MFM) [5], scanning tunneling microscopy (STM) [6], and magnetic x-ray microscopy [7]. Although these real space methods provide powerful images

B. Aktaş et al. (eds.), Nanostructured Magnetic Materials and their Applications, 327–343.

of different contrast and resolution, determining the magnetization vector, i.e. the angle and amplitude of the magnetization vector upon reversal in an external field is rarely possible. Two experimental techniques provide this information: vector-MOKE and polarized neutron reflectivity (PNR). Furthermore, by applying PNR to lateral structures it is possible to perform measurements at Bragg peaks generated by the artificial periodicity. Magnetic Bragg intensity does not take the statistical average but filters out correlation effects during reversal and is thus a powerful tool for analyzing domain interactions. In the following we will provide an introduction to vector-MOKE and PNR, followed by a discussion of the reversal mechanism of CoFe stripe arrays with the same periodicity but different width.

1. METHODS

In the following we describe the experimental methods of vector MOKE and of polarized neutron reflectivity (PNR). We start the discussion with simple examples for the reversal mechanism and distinguish between two main and ideal cases: reversal via coherent magnetization rotation and reversal via nucleation and domain formation. The sample preparation by lithographic means has been discussed at other places and shall not be repeated here [9].

1.1 **Vector magneto-optical Kerr effect**

Magnetic hysteresis measurements are usually performed with a MOKE setup in the longitudinal configuration. This implies the use of s-polarized light, a sample magnetization in the film plane, and a magnetic field applied in the scattering plane and parallel to the film plane. The resulting Kerr angle is then proportional to the component of the magnetization vector parallel to the field direction, $\theta_K^l \propto m_l$, where m_l is the longitudinal component of $\vec{M}$ projected parallel to $\vec{H}$. This kind of measurement can not distinguish between a magnetization reversal via domain rotation and/or via domain formation and wall motion. It is therefore advantageous to measure not only the component of the magnetization along the applied field, but also in the orthogonal direction in order to reconstruct the magnetization vector $\vec{M}$ from the measurement. The longitudinal MOKE can be used as a vector-magnetometer, if, in addition, the external magnetic field is applied perpendicular to the scattering plane and the sample is simultaneously rotated by $90°$ with respect to the scattering plane, keeping the rest of the setup constant. In this perpendicular configuration, MOKE detects the magnetization component parallel to the scattering plane and perpendicular to the magnetic field, $\theta_K^t \propto m_t$, as has been shown by [8].

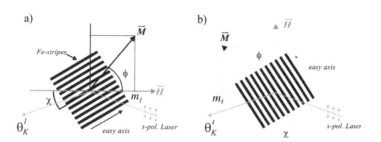

Figure 1. Definition of the sample rotation χ and the angle ϕ of the magnetization vector $\vec{M}$ for the case of the longitudinal setup (a) and the perpendicular setup (b). In order to measure the transverse magnetization component m_t the field and the sample are rotated by 90°, such that the angle χ is held constant, but the magnetization component m_t is in the scattering plane.

The geometry of the setup is sketched in Fig. 1. Both components, m_l and m_t, yield the vector sum for the average magnetization vector $\vec{M}$ sampled over the region, which is illuminated by the laser spot. This area is $\approx 1\,\mathrm{mm}^2$. The magnetization vector can be written as

$$\vec{M} = \begin{pmatrix} m_l \\ m_t \end{pmatrix} = |M| \begin{pmatrix} \cos\phi \\ \sin\phi \end{pmatrix}. \tag{1}$$

The proportionality constant between the Kerr angle θ_K and the two magnetization components is *a priori* unknown. However, there are good reason to assume that they are equal. In this case one can write:

$$\frac{m_l}{m_t} = \frac{\cos\phi}{\sin\phi} = \frac{\theta_K^l}{\theta_K^t}, \tag{2}$$

from which follows the rotation angle of the magnetization vector:

$$\phi = \arctan\left(\frac{\theta_K^t}{\theta_K^l}\right). \tag{3}$$

Furthermore one can express $|M|$, normalized to the saturation magnetization:

$$\frac{|M|}{|M|^{sat}} = \frac{\theta_K^l}{\theta_K^{l,sat}} \frac{1}{\cos\phi}.$$ (4)

Coherent rotation:

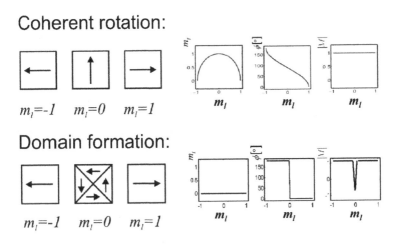

Domain formation:

Figure 2. Two limiting cases of magnetization reversal and the resulting vector-MOKE measurements. (a) shows the case of coherent rotation. The reversal is sketched and the transverse component, the angle and the magnitude of the magnetization are plotted as a function of the magnetization along the field. (b) depicts the case of domain formation and no rotation. The same quantities are plotted as in (a).

The results of vector-MOKE measurements provide important information which allow to distinguish between different magnetization reversals. Two limiting cases can easily be separated (see Fig. 2):

Coherent rotation (Fig. 2(a)) occurs when the length of the magnetization vector, $|\vec{M}|$, is constant during the reversal, while the magnetization vector $\vec{M}$ rotates from one direction into the other. Accordingly, the transverse component m_t, plotted as a function of the longitudinal component m_l, increases with decreasing longitudinal component, and exhibits a maximum at remanence.

Domain formation (Fig. 2(b)) is recognized by a magnetization vector, which remains aligned with the external field but changes its magnitude from negative to positive saturation and vice versa. In this case the transverse component stays zero for all field values. If no transverse component can be detected, no rotation of the magnetization takes place and the reversal is governed by domain processes.

1.2 **Polarized neutron reflectivity**

Neutron scattering from nanostructured magnetic arrays is a challenging task. Nevertheless it is worth pursuing this task because of the unique information neutron scattering can offer.

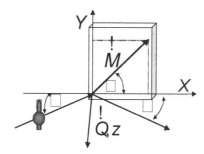

Figure 3. Scattering geometry for polarized neutron reflectivity studies. The Y-axis is the quantization axis for the neutrons and the non-spin-flip axis, the X-direction is the spin-flip axis. For specular reflectivity studies the scattering vector $\vec{Q}$ is parallel to the z-axis.

A few basic properties of polarized neutron reflectivity shall be recalled here for later use. For more details we refer to [10,11]. The experimental set-up is very similar to the MOKE set-up, as schematically shown in Fig. 3. The magnetization vector $\vec{M}$ may lie again in the sample plane and perpendicular to the scattering vector $\vec{Q}$. Furthermore, $\vec{M}$ shall make an angle θ against the x-axis. Next we assume that a monochromatic and polarized neutron beam is incident onto the sample at a scattering angle ϕ and that the magnetic moment of the incoming monochromatic neutron beam is aligned normal to the scattering plane (s-polarization) and parallel to the sample surface.

With polarized neutron reflectivity it is possible to measure independently the non spin-flip (NSF) reflectivities $R^{+,+}$, $R^{-,-}$ and the spin-flip (SF) reflectivities $R^{+,-}$, $R^{-,+}$. From the reflectivities, the nuclear and magnetic potential profile perpendicular to the sample plane can be retrieved:

$$V_{eff} = V_{nucl}(z) \pm V_{mag}(z), \tag{5}$$

where

$$V_{nucl}(z) = \frac{2\pi\hbar^2}{m} N_A b_{coh} \tag{6}$$

and

$$V_m(z) = -\mu \cdot \vec{B}_\parallel = \frac{2\pi\hbar^2}{m} N_A p_m. \tag{7}$$

Here μ is the neutron magnetic moment, $\vec{B}_\parallel = \mu_0(\vec{H} + \vec{M})$ is the effective magnetic induction in the sample plane, m is the neutron mass, N_A is the atomic number density, and b_{coh} is the coherent scattering length.

The difference of the specularly reflected NSF neutrons is proportional to the Y-component of the magnetic induction B_Y:

$$R^{+,+} - R^{-,-} \propto B_Y, \tag{8}$$

whereas the spin-flip reflectivities

$$R^{+,-} = R^{-,+} \tag{9}$$

are degenerate, and are proportional to the square of the X-component of the magnetic induction B_X:

$$R^{+,-}, R^{-,+} \propto B_X^2. \tag{10}$$

In the following we discuss three different reversal mechanisms and how they are expressed in neutron reflectivity measurements. Fig. 4 shows schematically in (a) the domain structure for nucleation and domain wall motion, in (b) for coherent rotation of the magnetization vector, and (c) reversal via domain formation. The neutron reflectivity is assumed to be taken at one particular value of the scattering vector, which is most sensitive to the magnetization direction, i.e. where the difference $R^{+,+} - R^{-,-}$ is largest. Starting from the "down" magnetization in saturation, the difference $R^{-,-} - R^{+,+} > 0$, but reverses sign at the coercive field. The cross over point with $R^{-,-} = R^{+,+}$ defines the coercive field H_c. For case (a), $R^{-,-} = R^{+,+} > 0$ and at the same time $R^{+,-} = R^{-,+} = 0$, as no x-component occurs during the magnetization reversal via nucleation and domain wall movement. In contrast, during reversal via coherent rotation the entire magnetization contributes to spin-flip scattering at the coercive field and both non-flip cross-sections drop to zero: $R^{-,-} = R^{+,+} = 0$. The third case of domain formation is in between the other two cases. At the coercive field

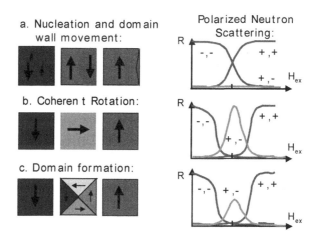

Figure 4. Sketch of three different possibilities for the magnetization reversal from negative saturation field to positive saturation field. This corresponds to only one branch of a magnetic hysteresis in an ascending field. Panel (a) shows schematically a reversal via domain nucleation and growth together with the respective specularly reflected intensity at a fixed scattering vector. In panel (b) a coherent rotation of the magnetization vector is assumed, and in panel (c) the reversal occurs via domain formation. For more details, see text.

spin-flip and non-flip scattering coexist. Reality in most cases is more complex. But the presented cases are good guidelines for the discussion of actual examples.

In comparison with vector-MOKE, PNR naturally provides a vector information of the magnetization via the simultaneously occurring NSF and SF reflectivities, without the need of resetting the sample. Furthermore, PNR can distinguish between case (a) nucleation and domain wall motion, and case (c) domain formation, as any transverse or x-component of the magnetization yields spin-flip scattering, while in MOKE experiments the transverse components compensate and do not give rise to a Kerr rotation.

In addition, PNR is depth sensitive and allows the analysis of the magnetization vector even for layers which are deeply buried under covers of other magnetic of non-magnetic materials. In contrast, MOKE analysis is limited to the near surface region within the skin depth of the particular material, which usually is on the order of 20-30 nm.

For laterally structured thin films with a defined periodicity, such as magnetic stripe arrays, additional reciprocal lattice streaks are generated, which occur in the x-y plane of the reciprocal space at regular spacing on the right and left side to the specular rod. Moreover, stripe arrays have, by design, an easy axis parallel to the stripe direction and a hard axis perpendicular to it. In order to study the reversal mechanism for different orientations of the field with respect to the stripe orientation, the sample needs to be rotated, while keeping the applied field perpendicular to the scattering plane and parallel to the polarization axis. Starting from a stripe arrangement perpendicular to the scattering plane (easy axis configuration), rotation of the stripe array implies an effective increase of the stripe separation with respect to the scattering plane, resulting in a shrinking separation between the reciprocal lattice rods and an increasing hard axis behavior. In Fig. 5 the scattering geometry is shown together with the definition of the angles, which are used later on for the discussion of examples. The easy axis corresponds to $\chi = 0°$ and the hard axis to $\chi = 90°$.

The analysis of the magnetization vector from the intensity of Bragg-reflection is different as compared to the analysis of the polarized reflectivity. This is due to the fact that Bragg reflections are the result of real space correlations, here of real space magnetic correlations between the stripes. Accordingly the magnetic form factor at the position of the first order Bragg peak is approximately proportional to:

$$F_m \propto \sqrt{R^+} - \sqrt{R^-}, \tag{11}$$

and the intensity of the Bragg peak follows from $I_{Bragg} \propto F_m^2$.

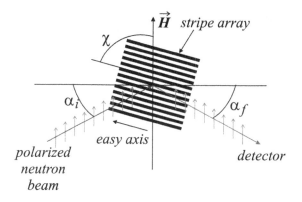

Figure 5. Sketch of the neutron scattering geometry. χ is the angle of the sample rotation with respect to the applied field (same definition as in Fig. 1). The magnetic field $\vec{H}$ is applied perpendicular to the scattering plane. α_i and α_f refer to the incident and exit angles of the neutrons to the sample surface.

2. EXPERIMENTAL EXAMPLES

In the following we discuss two examples for the analysis of the magnetization reversal of a stripe pattern using vector-MOKE and polarized neutron scattering. In both cases the stripes consist of a 20 x 10 mm^2 and 90 nm thick polycrystalline $Co_{0.7}Fe_{0.3}$ film grown by DC magnetron sputtering and patterned subsequently by lithographic means. Details of the sample preparation are provided in Ref. [9]. The films have no further intentionally induced anisotropy, such that their anisotropy is dominated by the shape anisotropy. The thin $Co_{0.7}Fe_{0.3}$ stripes have a width of 1.2 μm and a grating parameter of $d = 3\mu$m as can be seen from the AFM topograph reproduced in Fig. 6. Due to the high aspect ratio, the easy axis in this pattern is aligned parallel to the stripes and in remanence the stripes are in a single domain state. The second sample has the same grating parameter but twice the width of the thin stripes.

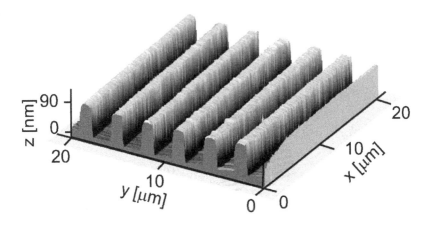

Figure 6. Surface topography of the array of Co $_{0.7}$ Fe $_{0.3}$ stripes obtained with an atomic force microscope shown in a 3-dimensional surface view. The displayed area is 20 x 20 μ m ^{2} .

2.1   Vector-MOKE results from thin FeCo-stripes

The left row of Fig. 7 shows four typical longitudinal MOKE hysteresis loops taken from the CoFe stripes with different in-plane angles χ. The hysteresis loop in (a) corresponds to an external magnetic field oriented parallel to the stripes. In this case, we find an almost square hysteresis loop, which represents the typical behavior of a sample when magnetically saturated parallel to the easy axis of the magnetization. The coercive field is $H_c = 140$ Oe. The coercive field increases to $H_c = 200$ Oe and $H_c = 320$ Oe for intermediate angles of rotation of $45°$ (c) and $63°$ (e), respectively. Fig. 7 (g) shows the corresponding hysteresis loop for the CoFe stripe array oriented perpendicular to both the external field and the plane of incidence. Here, a typical hard axis hysteresis loop is obtained. The saturation field measured with MOKE in the hard axis configuration exceeds 1000 Oe.

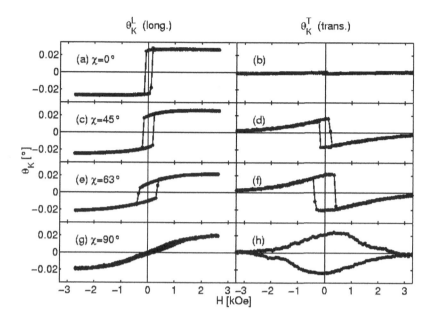

Figure 7. MOKE hysteresis loops measured in the longitudinal configuration (left figures) and in the transverse configuration (right figures). For both configurations in (a) and (b) the external magnetic field is oriented parallel to the stripes (easy axis configuration); in (c) and (d) the stripes are rotated by $45°$ and in (e) and (f) by $63°$ with respect to the direction of the external field; in (g) and (h) the magnetic field is oriented perpendicular to the stripes, (hard axis configuration).

The transverse component of the magnetization vector is reproduced in the right column of Fig. 7. The hysteresis loop reproduced in (b) was determined for a magnetic field parallel to the CoFe stripes, i.e. parallel to the easy axis. Ideally, within this configuration the measured Kerr rotation should remain zero unless components of the magnetization lie in the plane of incidence during the magnetization reversal process. As can be seen from Fig.7(b), the measured Kerr rotation is indeed almost zero, thus no rotation of the magnetization occurs in this configuration. Fig.7 (d) and (f) show the corresponding transverse component of the Kerr rotation measured for higher angles of rotation. Finally Fig. 7 (h) reproduces the hysteresis loop for a magnetic field parallel to the hard axis direction, i.e. in a direction perpendicular to the stripes and the plane of incidence. In this configuration the the transverse magnetization component always reflects a rotation of the magnetization away from the magnetic field.

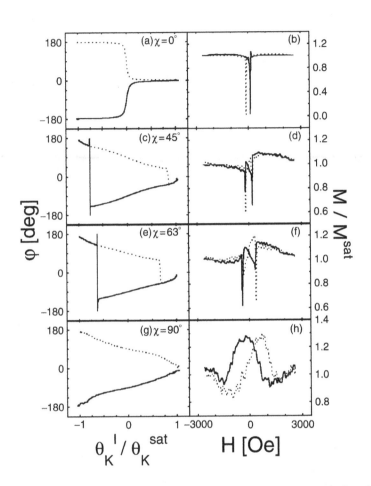

Figure 8. Plot of the angle of rotation (left panel) and magnitude of the magnetization vector (right panel) evaluated from the transverse and longitudinal components of the MOKE signal and for different sample rotation angles χ according to the measurements shown in Fig.7. Solid lines are for ascending magnetic fields and dotted lines for descending field.

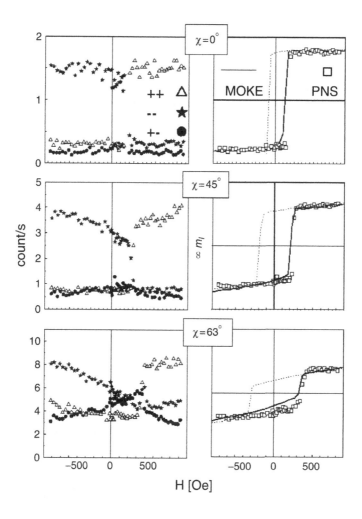

Figure 9. Magnetization reversal measurements performed at the first order Bragg peak for three different cross sections (left column) and calculated curves from the polarized neutron measurements (right column). The calculated curves are compared to longitudinal MOKE hysteresis loops, reproduced as solid and dotted lines. The top row depicts measurements at a sample rotation of $\chi = 0°$, the middle row measurements at a sample rotation of $\chi = 45°$ and the bottom row measurements at a sample rotation of $\chi = 63°$.

From the longitudinal and transverse components the magnetization vector can be reconstructed as discussed in the introduction. This is shown in Fig. 8, where the angle of the magnetization vector and the magnitude is plotted as a function of the external field for all four stripe orientations. The analysis confirms that for a magnetic field direction parallel to the stripes the reversal proceeds by nucleation and domain wall motion within a very narrow region around the coercive field, H_c (see Fig. 8(a,b)). For field directions $0 < \chi < 90°$ the reversal is dominated by a coherent rotation up to the coercive field, where some domains are formed (see Fig. 8(c-f)). For $\chi = 90°$ the reversal mechanism appears to be entirely dominated via a coherent rotation of the magnetization vector and no switching takes place. In the hard axis direction the magnetization vector describes a complete 360° rotation during the full magnetization cycle without any discontinuity see Fig. 8(g,h). For other directions a switching of the magnetization of 180° is observed at H_c, which can be viewed as a head-to-head domain wall movement through the stripe. These conclusions have been confirmed by imaging the domains within the stripe array via Kerr microscopy [9].

2.2   Polarized neutron scattering from thin FeCo-stripes

Magnetization reversal measurements have been carried out using polarized neutron scattering at small angles [9]. The experiments have been carried out using the polarized neutron reflectometer ADAM at the Institut Laue-Langevin, Grenoble, France [12]. In particular, the first order Bragg peak of the stripe pattern was scanned for different sample rotation angles χ and the magnetic contribution to the Bragg peak was determined from the eq.? for the magnetic form factor.

The left column of Fig. 9 shows hysteresis measurements performed at the first order Bragg peak via polarized neutron scattering. The top row in Fig. 9 for $\chi = 0°$ exhibits a large splitting of the $(+,+)$ and $(-,-)$ intensities, which reverses suddenly at the coercive field H_c. Almost no spin flip scattering occurs over the entire field range, indicating that the magnetization reversal for the easy axis configuration takes place in form of nucleation and fast domain wall movements at H_c. At $\chi = 45°$ the $(+,+)$ - $(-,-)$ splitting starts slopes down starting from field values far above and below H_c, while the spin flip scattering gently increases towards the coercive field. This behavior indicates a coherent rotation of the magnetization vector into the field direction. The situation is more pronounced for the sample rotation of $\chi = 63°$. The $(-,-)$ intensity continuously decreases with increasing field from negative to positive field values, while the $(+,+)$ intensity is more or less constant for most of the field values and changes suddenly at the coercive field. At the same time the spin flip intensity increases again towards the coercive field, however with a larger slope. All four cross sections are again consistent with a coherent rotation of the magnetization vector away from the easy axis.

A quantitative analysis can be provided by evaluating the magnetic form factor F_m, as discussed previously. By proper normalizing, the neutron hysteresis curves (open squares) are compared to the respective longitudinal MOKE hysteresis loops (solid and dotted lines) for the same sample rotation, as shown in the bottom graphs of Fig. 9. The overall agreement is rather good for all three sample rotation angles. Small deviations are only visible for $\chi = 63°$ close to the coercive field.

The neutron scattering results confirm the conclusions drawn from vector-MOKE experiments. In case that the field is applied parallel to the stripe axis (easy axis), a domain nucleation and domain wall movement occurs within a narrow field range at the coercive field. For all other sample orientations a coherent magnetization rotation with increasing field is observed with some domain nucleation occurring just around H_c. However, for a stripe orientation perpendicular to the applied field the domain rotation is complete without nucleation processes.

2.3 Magnetization reversal of thick CoFe-stripes

Similar experiments as for the thin CoFe-stripes were also carried at CoFe-stripes with twice the thickness (2.4 μm), while keeping the periodicity constant. Vector-MOKE experiments yield for the easy axis essentially the same results as for the thin stripes. However, with increasing angle between external field and stripe direction differences occur. At an angle of $\chi = 90°$ the transverse component m_t is not completely recovered, indicating that the hard axis reversal is characterized by domain formation instead of coherent rotation as observed for the thinner stripes.

Polarized neutron scattering experiments have been carried out in specular and in off-specular Bragg mode [13]. The specular PNR and vector-MOKE data are in good agreement, confirming that domain formation plays a more important role for the thick stripes as compared to the thinner stripes. For most of the χ values coherent rotation together with the formation of ripple domains close to coercivity controls the reversal process. The ripple domains give rise to diffuse scattering in the reciprocal space between the specular rod and the Bragg reflections.

Important difference is recognized between the magnetic hysteresis derived from the specular reflectivity as compared to the hysteresis which follows from the first Bragg peak. Accordingly the agreement between the Bragg derived hysteresis and the vector-MOKE data is less pronounced. This difference most likely is due to correlation effects. The magnetic Bragg peaks are very sensitive to correlation effects during the magnetization reversal. Dipole-dipole interactions, which likely plays an increased role for the thicker stripes due to the presence of stray fields, may enhance the coordinated domain reversal in adjacent stripes.

3.   DISCUSSION AND SUMMARY

We have described experimental methods to determine the vector components of the magnetization in stripe arrays and compared results obtained via vector-MOKE and polarized neutron scattering. From both methods the x- and y-components of $\vec{M}$ can be retrieved, which allows a complete analysis with respect to orientation and magnitude of the magnetization vector with respect to the applied field. From this analysis, in turn, the magnetization reversal can be characterized as being dominated by domain formation, by nucleation and domain wall motion, or by coherent rotation. Nevertheless, both methods, vector-MOKE and polarized neutron scattering, do not measure exactly the same properties. First, MOKE is sensitive to intraband transitions of the ferromagnetic film, while polarized neutron scattering at small angles senses the magnetic induction in the sample plane. In the presence of stray fields, this difference may become important. Effects of stray fields have been seen for the thick stripe array and a magnetization parallel to the hard axis. Then the magnetic field lines emanate from the sides of the stripes and fill up the gap from one stripe to the next. Polarized neutrons sense these stray fields and react by spin flip.

There is another important difference between MOKE and PNR. Vector-MOKE experiments, as presented here, are taken in a specular configuration, while the polarized neutron intensity has been evaluated at Bragg reflections. The difference is one of analyzing the average as compared to correlations. In the language of scattering physics this relates to the self correlation function in case of specular vector-MOKE measurements as compared to pair-correlation in case of polarized neutron scattering. Differences, which have been observed between MOKE and neutron results may be due to this effect. One could circumvent this difference by combining vector-MOKE with Bragg-MOKE, i.e. the MOKE effect at high orders of interference from the stripe pattern [14]. However, this type of experiment has not been performed up to now.

To summarize, for the analysis of the reversal mechanism of laterally patterned media, vector-MOKE is a very good, fast and easy to use method. However, polarized neutron scattering and reflectivity, although more elaborate and costly, offers a more complete analysis of various processes, as the magnetization vector in different layers can be analyzed independently, correlation effects can be filtered out, and magnetic coherence lengths can be determined.

ACKNOWLEDGMENTS

We are grateful to K. Rott and H. Brückel, Universität Bielefeld, for the sample preparation, and to J. McCord, IFW Dresden, for carrying out MOKE-microscopy. This work was supported by *SFB 491* of the Deutsche Forschungsgemeinschaft: "Magnetic Heterostructures: Structure and Electronic Transport".

REFERENCES

[1] A. Hubert and R. Schäfer, *Magnetic Domains: The Analysis of Magnetic Microstructures*, Springer, Heidelberg-Heidelberg-New York, 1998

[2] J.N. Chapman, A. Johnston, L.J. Heydermann, S. McVitie, W.A.P. Nicholson, and B. Bormans, *IEEE Trans. Mag.*, **30**, 4479 (1994).

[3] M. R. Scheinfein, J. Unguris, M. H. Kelley, D. T. Pierce, and R. J. Celotta, Rev. Sci. Instr. **61**, 2501 (1990).

[4] J. Stöhr, A. Scholl, T. J. Regan, S. Anders, J. Lьning, M. R. Scheinfein, H. A. Padmore, and R. L. White, Phys. Rev. Let. **83**, 1862 (1999).

[5] U. Memmert, P. Leinenbach, J. Lösch, and U. Hartmann, J. Magn. Magn. Mater. **190**, 124 (1998)

[6] A. Wachowiak, J. Wiebe, M. Bode, O. Pietzsch, M. Morgenstern, R. Wiesendanger. Science **298**, 577 (2002)

[7] P. Fischer, T. Eimuller, G. Schütz, G. Schmahl, P. Guttmann, and G. Bayreuther, *J. Mag. and Magn. Mat.*, **198**, 624 (1999).

[8] C. Daboo, R. J. Hicken, E. Gu, M. Gester, S. J. Gray, D. E. P. Eley, E. Ahmad, J. A. C. Bland, R. Poessl, J. N.Chapman, *Phys. Rev. B*, **51**, 15964 (1995).

[9] Theis-Bröhl K, Schmitte T, Leiner V, Zabel, H Rott K, Brückl H and McCord J 2003 Phys. Rev. B 67, 184415

[10] Majkrzak C F 1989 Physica B **156** & **157** 619

[11] H. Zabel and K. Theis-Bröhl J. Phys.: Condens. Matter 15, S505 (2003)

[12] Schreyer A, Siebrecht R, Englisch U, Pietsch U and Zabel H 1998 Physica B **248** 349

[13] Theis-Bröhl K, 2003, Adv. Sol. State Phys. 48, in print

[14] Till Schmitte, Kurt Westerholt, Hartmut Zabel J. Appl. Phys. 92 (8), 4524 (2002)

[15] Temst K Van-Bael and M J Fritzsche H 2001 Appl. Phys. Lett. **79** 991

[16] Temst K, Van-Bael M J, Moshchalkov V V, Bruynseraede Y, Fritzsche H and Jonckheere R 2002 Appl. Phys. A in print

FMR STUDIES OF Co NANOWIRE ARRAYS

O. Yalçın [1,2] , F. Yıldız [1], B.Z. Rameev [1,3], M.T. Tuominen [4], M. Bal [4], M. Özdemir [5], B. Aktaş [1]

[1] Gebze Institute of Technology, 41400 Gebze-Kocaeli, Turkey
[2] Gaziosmanpaşa University, Department of Physics, 60110 Tokat, Turkey
[3] Kazan Physical-Technical Institute, 420029 Kazan, Russia
[4] University of Massachusetts, Department of Physics, Amherst, MA 01003, USA
[5] Marmara University, Department of Physics, 81040, Göztepe - İstanbul, Turkey

Abstract: Magnetic properties of ultrahigh density Co nanowire arrays fabricated by a nanoporous template have been studied by ferromagnetic resonance (FMR) technique. Diblock copolymer P(S-b-MMA) composed of polystyrene and polymethylmethacrylate has been used to prepare the template for nanowire deposition. Two samples with the same wire length of 1 µm and different wire diameters of 24 and 12 nm have been grown by electrodeposition technique in the pores of the P(S-b-MMA) template. The fabricated nanowires were perpendicularly and hexagonally arrayed in the film plane. FMR signal has been studied as a function of orientation of the nanowire arrays with respect to the applied magnetic field. Very broad (a few kOe) FMR absorption lines have been observed. Linewidth, amplitude and resonance field of FMR spectra strongly depend on the orientation of the applied field. The magnetic parameters have been obtained from experimental data by means of a theoretical model. The 24 nm thick nanowire arrays show a uniaxial anisotropy with the hard axis direction perpendicular to the sample plane (parallel to the wire axes).

Key words: nanowire arrays; ferromagnetic resonance; dipolar interaction

1.   INTRODUCTION

The fabrication and study of ordered nanowire arrays have attracted a continuous interest in recent years. It is due to the fact that the nanowire arrays are expected to have broad applications in many areas of nanotechnology, such as magnetic storage media, magneto-electronic

B. Aktaş et al. (eds.), Nanostructured Magnetic Materials and their Applications, 345–356.

devices, thermoelectric nanowires, field emission sources, etc. Diverse preparation techniques have been applied to fabricate the nanowires, which are shadow masks, self-assembled structures, nanoimprint, radiation damage, laser interference lithography, electron beam lithography, focused ion beam milling, scanning probe lithography, X-ray lithography, and copolymer nanolithography. The successful fabrication procedures and characterization of nanowires by suitable parameters have been summarized by Martín [1], and Skomski [2]. The magnetic properties of nanowires have been investigated by using the magneto-optical Kerr effect, neutron diffraction, alternating gradient magnetometry, superconducting quantum interference device, Brillouin light scattering, scanning magnetoresistance, scanning Hall microscopy, torque magnetometry, vibrating sample magnetometry, magnetic force microscopy (MFM), spin- polarised scanning tunnelling microscopy and ferromagnetic resonance (FMR) techniques [1-10]. Magnetic behaviour of the Co and Ni electrodeposited nanowire arrays with diameters of 200 and 400 nm have been studied in Ref. [11]. The Ni nanowire arrays with diameter from 30 to 55 nm have been studied by magnetometry and MFM techniques [12,13]. Magnetic hysteresis loop of ultrahigh-packed Ni nanowire arrays with diameter 10 nm have been measures and modelled in Ref. [14]. Magnetization reversal process in Ni nanowire arrays has been studied as a function of temperature in the 150-350 K range [15]. Remanent magnetization states of Ni, Co, CoP and CoNi arrays with diameters of 57-180 nm have been modelled by micromagnetic calculations [16]. FMR experiments have been performed for some nanowire arrays with wire diameters ranging from 35nm to 500nm [17-20].

In this study we investigate by FMR technique two arrays of Co nanowires with diameters of 24 and 12 nm. The nanowires are 1μm in length, and distances between the axes of nearest- neighbour wires are 48 and 24 nm, respectively. The nanowires are perpendicularly ordered in hexagonal symmetry in lateral dimensions of polymethylmethacrylate (PMMA) film. The magnetic anisotropies and inter-wire dipole-dipole interactions of these nanowire arrays have been studied.

2.   EXPERIMENTAL TECHNIQUE

The ultrahigh-density nanowire arrays have been obtained by DC electro-deposition method using the copolymer template, as described in [21,22]. This nanotemplate technique utilizes diblock copolymer, P(S-b-MMA), composed of two chemically different (polystyrene and polymethylmethacrylate) compounds that form polymer chains joined by acovalent bond. The produced templates were used to electrodeposit the arrays of Co nanowires with high vertical aspect ratios and high packing density. Two arrays of Co nanowire arrays with diameters of $d = 24$ (12) nm and distances between the axes of nearest-neighbour wires $r = 48$ (24) nm have been fabricated. The nanowire length was (L) 1μm for the both arrays.

FMR measurements have been performed at room temperature using X-band (9.5 GHz) commercial Bruker EMX spectrometer. The anisotropic FMR signal has been observed and recorded at different angles for both in-plane geometry (IPG, the applied field is in the film plane) and out-of-plane geometry (OPG, the applied field is rotating from the film plane to the film normal). A goniometer has been employed to rotate the sample around the vertical axis. The sample geometry, relative orientation of the equilibrium magnetization M, the applied magnetic field H and experimental axis systems are shown in Fig 1. A specially designed computer programme has been used to model the FMR spectra of the Co nanowires.

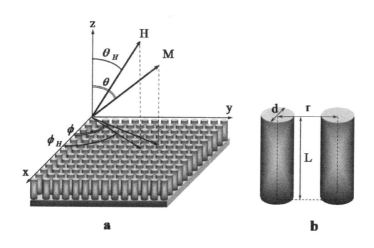

Figure 1. The geometry of FMR measurements. Orientations of equilibrium magnetization vector M and the applied magnetic field H with respect to the sample plane are shown. Definition of Co nanowire array parameters (d=12/24 nm, r=24/48 nm, L=1mm;) is also presented at the right part of the figure.

3.   THEORETICAL MODEL

The ferromagnetic resonance condition of nanowire film can be obtained from the total free energy,

$$E = -MH\left[\sin\theta\sin\theta_H\cos(\phi-\phi_H)+\cos\theta\cos\theta_H\right]+K_{eff}\sin^2\theta, (1)$$

where the first term represents the Zeeman energy of the film in the applied magnetic field (H) that applied at the spherical angles θ_H and ϕ_H, the second term is the effective anisotropy energy of the nanowire film, characterized by the effective uniaxial anisotropy parameter,

$$K_{eff} = \pi M^2(1-3P)+K_u, \tag{2}$$

where the first term is due to the magnetostatic energy of a perpendicularly-arrayed nanowire array [18,19,23] and constant, K_u, takes into account some additional second-order uniaxial anisotropy [17] with the symmetry axis along wire direction. Parameter P is the packing density (porosity) of wires, determined for a perfectly ordered nanowire array as

$$P = \frac{\pi}{2\sqrt{3}}\frac{d^2}{r^2}. \tag{3}$$

In Eq. (1) the spherical polar angles θ and ϕ for the saturation magnetization vector M are determined by the static equilibrium conditions as

$$dE/d\theta = MH_{eff}\sin 2\theta - MH(\cos\theta\sin\theta_H - \sin\theta\cos\theta_H) = 0, \quad \text{(4a)}$$
$$dE/d\varphi = MH\sin\theta\sin\theta_H\sin(\phi-\phi_H) = 0. \tag{4b}$$

Neglecting the damping term one can write the equation of motion for the magnetization vector M as

$$\frac{1}{\gamma}\frac{d\mathbf{M}}{dt} = \mathbf{M}\times\mathbf{H}_{eff}. \tag{5}$$

Here the γ is the gyromagnetic ratio and H_{eff} is the effective magnetic field that includes the applied magnetic field and the internal field due to the anisotropy energy.

The dynamic equation of motion for magnetization with the Bloch-Bloembergen damping term is given as

$$\frac{1}{\gamma}\frac{d\mathbf{M}}{dt} = \mathbf{M} \times \mathbf{H}_{eff} - \frac{\mathbf{M} - \boldsymbol{\delta}_{iz}M_0}{\mathbf{T}} \tag{6}$$

Here, $\mathbf{T} = (T_2, T_2, T_1)$ represents both transverse (for M_x and M_y components) and the longitudinal (for M_z component) relaxation times of the magnetization. That is, T_1 is the spin-lattice relaxation time, T_2 is the spin-spin relaxation time, and $\boldsymbol{\delta}_{iz} = (0,0,1)$ for (x,y,z) projections of the magnetization. In spherical coordinates the Bloch-Bloembergen equation is can be written as

$$\frac{1}{\gamma}\frac{d\mathbf{M}}{dt} = \frac{\mathbf{M}}{|M|} \times \nabla E - \frac{\mathbf{M}_{\theta,\phi}}{\gamma T_2} - \frac{\mathbf{M}_z - M_0}{\gamma T_1}, \tag{7}$$

where, the torque is obtained from the energy density through the expression

$$\nabla E = -\left(\frac{\partial E}{\partial \theta}\right)\hat{e}_\phi + \frac{1}{\sin\theta}\left(\frac{\partial E}{\partial \phi}\right)\hat{e}_\theta. \tag{8}$$

For a small deviation from the equilibrium orientation, the magnetization vector M can be approximated by

$$\mathbf{M} = M_s\hat{e}_r + m_\theta\hat{e}_\theta + m_\phi\hat{e}_\phi, \tag{9}$$

where the dynamic transverse components are assumed to be sufficiently small and can be given as

$$\begin{aligned} m_\theta(z,t) &= m_\theta^0 \exp i(\omega t \pm kz) \\ m_\phi(z,t) &= m_\phi^0 \exp i(\omega t \pm kz) \end{aligned}, \tag{10}$$

Using these solutions in Eqs. (7) and (8) one can derive the dispersion relation

$$\left(\frac{\omega_0}{\gamma}\right)^2 = \frac{1}{M^2\sin^2\theta}\left(E_{\theta\theta}E_{\phi\phi} - E_{\theta\phi}^2\right) + \left(\frac{1}{\gamma T_2}\right)^2. \tag{11}$$

The power absorption from *rf* field in a unit volume of a sample is given by

$$P = \frac{1}{2}\omega\chi_2 h_1^2, \tag{12}$$

where ω is the microwave frequency, h_1 is the amplitude of the magnetic field component and χ_2 is the imaginary part of the high-frequency susceptibility. The field-derivative ESR absorption spectrum is proportional to $d\chi_2 / dH$ and the magnetic susceptibility $\chi(=\chi_1 - i\chi_2)$ is given as

$$\chi = \frac{m_x}{h_x} = \left(\frac{m_\phi}{h_\phi}\right)_{\phi=0}. \tag{13}$$

The theoretical absorption curves are obtained by using the imaginary part of high frequency magnetic susceptibility as a function of applied field [24-27].

$$\chi = \frac{4\pi M_s \left(\dfrac{1}{M_s}\dfrac{\partial^2 E}{\partial \theta^2}\right)\left[\left(\dfrac{\omega_0}{\gamma}\right)^2 - \left(\dfrac{\omega}{\gamma}\right)^2 + \dfrac{2i\omega}{\gamma^2 T_2}\right]}{\left[\left(\dfrac{\omega_0}{\gamma}\right)^2 - \left(\dfrac{\omega}{\gamma}\right)^2\right]^2 + \left(\dfrac{2\omega}{\gamma^2 T_2}\right)^2}. \tag{14}$$

The dispersion relation can be derived by substituting Eq. (1) into Eq. (11) as

$$\left(\frac{\omega_0}{\gamma}\right)^2 = \left[H\cos(\theta - \theta_H) + H_{eff}\cos 2\theta\right]$$
$$\times \left[H\cos(\theta - \theta_H) + H_{eff}\cos^2\theta\right] + \left(\frac{1}{\gamma T_2}\right)^2, \tag{15}$$

where $H_{eff} = 2\pi M_s(1 - 3P) + \dfrac{2K_u}{M_s} \tag{16}$

is the effective anisotropy field derived from the anisotropy energy of nanowire array (Eqs. 1 and 2).

4.   RESULT AND DISCUSSION

The FMR spectra for various orientations of nanowire arrays with respect to the applied magnetic field are given in Figs. 2, 3 and 4. Fig. 2 and Fig. 3 present out-of-plane FMR spectra for the nanowire arrays with diameter of 24 nm and 12 nm, respectively. The in-plane FMR spectra for the 12 nm diameter nanowire array measured in the in-plane geometry are presented in Fig.4.

As it can be seen from the figures, the observed FMR signals are very broad and highly asymmetric with respect to the baseline. Broadening of FMR signal can be related to the dipolar fields, which are expected to be highly inhomogeneous. The dipolar coupling between the wires favours to the "antiferromagnetic-like" magnetic configuration at zero applied magnetic field, i.e. the nearest-neighbour wires are magnetized in antiparallel directions. As a result, the overall magnetization of a nanowire array is under-saturated at relatively low (~1 kOe) applied magnetic fields [6,28]. Therefore, the local magnetization and dipolar field strongly vary in the lateral dimension of the array (from the central axis of one wire to the next one) as well as along a wire axis. Thus, broadening of the FMR lines reflects a high dispersion of the local effective fields or, equivalently, of the FMR resonance fields, while asymmetric shape of FMR signals is entirely due to under-saturation effects in low magnetic fields.

In order to deduce magnetic parameters, the above theoretical model has been used to analyze the experimental data. Computer modeling of the FMR spectra for the 24 nm diameter nanowire array has been performed (dashed lines in Fig.2). As seen from the figure there is rather good agreements between the experimental and theoretical spectra, except the region of low magnetic fields where the experimental FMR absorption signal is distorted due to under-saturation effects. The angular dependencies of the theoretical and experimental spectra are in good agreement as well.

Using the packing density factor for the nanowire arrays (about 20%) and the values of the saturation magnetization (1400 G) and ω/γ (3150 Oe) appropriate for the bulk Co, the values of the uniaxial anisotropy constant, $K_u = -3.8\times10^6$ erg/cm^3 and the transverse relaxation time, $T_2 = 0.20\times10^{-8}$ s have been obtained for the 24 nm thick nanowire array. It is seen that the additional magnetic anisotropy (due to surface or magnetocrystalline anisotropy) appeared to be uniaxial with the hard axis direction along the wire axis (the easy axis is perpendicular to wire, i.e. lies in the sample plane). The deduced value of the anisotropy corresponds to the *effective anisotropy field* of about 2 kOe.

Since splitting of the FMR signal into two modes has been observed for the 12nm thick wire array, the simple model outlined above could not be directly applied for this sample. However, the effective anisotropy has been estimated to be much larger than the obtained value for the thicker wire.

Comparing the out-of-plane angular dependence of the FMR spectra for nanowire arrays with diameter of 12 nm (Fig. 3) and 24 nm (Fig. 2) one can note that there is a second mode for the both samples. For the 24 nm thick wire array this mode is observed as high-field ($\sim$ 5 kOe) shoulder of FMR signal for the orientation where the magnetic field applied in the sample plane. This mode is much more clearly seen in Fig. 3 for the thinner Co nanowires. We can attribute this mode to the surface spin-wave mode (SWM). It is in correspondence with the angular dependence of the mode, it is known that SWM depends on angle more sensitively than the main mode. We suppose that the surface mode originates from the surface anisotropy of individual wire, and the hard axis of the surface anisotropy energy is parallel to the wire axis (i.e. the easy axis is in radial direction to wire). Other probable reason is the additional "coupled" exchange-magnetostatic spin-wave modes, as it has been demonstrated in Ref. [28], where FMR measurements on a square array of submicron size circular permalloy disks have been performed. It has been shown that the intensity and resonance field of these modes depend on the direction of the applied magnetic field with respect to the square array. Resonance frequency of this mode depends on the dipolar interactions, which in turn reflect a symmetry of the array lattice. Therefore, the observed strong dependence of the second FMR mode on the sample orientation may be related to excitation of such "coupled" exchange-magnetostatic spin wave modes.

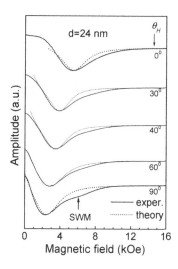

Figure 2. Out-of-plane FMR spectra of the 24 nm diameter Co nanowire array. The angles of the film plane with respect to the applied field are labeled. The experimental and modeling results are presented by solid and dashed lines, respectively.

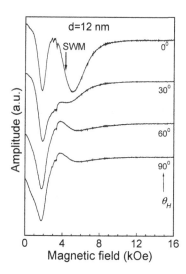

Figure 3. Out-of-plane FMR spectra of the 12 nm diameter Co nanowire arrays. The angles of the film plane with respect to the applied field are labeled.

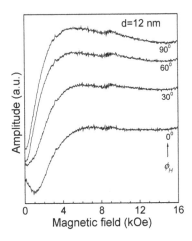

Figure 4. In-plane FMR spectra of the 12nm diameter
Co nanowire array. Azimuthal angles in the film plane
with respect to the applied field are labeled.

The in plane angular variations of the FMR spectra are given in Fig. 4 for
12 nm thick nanowire arrays. In this case magnetic field is rotated in the
sample plane, i.e. the field is kept always perpendicular to the wire axes.
Normally one does not expect any anisotropic behaviour for this angular
study. However, one can see from the Fig. 4, there is noticeable anisotropy
in the resonance fields. One of the possible reasons for in-plane anisotropy is
a tilt of the wire axes with respect to the film normal. This tilt can arise from
the tilt of the prepared polymer molecule due to a misalignment of the DC
electric field that was applied during the procedure of template preparation.
As result, the tilted demagnetizing field contribution to the axial anisotropy
in the film plane appears.

In conclusion, Co nanowire arrays, which were perpendicularly and
hexagonally arrayed in the film plane, have been grown by electrodeposition
technique in the pores of the P(S-B-MMA) template. Two samples with wire
diameters of 24 and 12nm have been studied by ferromagnetic resonance
(FMR) technique at various orientations of the nanowires with respect to the
applied magnetic field. Very broad and asymmetric FMR absorption lines,
which were strongly dependent on the orientation in the applied field, have
been observed. Computer modeling of FMR spectra from the 24nm thick
nanowire arrays have been performed and magnetic parameters have been
obtained. A uniaxial anisotropy with easy axis direction in the sample plane
(perpendicular to the wire axes) has been determined.

REFERENCES

1. J. I. Martín, J. Nogués, Kai Liu, J.L. Vicent, Ivan K. Schuller. J. Magn. Magn. Mater. 256 (2003) 449-501, and references therein.
2. R. Skomski. J. Phys.: Condens. Matter 15 (2003) R841-R896, and references therein.
3. J. Rivasa, A. Kazadi Mukenga Bantu, G. Zaragoza, M.C. Blanco, M.A. Lopez-Quintela. J. Magn. Magn. Mater. 249 (2002) 220.
4. A. Fert, L. Piraux. J. Magn. Magn. Mater. 200 (2000) 338.
5. P. M. Paulus, F. Luis, M. Kröll, G. Schmid, L. J. de Jongh. J. Magn. Magn. Mater. 224 (2001) 180.
6. L. Piraux, S. Dubois, E. Ferain, R. Legras, K. Ounadjela, J. M. George, J. L. Maurice, A. Fert. J. Magn. Magn. Mater. 165 (1997) 352.
7. R. Ferrè, K. Ounadjela, J. M. George, L. Piraux, S. Dubois. Phys. Rev. B 56 (1997) 14066.
8. Y. Jaccard, Ph. Guittienne, D. Kelly, J-E Wegrove, J-Ph. Ansermet. Phys. Rev. B 62 (2000) 1141.
9. J. F. Cochran, J. M. Rudd, M. From, B. Heinrich, B. Bennet, W. Scwarzacher, W. F. Egelhoff. Phys. Rev. B. 45 (1992) 4676.
10. O. Yalçın, F. Yıldız, M. Özdemir, B. Aktaş, M.T. Tuominen. Abst. International Conference on Magnetism (ICM 2003), Rome, Italy, July 27- August 1, 2003, p. 408.
11. J. Rivasa, A. Kazadi Mukenga Bantu, G. Zaragoza, M.C. Blanco, M.A. López-Quintela. J. Magn. Magn. Mater. 249 (2002) 220–227.
12. K. Nielsch, R. B. Wehrspohn, J. Barthel, J. Kirschner, U. Gösele, S. F. Fischer, H. Kronmüller. Appl. Phys. Lett. 79 (2001) 1360.
13. K. Nielsch, R. B. Wehrspohn, J. Barthel, J. Kirschner, S. F. Fischer, H. Kronmüller, T. Schweinböck, D. Weiss, U. Gösele. J. Magn. Magn. Mater. 249 (2002) 234–240.
14. M. Zheng, R. Skomski, Y. Liu, D. J. Sellmyer. J. Phys.: Condens. Matter 12 (2000) L497-L503.
15. A. Pérez-Junquera, J. I. Martín, M. Vélez, J. M. Alameda, J. L. Vicent. Nanotechnology 14 (2003) 294- 298, and references therein.
16. C. A. Ross, M. Hwang, M. Shima, J. Y. Cheng, M. Farhoud, T. A. Savas, Henry I. Smith, W. Schwarzacher, F. M. Ross, M. Redjdal, F. B. Humphrey. Phys. Rev. B 65 (2002) 144417-1.
17. U. Ebels, J.-L. Duvail, P. E. Wigen, L. Piraux, L. D. Buda, K. Ounadjela. Phys. Rev. B 64 (2001) 144421.
18. M. Demand, A. Encinas-Oropesa, S. Kenane, U. Ebels, I. Huynen, L. Pirax. J. Magn. Magn. Mater. 249 (2002) 228.
19. A. Encinas-Oropesa, M. Demand, L. Piraux, I. Huynen, U. Ebels. Phys. Rev. B 63 (2001) 104415.
20. A. Encinas-Oropesa, M. Demand, L. Piraux, U. Ebels, I. Huynen. J.Appl.Phys.89 (2001) 6704.
21. T. Thurn-Albrecht, J. Schotter, G. A. Kästle, N. Emley, T. Shibauchi, L. Krusin-Elbaum, K. Guarini, C. T. Black. M. T. Tuominen, T. P. Russell. Science, 290 (2000) 2126.
22. A. Ursache, M. Bal, J. T. Goldbach, R. L. Sandsrom, C. T. Black, T. P. Russell, M. T. Tuominen. Mat. Res. Soc. Symp.Proc. Vol. 721.
23. J. Dubowik. Phys. Rev. B 54 (1996) 1088.
24. B. Aktaş, M. Özdemir. Physica B 119 (1994) 125.
25. B. Aktaş. Solid State Commun. 87 (1993) 1067.
26. L. J. Maksymowicz, D. Sendorek. J. Magn. Magn. Mater. 37 (1983) 177.
27. P.E. Wigen. Thin Solid Films 114 (1984) 135.

28. S. Jung, B. Watkins, L. DeLong, J. B. Ketterson, V. Chandrasekhar. Phys. Rev. B 66 (2002) 132401.

FINITE-TEMPERATURE MICROMAGNETICS OF HYSTERESIS FOR MISALIGNED SINGLE IRON NANOPILLARS

M. A. Novotny [1,2], S. M. Stinnett [1,2], G. Brown [3,4], P. A. Rikvold [4,5]

[1]*ERC Center for Computational Sciences, Mississippi State Univ., Mississippi State, MS, USA*
[2]*Department of Physics and Astronomy, Mississippi State Univ., Mississippi State, MS, USA*
[3]*Center for Computational Sciences, Oak Ridge National Lab, Oak Ridge, TN, USA*
[4]*School for Computational Science and Information Technology, Florida State University, Tallahassee, FL, USA*
[5]*Center for Materials Research and Technology and Department of Physics, Florida State University, Tallahassee, FL, USA*

Abstract: We present micromagnetic results for the hysteresis of a single magnetic nanopillar that is misaligned with respect to the applied magnetic field. We provide results for both a one-dimensional stack of magnetic rotors and of full micromagnetic simulations. The results are compared with the Stoner-Wohlfarth model.

Keywords: hysteresis, micromagnetics, nanomagnets

1.   INTRODUCTION

Although hysteresis in single-domain nanomagnets has been known for many decades, there is currently much interest in looking anew at this phenomenon. This is partly driven by recent experiments on single-domain nanomagnets, in which the hysteresis and magnetization reversal of a single single-domain nanomagnets can be measured. It is also driven by the applications of nanomagnets, in particular to magnetic recording. Finally, one is no longer confined to pencil-and-paper calculations to understand

B. Aktaş et al. (eds.), Nanostructured Magnetic Materials and their Applications, 357–363.

physical phenomena. Available computer resources allow calculations which are much more complicated. Here we present large-scale computer simulations of hysteresis for two different model systems of single-domain nanoscale Fe pillars. We focus on the hysteresis when the long axis of the pillars is misaligned with the applied magnetic field.

2.   METHODS AND MODELS

2.1   Coherent Rotation

Given a single-domain particle with uniaxial anisotropy, it is possible to find the quasi-static equilibrium position of the magnetization when a magnetic field is applied at some angle to the easy axis. It is assumed that the magnetization can be represented by a single vector, M, with constant amplitude, M_S. The energy density is then

$$E = K \sin^2 \theta - M_S H \cos(\phi - \theta), \tag{1}$$

where K is the uniaxial anisotropy constant, H is the magnetic field applied at an angle ϕ to the easy axis and θ is the angle the magnetization makes $M(r_i)$ with the easy axis. Stoner and Wohlfarth showed that the critical transition curve for the coherent reversal of the magnetization is given by, [1]

$$h_{AX}^{2/3} + h_{AY}^{2/3} = 1 \tag{2}$$

where h_{AX} and h_{AY} are the components of the magnetic field along the easy and hard axes respectively. Equation (2) is the well-known equation of a hypocycloid of four cusps, also known as an astroid.

2.2   Micromagnetics

For systems where the spins are not aligned and/or where the field is changing too rapidly for the magnetization to reach its quasi-static equilibrium position, it is usually necessary to use micromagnetics to determine the reversal process. The basic approach is to discretize the system into a series of coarse-grained magnetization vectors,, where r_i is the position of the i-th magnetization vector. Each spin is assumed to have uniform magnetization, M_S, corresponding to the saturation magnetization of the bulk material, a valid assumption for temperatures well below the Curie temperature [2]. The time evolution of each spin is given by the Landau-Lifshitz-Gilbert (LLG) equation [3,4],

$$\frac{d\mathbf{M}(\mathbf{r_i})}{dt} = \frac{\gamma_0}{1+\alpha^2}\left(\mathbf{M}(\mathbf{r_i})\times\left[\mathbf{H_T}(\mathbf{r_i})-\frac{\alpha}{M_S}(\mathbf{M}(\mathbf{r_i})\times\mathbf{H_T}(\mathbf{r_i}))\right]\right) \quad (5)$$

where $\mathbf{H_T}(\mathbf{r_i})$ is the total local field at the i-th position, γ_0 is the gyromagnetic ratio (1.76×10^7 rad/Oe-s), and α is a dimensionless phenomenological damping term which determines the rate of energy dissipation. The first term represents the precession of each spin around the local field, while the second term is a dissipative term that drives the motion of the magnetization towards equilibrium. For the sign of the undamped precession term, we follow the convention of Brown [3].

The total local field, $\mathbf{H_T}(\mathbf{r_i})$, may include contributions from the applied field (Zeeman term), the crystalline anisotropy (set to zero in our model), the dipole field, and exchange interactions. At nonzero temperatures, thermal fluctuations also contribute a term to the local field in the form of a stochastic field which is assumed to fluctuate independently for each spin. The fluctuations are assumed Gaussian, with zero mean and (co)variance given by the fluctuation-dissipation theorem [5].

While the stochastic thermal field requires careful treatment of the numerical integration in time, the most computationally intensive part of the calculation involves the dipole term. For systems with more than a few hundred spins, it is necessary to use a more advanced algorithm. We use the Fast Multipole Method (FMM), the implementation of which is discussed elsewhere [5].

In this paper, we examine two model systems. The first is a nanopillar with dimensions of 5.2 nm x 5.2 nm x 88.4 nm. The cross-sectional dimensions are small enough (about 2 exchange lengths) that the assumption is made that the only significant inhomogeneities in the magnetization occur along the long axis [6] (z-direction). The particles in this model, discussed previously [5], are therefore discretized into a linear chain of 17 spins along the long axis of the pillar.

The second model system consists of a single nanopillar with dimensions 9 nm x 9 nm x 150 nm. The dimensions were chosen to correspond to arrays of Fe nanopillars fabricated by Wirth, *et al.*[7]. In this model, the system is discretized into 4949 sites (7 x 7 x 101) on the computational lattice.

Material properties in both systems were chosen to correspond to bulk Fe. The saturation magnetization is 1700 emu/cm^3 and the exchange length (the length over which the magnetization can change appreciably) is 3.6 nm. We take $\alpha = 0.1$ to represent the underdamped behavior usually assumed to be present in nanoscale magnets.

3. RESULTS AND ANALYSIS

In Fig. 1 we present hysteresis loops at $T = 100$ K for the second model for the field at $0°$ and $45°$ to the long axis of the particle. The loops were calculated using a sinusoidal field of period 15 ns which started at the maximum value of 5000 Oe. In all loops shown here, the reported magnetization is the component along the long axis of the particle. At $45°$, the magnetization vector is initially pulled away from the easy axis by the large magnetic field. As the field is reduced to zero, the magnetization relaxes towards the easy axis, reaching essentially saturation at zero field.

Figure 2 shows the *z*-component of the magnetization in these pillars for various times during the switching process with the field at $45°$ to the long axis of the pillar, under the same conditions as Fig. 1. Note that the magnetic end-caps are the sources of the finite-temperature nucleation that leads to the reversal of the hysteresis. Furthermore, note that these particles do not have a uniform magnetization vector, even though they are single-domain particles.

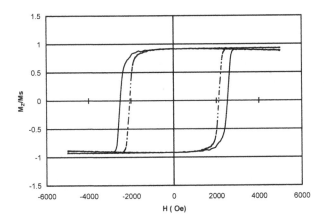

Figure 1. Hysterisis Loops at T= 100 K for the field aligned (dashed) and misaligned at $45°$ (solid). The simulation was performed over one half of the loop and the results reflected to generate the entire loop.

Figure 3a presents hysteresis loops for the first model for the field misaligned at various angles. The applied field is again sinusoidal with a period of 200 ns. Note that when the applied field is perpendicular to the long axis of the pillar, the hysteresis loop has a bubble shape qualitatively consistent with both the Stoner-Wohlfarth (SW) model (Fig. 3b) and the experimental results of Wirth *et al.*[7]. Quantitatively, however, the field at which nonzero magnetization appears is significantly lower than both what is expected from SW and from the experimental value.

Figure 2. Snapshots of the z-component of the magnetization at (from left to right) 0.00 ns, 4.875 ns, 5.000 ns, 5.075 ns, and 5.100 ns. One quarter of the pillar has been removed from the illustration to show the behavior of M_Z inside the pillar.

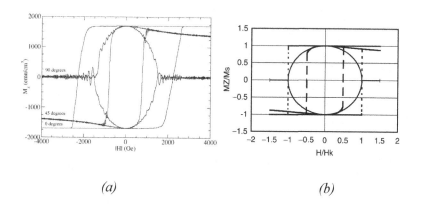

       (a)   *(b)*

Figure 3. Hysteresis Loops for 0°, 45°, and 90° misalignment of the field with the long axis of (a) small pillars and (b) the SW model.

4.   DISCUSSION AND CONCLUSIONS

In this brief paper, we presented finite-temperature micromagnetic simulations of hysteresis for a single Fe nanopillar. Two different micromagnetic models were simulated, both of which used fast periods of the sinusoidal applied field. The results were compared with the quasi-static, zero-temperature predictions of the Stoner-Wohlfarth model. We observed that when the field is aligned with the long axis of the particle, all models have qualitatively similar hysteresis loops, although the micromagnetic loops for the aligned case are more rounded than that of the SW model. When the applied field is at 45° to the long axis of the particle, the SW and small pillar models give switching fields which are lower than when the field is aligned, while the large pillar model gives a switching field which is larger than the aligned case, more consistent with experimental results [7]. When the field is perpendicular the long axis, the hysteresis loop has a bubble shape. However, the SW model gives a switching field which is equal to the aligned case, while the small and large pillar models give switching fields that are smaller and larger than the aligned case, respectively.

These results demonstrate that although simple models of magnetization switching in magnetic nanopillars can yield qualitatively correct results, to obtain quantitatively correct results, large-scale micromagnetic calculations are required. Only such detailed calculations will enable quantitative comparisons with finite-temperature, fast-period hysteresis loops.

ACKNOWLEDGEMENTS

The authors gratefully acknowledge support from NSF grant DMR0120310.

REFERENCES

1. E. C. Stoner and E. P. Wohlfarth, *Phil. Trans. Roy. Soc.* **A240**, 599 (1948).
2. D. A. Garanin, Phys. Rev. B **55**, 3050 (1997).
3. W. F. Brown, *Micromagnetics*, (Wiley, 1963).

4. U. Nowak, Annual Reviews of Computational Physics IX , Ed. by D. Stauffer, World Scientific, Singapore (2001), p. 105.
5. G. Brown, M. A. Novotny, and P. A. Rikvold, Phys. Rev. B **64**, 134432 (2001).
6. E. D. Boerner and H. N. Bertram, IEEE Trans. Magn. **33**, 3052 (1997).
7. S. Wirth, M. Field, D. D. Awschalom, and S. von Molnár, Phys. Rev. B **57**, R14028 (1998).

PART 7: MAGNETIC NANOCONTACTS

MAGNETORESISTANCE AND MAGNETOSTRICTION IN MAGNETIC CONTACTS

N. García, H. Wang, H. Cheng, C. Guerrero, N. D. Nikolic and
A. C. Papageorgopoulos
Laboratorio de Física de Sistemas Pequeños y Nanotecnología
Consejo Superior de Investigaciones Científicas
Serrano 144, Madrid 28006, Spain,
nicolas.garcia@fsp.csic.es

Abstract: In this paper we describe the magnetoresitive behaviour of ballistic magnetic contacts electrodeposited in nonmagnetic electrodes. We show large magnteoresistive values, up to 400% at room temperature at low applied magnetic fields (25Oe), with many repetitive magnetoresistive loops that could be used for future large integration (Terabits/inch^2) in magnetic recording processes. In addition, we analyse the variation of the resistance of the contact under an applied magnetic field from the point of view of possible magnetostrictive effects that may modify the contact. Moreover, it is shown that the magnetic field applied to the systems during electrodeposition can change the morphology and structure of the deposits in a subtle way. In particular, the magnetoresistance of the magnetic deposits gives rise to distinct arboreous-dendritic morphologies. This phenomenon parallels many processes found in nature.

Key words: Ballistic Magnetoresistance, Electrodeposition, Magnetostriction

1.   INTRODUCTION

The information presented in this article is a synthesis of studies in ballistic magnetoresistance, electrodeposition and local probe magnetostriction measurements, combined essentially into one complete experimental effort to better understand the magnetoresistive behaviour of nanocontacts. Each area shall be expounded upon distinctly to prevent confusion in the introduction as well as in the main text. In order to

B. Aktaş et al. (eds.), Nanostructured Magnetic Materials and their Applications, 367–381.
© 2004 *Kluwer Academic Publishers. Printed in the Netherlands.*

understand this experimental effort, and the results gained by it, some background is necessary in each of these three areas of study.

Regarding the phenomenon of ballistic magnetoresistance, as electronic devices shrink in size the wave nature of electrons becomes increasingly important, while this results in many interesting quantum phenomena. One of these, and an important example itself, is the Giant Magnetoresistance (GMR) effect discovered in 1988, in which the height limit (Z-direction) of a multilayered sample is in the quantum regime, while the dimensions of the X-Y plane is far beyond the region of quantum characteristic size. The present theoretical model mostly focuses on the transport mechanism through a coupled spin-chain. The question that concerns us here is the following: What kind of spin-related transport mechanism could we expect if the X-Y plane dimensions of a specimen as well as the Z dimension shrink to quantum characteristic size, and a small contact is formed? The need to answer this question prompted our original proposal to investigate the transport properties of nanocontacts. Thanks to the cooperation with other groups, we have established some basic rules for the transport mechanism of nanocontacts in theory, which is the so-called Ballistic Transport mechanism. The related spin-dependent Ballistic Transport phenomenon is defined as Ballistic Magnetoresistance (BMR). In experiments, we have discovered not only large values of BMR, but also significantly increased the stability of BMR. In our initial experiments, we had mainly fabricated ferromagnetic nanocontacts (Ni, Fe) on ferromagnetic electrodes (Ni, Fe, NiFe *etc.*). In this configuration, however, the contribution of macro-electrodes on magnetic transport properties could not be simply neglected. The investigation described in this contribution was extensively carried out in the last months, and shows a small contribution from macro-electrodes. Another approach is using nonmagnetic electrodes instead of ferromagnetic electrodes. A large and repeatable BMR effect can be also observed in this system. It is clear that BMR originates only from the local properties of nanocontact itself. In order to understand the contribution of macro-electrodes it is necdessary to study in more detail the electrochemical processes of how these macro-electrodes are fabricated. Hence, the field of electrochemistry and the experimental methods of electrodeposition have come to play a large role in our BMR research.

As mentioned, therefore, the nanocontacts used in this study (Ni, Fe and NiFe) were obtained by electrochemical deposition from corresponding acid sulfate solutions. The electrodepositions were performed on both nonmagnetic (Cu) and magnetic (NiFe and Ni) electrodes. The advantage of the fabrication of nanocontacts by electrochemical deposition, compared with other methods of depositing metals, is that the former electrochemical method is the easiest way to produce less coarse, more compact and more uniform and evenly distributed deposits *at the nanocontact zone.* In addition, it is known that morphologies of electrochemically obtained metal deposits can be strongly changed if electrodeposition is performed in the presence of a magnetic field. Changes in morphology of metal deposits are ascribed to the Lorentz force [1]. During the electrolysis, this force acts on the migration of ions and induces convective flows of the electrolyte close to the electrode surface. This effect on the electrodepositing process is known as the *magnetohydrodynamic* (MHD) effect.

The largest effect of this force, and consequently, the largest effect on convective mass transport of the electrolyte, can be realized when the magnetic field *B* is applied parallel to the electrode surface (*i.e.* an external magnetic field is oriented perpendicular to the direction of the ion flux). On the other hand, when the magnetic field is applied perpendicular to the electrode surface (*i.e.* the applied magnetic field is parallel to the direction of the ion flux), no drastic changes on the growth are expected *a priori,* except through the effects associated with gradients and the gravity-induced convection [1].

Finally, the study and measurement of magnetically induced strains on nanocontacts composed of magnetic materials either through the macro-electrode substance or deposit, was imperative in order to solve problems posed by the alleged magnetostrictive motion of the specimens used for ballistic magnetoresistance measurements. This necessity led us to devise a novel method for measuring these strains on the nanometer scale directly using an atomic force microscope. The method itself, and the results obtained are described in detail further in the text, whereas, here a brief description of the phenomenon is appropriate.

In essence, magnetostriction is the phenomenon whereby the shape of a ferromagnetic specimen changes during the process of magnetization. The deformation of the specimen, $\Delta l/l$ induced by its exposure to the magnetic field is usually in the range 10^{-5} to 10^{-6}. [2]. The degree of magnetostrictive strain is dependent on the field strength *H*, and increases with the latter to a saturation value l. This effect is caused by the shape change of the crystal lattice in each domain in the direction of magnetization, while this is accompanied by a subsequent rotation of the axis of spontaneous strain. The inevitable outcome is that the shape of the whole specimen changes. In terms of the magnetostrictive effect, materials can be divided in those with positive magnetostriction, where the material expands when magnetized, and those with negative magnetostriction, where it contracts. Not only does the magnetic field induce a strain on the materials in mention, but applied unidirectional stress tends to change the materials' magnetization. Thus, materials exhibiting positive magnetostriction have their magnetization increased with the application of stress, while those exhibiting negative magnetostriction show a decrease in their magnetization [3]. Different methods are used to measure the strain caused by the application of a magnetic field, the most popular of which are the strain gauge and capacitance methods [2]. Never before scanning probe microscope technology been utilized to measure and directly observe the lateral magnetostriction of a wire or configuration of wires under an applied magnetic field. This technique proved to be an ideal way to measure the movements or lack thereof, of specimens used in ballistic magnetoresistance measurements and allowed for the possibility for the observation and measurement of magnetostrictive strains at the same time that ballistic magnetoresistance measurements take place. Given the contribution of this method to the deepening understanding of the ballistic magnetoresistance phenomena and its greater applications, we thought it appropriate to describe our results involving the magnetoresistance of contact-configured wires in greater detail in the text.

2.   BALLISTIC MAGNETORESISTANCE IN ATOMIC AND NANOMETER SIZE ELECTRODEPOSITED CONTACTS

Ballistic magnetoresitance (BMR) in point atomic contacts with resistances (R) larger than say 1000Ω exhibits very large values, up to 500% [4-6], and the general trend is that BMR increases when R decreases. Ballistic magnetoresistance is the resistance that the electric current manifests through a point contact when a magnetic field is applied. The reason for this phenomenon is explained in terms of domain wall (DW) scattering in magnetic materials [7, 8] assuming that the DW width formed at the contact is of the order of the contact size. If the DW is narrow, strong non-adiabatic scattering with spin conservation at both sides of the contact is predicted by an old theory of Cabrera and Falicov [8], and applied to atomic contacts by Tatara *et al* [7]. The BMR can be expressed by

$$BMR(\%)=\pi^2/4(\zeta^2/1-\zeta^2)F(\xi). \tag{1}$$

In the above equation, the factor F is a dynamic term that reduces BMR according to the DW width and shape, and describes the adiabaticity in the electron spin transfer from one side to the other of the contact. The other term in Eq. 1 is the ratio of the difference in the density of states between majority and minority spins at the Fermi level in the electrodes at both sides of the contact. This is basically the Julliere formula [9] for two electrodes made of the same metal. The resistance is spin-ballistic if the contact is small and the spin mean free path (l) is larger than the contact size.

The resistance can be estimated assuming that one atom at the contact provides a conducting channel, and that that one atom takes up a surface area of approximately 0.1 nm^2 by

$$R(\Omega)=12900\Omega/(10a^2(nm^2)) \tag{2}$$

where 12900 Ω is the quantum of resistance, and a is the size of the contact in nm. The above formula is basically the Sharvin´s formula for ballistic transport [10]. The typical values of l for the spin mean free path are 30-50nm. It can, therefore, be observed from the above formula that the limits for ballistic transport are values of R larger than 1 Ω. For much smaller values of R the classical Maxwell formula controls the resistance that becomes non-ballistic. Fig. 1 shows BMR values for Ni, Fe, and Co as a function of the conductance expressed in quantum units. The resistance of the contact is obtained simply by dividing the quantum of the 12900 Ω resistance by the number of conducting channels (given in the upper x-axis). Fig.1 indicates that for 20 channels and 2 nm^2 of contact area, (corresponding to 600 Ω!), the BMR is largely reduced to a 10% value. This is in accordance with theory, as indicated by the good agreement between theory (lines) and experiments (dots) in the Fig 1. While this observation may produce very interesting physics from the theoretical and experimental points of view, it does not qualify for having technological applications because the contacts are unstable, lasting only a few minutes.

To overcome this problem and improve contact stability, we have electrodeposited material in magnetic and nonmagnetic wires and films forming a "T" configuration to make contacts. To our surprise, we have observed even larger values of MR, up to 700% in the Ni contacts deposited on small gaps between Ni wires [11] with typical R values of 5-20 Ω. These MR values clearly cannot be explained by DW scattering [12]. This is because the contact sizes are of the order of 10nm and then the DW scattering [7, 8 and 12] yield BMR values not larger than 10%. Recently, much larger values (many thousands per hundred!) of MR have been claimed as well [13, 14]. This effect of large BMR values in 10nm size contacts may be assigned to the formation of a very thin "dead layer", or non-stoichiometric compound (oxide, sulfite, etc.) at the nanocontact region, which may reconfigure the spin density of states defining the electron transport [14, 15]. While the very large values are very difficult to reproduce and to stabilize, we have been able to recently [14] to obtain very stable contacts with high reproducibility R(H) curves; i.e., variations of resistance versus applied magnetic field H.

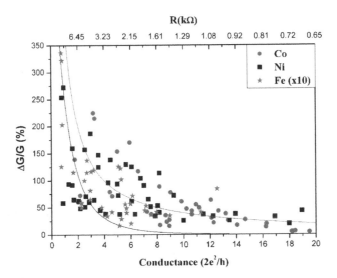

Figure 1. Magnetoconductance as a function of the contact conductance (bottom x-axis) and the contact resistance (top axis) for Ni, Co and Fe contacts. Black and red are the theory lines approximations in the limits of small and large number of conducting channels, respectively.

These recent magnetic contacts are formed using two films of Cu as a substrate that are separated by a gap that is between 2 and 100 μm wide. Once we have the desired gap, a contact can be generated with the appropriate electrodepositing procedures (Fig. 2a), using a magnetic material as the deposit substance. This material can be an element such as Ni, Fe or Co, but it can also be a binary compound like Permalloy $Ni_{85}Fe_{15}$, which is very soft and has a small coercive field. Utilizing this procedure, we have obtained extremely stable and reproducible R(H) curves that can last for thousands of loops. As an example, Fig. 2b shows the reproducibility of the R(H) curves (in the case of NiFe nanocontact), depicting only some tens of curves out of the hundreds that have been measured. The reproducibility is astonishing, and gives hope for the development of highly sensitive sensors as magnetic reading heads. At present we can obtain these kinds of results with a yield of 60% of the thousands of samples we have studied. The BMR values obtained with this method exhibited a reproducibility range between 20 and 300% for contacts in the 10 Ω range that, in turn, correspond to contacts of 10nm in size.

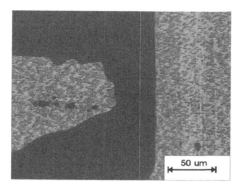

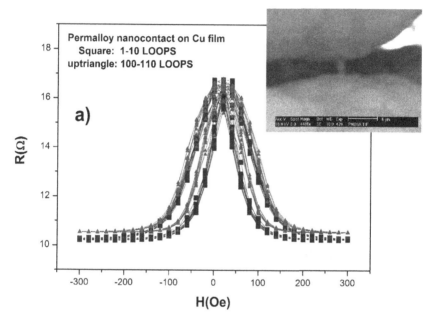

Figure 2. a) Optical microscope photograph taken near the contact area of the "T" configured Cu film arrangement. b) BMR curves of deposited permalloy contacts on Cu film. We measured the sample twice, taking one hundred loops per measurement. Indicated here are the first ten loops of each measurement: the upward-pointing triangular data points depict loops 1 through 10, while the square ones loops 100 through 110. The inserts at the top right of the curve correspondingly depict SEM scans of the contacts in mention with the scales shown at the bottom of the photos.

3.   MAGNETORESISTANCE OF THE DEPOSIT CONTROLS ITS MORPHOLOGY AND STRUCTURE

To prove the magnetoresistive effect we have performed electrochemical experiments applying a magnetic field while electrodeposition is taking place. The procedure is as outlined below:

We grow Ni by electrodepositing with zero applied field at several cathodic potentials.

Then we perform the same experiments with an applied magnetic field, using a weak field up to 500 Oe. This field is approximately the coercive field of the Ni electrodeposits as we observe in BMR experiments of nanocontacts (see Ref. [11], not to be confused with Fig. 2b where the coercive fields are much smaller, since in this case the deposit is permalloy with a 25 Oe coercive field).

The field is applied in the direction perpendicular to the plane of the deposit in such a way that the cross product $H \times v = 0$ (H and v being the applied field and the velocity of the ions in the electrochemical solution); i.e no Lorentz force exists, so no surface smoothing effects are expected [16, 17]. In fact, the only gradient effects expected should be weak because of the small fields applied (500 Oe maximum).

Under these conditions, although no important results were expected [16, 17], we found very large effects due to subtle small fields. We assign these to the resistance of the deposits with and without an applied field. This resistance should be small for an applied field due to magnetoresistive effects. When the field is applied the DW are removed, as discussed previously, and the resistance decreases.

In Fig. 3a we depict experiments showing SEM micrographs of the deposits of Ni on Cu for a cathodic potential of –1300 mV/SCE with no field, where the observed structure is rough, with clearly visible nickel clustered features. Here, every nickel cluster consisted of very small clusters, nano-sized dimensions. However, when a field H = 500 Oe is applied in the direction where the ions in the solution move, we observe an arboreous bead-dendritic structure with long branches, exhibiting a formation similar to that of sea algae (Fig. 3b), which is completely different to that when H = 0 (Fig. 3a). The structure of this deposit consisted of thin nickel branches, which terminated in flower aggregates of nickel, where one flower aggregate of nickel consists of thin nickel branches (or filaments) formed by small nano-sized nickel clusters, like a rosary (Fig. 3c). It is noteworthy that these beads of clusters making up the branches, are connected by nanocontact size junctions (as illustrated in Fig. 3d), since

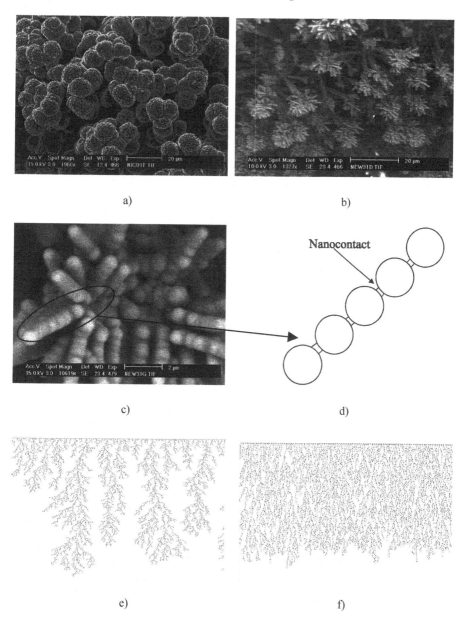

Figure 3. SEM micrographs of nickel deposits obtained at a cathodic potential of -1300 mV/SCE from Watt plating solution with addition of coumarin: a) without, and b) and c) with a perpendicular oriented magnetic field of 500 Oe. d) Schematic illustration of the part of a surface from Fig. 3c with a clearly denoted place of nanocontact, and the result of simulation of a growth on the surface: e) the filaments resistance is zero, and f) the filaments resistance is taken into account.

these may be the cause of the very strong magnetorersitance that we have already observed in electrodeposited magnetic nanocontacts [4, 8, 11].

The changes observed in nickel morphologies are due to the fact that when the field is 500 Oe the resistance is smaller than that of the case for zero field because the DW have been removed, whereas in the latter case the DW are present and contribute to the resistance. The effective cathodic potential between the end of the filamentous deposits and the saturated calomel reference electrode is smaller for zero fields due to the larger resistance of the filaments. For this reason these are not allowed to grow.

This can be explained in more detail in the following way: in order to grow branched structures, it is necessary that the effective potential at the end of the branches (or filaments) is the same as the applied one. If these filaments, however, are very resistive then this potential will be reduced to one smaller than that needed for electrodeposition of the Ni. In the case of magnetic deposits, with or without an applied field, the effective potential at the end of the filaments will be:

$$U_{eff} = U_c - IR\,(B=0,\,l) \tag{1a}$$

for zero field, and

$$U_{eff} = U_c - IR\,(B=B_0,\,l) \tag{1b}$$

for a field $B = B_0$, where U_{eff} and U_c are the effective and the applied cathodic potentials respectively, I is the current of deposition and R the resistance of the filament l wether or not it is under an applied magnetic field, yet which is field-dependent in the case of magnetic materials; i.e., materials that are magnetoresistive. This resistance for magnetic materials may be very different without or with applied magnetic field and, in fact, it was shown that when nanocontacts are formed, the differences due to domain wall scattering when no field is applied, can be factors up to 10 [4, 8, 11]. Therefore, the resistance for zero field is too large and the U_{eff} is much smaller in magnitude than − 1300 mV/SCE, and very close to minimum potential for a nickel electrodeposition of about − 700 mV/SCE. Hence, this resistance generates a potential drop of about −600 mV, whereupon no branched structure can grow, and, instead, a rough clustered structure is obtained (Fig. 3a). As mentioned earlier, when a sufficiently large magnetic field is applied (500 Oe, fields smaller than 300 Oe have much smaller effect), DW are erased in the filaments and the resistance is much smaller. In that case, the effective potential is high enough to keep the arboreous bead - dendritic structure from growing. It is noteworthy that the growth reported here is not formed by needle-like structures [18], but is conformed by a bead-arboreous-dendritic structure; i.e., the filaments have the form of a rosary made up by beads, and do not have any special orientation depending of the applied field. This is at variance with the needle-like structure observed in the deposits of Fe at a 2000 Oe field [18]. The field, furthermore, does not affect to the growth of Cu since this material is non-magnetic. The copper deposits obtained with and without a perpendicular oriented magnetic field of 500 Oe always formed an arboreous bead-dendritic structure.

This idea is illustrated by a computer simulation using the diffusion limited aggregation (DLA) model [19, 20]. Fig. 3e is a simulation considering that the filaments resistance is zero, where an arboreous-dendritic structure is obtained for the deposit, in agreement with Fig. 3b and the discussion above. However, when the resistance of the wires is taken into account, the structure is more compact as described in Fig. 3f, which correlates well with Fig. 3a. These experiments, therefore, show how subtle magnetic fields can provide different kinds of aesthetically impressive structures. As many processes in nature are mostly electrochemical, it may be that the formation of natural structures (organic and otherwise) may be influenced by subtleties of stray environmental magnetic fields.

4. MAGNETOSTRICTION MEASURED WITH AN ATOMIC FORCE MICROSCOPE

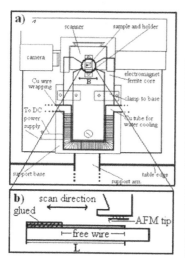
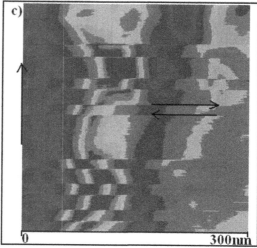

Figure 4. a) Top-view schematic of our AFM magnetostriction measurement setup. The sample is depicted in the bold circle between the two poles of the electromagnet and under the scanner. The double arrow marked B indicates the direction of the magnetic field. b) Side-

view zoom of the sample configuration, including the AFM tip situated at the bottom of the scanner. The samples were metallic wires, glued on one end, while scanning occurred as close to the free tip of the wire as possible. The tip scans along the length of the wire resulting in a topographic image similar to c). Part c) shows a typical scan whose area is 300 nm x 300 nm (1 nm ~ 1/1000000 mm). The corresponding wire length is from left (toward glued portion) to right on the scan, while the scan itself is moving in the direction of the vertical arrow (bottom to top). The material used is permalloy (an alloy of nickel and iron), and the length of the wire is about 10 mm. Field application causes a linear shift in the whole picture to the left, indicating contraction, whereas when the field is removed the image moves to the right, and the wire expands to its original position (see corresponding horizontal arrows). The shift itself was measured at 11 nm for the field strength applied.

The method is a simple and direct depiction of a phenomenon previously measured only indirectly, and its application potential vast, ranging from nanotechnology and nanocircuitry to the exploration of new magnetic materials with possible extensions in biology. Besides technological applications the direct observation of magnetostriction holds the potential to expand the understanding of the relationship between mechanical and magnetic properties of materials, and may even lead to the observation of magneto-mechanical phenomena on the atomic level.

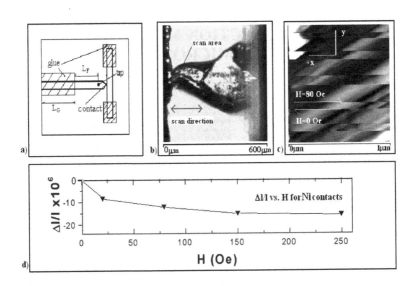

Figure 5. a) Schematic of the Ni "T"-contact. Two wires, glued perpendicular to each other with the gap between them closed by electro-deposition of the same material. The free wire length L_F is considered the distance between the glue boundary at one end of the wire and the AFM tip at the other. b) AFM video camera image (600 μm range) of the contact formation

between the two perpendicular wires. Scan position and scan direction are shown, while the contact has been measured to be 30 μm across. c) Shows a 1 μm scan of the Ni surface, scanned at the contact area of fig. 5b, with a magnetic field applied and removed five times during the scan. The scan direction is from top to bottom and H = 80 Oe. d) Plot of the deformation ($\Delta l/l \times 10^6$) of the "T"-contact vs. increasing magnetic field strength H (Oe), up to 250 Oe in a direction parallel to its length.

With this technique of measuring the lateral strain of magnetic materials, we have investigated the possibility of displacements or motions of contacts that give rise to the large BMR depicted in Fig. 2b with large reproducibility. After a multitude of measurements on free wires, problems such as sample clamping and strain due to the magnetization of the specimen as a whole were resolved, and the validity and reliability of this lateral (with respect to the sample and scan plane) magnetoelastic strain measurement technique was established [21]. We first measured contacts formed by electrodepositing Ni and permalloy ($Ni_{79}Fe_{21}$) on corresponding wires of the same materials. These contacts were used in BMR measurements, but had showed little or no magnetoresistance. We shall describe below the case for Ni deposited on Ni wires of 0.25 mm diameter, since Ni has a magnetostrictive strain over 15 times that of permalloy $Ni_{79}Fe_{21}$, and results are more striking.

A schematic of the Ni "T"-contact is shown in Fig. 5a. Two wires are firmly glued perpendicular to each other and the gap between them is closed by electro-deposition of the same material. The free wire length L_F considered is the distance between the glue boundary at one end of the wire and the AFM tip at the other. Fig. 5b shows an AFM video camera image magnified to the 600μm range of the contact formation between the two perpendicular wires. The precise scan position and scan direction are also shown, while the contact has been measured to be 30 μm across. Due to the increased clamping of the "T"-configured wires onto the sample base, the contact wire parallel to the direction of the applied magnetic field exhibits magnetoelastic shifts about half those of the free wire. Figure 5c shows a 1 μm scan of the Ni surface, scanned at the contact area of Fig. 5b, with a magnetic field applied and removed five times during the scan. The scan direction is from top to bottom. It is evident from the scan that the shift is in the negative x-direction, in agreement with literature on magnetostriction for Ni wires under a field parallel to their lengths [2, 3]. Measurements have also been performed at different positions along the length of the measured wire, and it is clear that the shift increases as L_F also increases. The above is an indication that the shifts are due to contraction of the wire under the influence of the magnetic field, and not to a motion of the wire configuration within the clamping (i.e. a parallel shift of the whole wire configuration). The contact resistance was monitored throughout all the magnetic field applications involving "T"-configured contacts. The value of this resistance was measured at 1.5 Ω, which represents the effective conductive portion of the contact. From previous studies we have found this particular value to correspond to conducting contact of 30 nm [22], in contrast with the total contact geometry (30 μm diameter area). Most of the total contact area is actually comprised of non-conductive oxides.

In Fig. 5d the graph represents the deformation values of the "T"-contact vs. increasing magnetic field, up to 250 Oe in a direction parallel to its length. The observed deformation at 250 Oe is about half the value of the free wire (-16 nm/mm), while it would be expected that a 100 nm contraction (over the 6 mm free length) would break the contact. In fact, the monitored resistance remained stable during multiple consecutive field applications. In other words, magnetoelastic strain does not alter the resistance across the contact, and more importantly *does not automatically imply large magnetoresistance.*

An even more striking result involves the measurement of the magnetostricion of a specimen that has exihibited a 200% BMR response. The sample was a contact formed by the electrodeposition of permalloy to bridge the 30 μm gap between two pure Cu wires aligned tip-to-tip. The applied field was H = 850 Oe, and no shift was observed in the scan upon field application, meaning that if there was a strain present, it was $< 5 \cdot 10^{-8}$. Thus, we may conclude that not only does magnetostriction not imply BMR, but that also the presence of BMR does not mean an observable magnetostriction is present. The two effects are thus not related to each other. This tends to indicate that the observed values attained in the BMR measurements truly correspond to magnetoresistance with a high degree of reproducibility. If this is the case, these contact devices open a new ground for very sensitive sensors and especially as reading heads for reading magnetic information compacted in the Terabit/inch². At present the research is going on in miniaturization of samples and stabilization of the electrodeposited contacts for periods of weeks.

5.   ACKNOWLEDGEMENTS

This work has been supported by the Spanish DGICyT.

6.   REFERENCES

1. O.Devos, A.Oliver, J.P.Chopart, O.Aaboubi and G.Maurin, J. Electrochem. Soc., 145, 401 (1998).

2. S. Chikazumi, *Physics of Ferromagnetism*, Oxford University Press (1999).
3. R. M. Bozorth, *"Ferromagnetism"*, D. Van Nostrand Co. Inc, New York, NY (1951).
4. N. García, M. Muñoz and Y-W Zhao, Phys. Rev. Lett. 82, 2923(1999).
5. S. H. Chung, M. Muñoz, N. García, W. F. Egelhoff and R. D. Gomez, Phys. Rev. Lett. 89, 287203(2002).
6. J. J. Verluijs, A. M. Bari and J. M. D. Coey, Phys. Rev. Lett. 87, 26601 (2001).
7. G.Tatara, Y-W Zhao, M. Muñoz and N. García, Phys. Rev. Lett. 83, 2030(1999).
8. G. G. Cabrera and L. M. Falicov, Phys. Status Solidi, B 61, 539 (1974).
9. M. Julliere, Phys. Lett. 54A, 225(1975).
10. Yu. V. Sharvin,Zh. Eksp. Teor. Fiz. 48, 984(1965). [Sov.Phys. JETP 21, 655(1965)].
11. N. García, M. Muñoz, G. G. Qian, H. Rohrer, I. G. Saveliev and Y.-W. Zhao, Appl. Phys. Lett. 79, 4550 (2001).
12. V. A. Molyneux, V.V. Osipov and E. V. Ponizowskaia, Phys. Rev. B 65,184425 (2002).
13. H. D. Chopra and S. Z Hua, Phys. Rev. B 66, 0204403(2002).
14. Hai Wang, Hao Cheng and N. García arXiv:cond-mat/0207516v1 We should mention here that very large variation of resistance with magnetic field up to practically infinity have been obtained with Ni wires, but this needs further understanding in the view of magnetoelastic deformations. The possibilities of magnestostriction effects are discussed below.
15. N. Papanikolaou, arXiv:cond-mat/0210551v1.
16. R. Aogaki, J. Electrochm. Soc. 142, 2954(1995).
17. A. Oliver, J.P.Chopart and J. Douglade, J. Electroanal. Chem., 217, 443 (1987).
18. S. Bodea, L. Vignon, R. Ballou and P. Molho, Phys. Rev. Lett., 83, 216 (1999).
19. T.A. Witten and L.M.Sander, Phys. Rev. Lett. 47, 1400 (1981).
20. E. Louis, F. Guinea and L.M Sander, Phys. Rev. Lett. 68, 209 (1992).
21. A.C. Papageorgopoulos, Hai Wang, C. A. Guerrero and N. Garcia, *in press in J. Mag. Mag. Mat.*
22. N. Garcia, G.G. Qian, and I. G. Saveliev, *Appl. Phys. Lett.* 80, 1785 (2002).

CONDUCTANCE QUANTIZATION IN MAGNETIC AND NONMAGNETIC METALLIC NANOSTRUCTURES

W. Nawrocki and M. Wawrzyniak

Poznan University of Technology, ul. Piotrowo 3A, 60-965 Poznan, POLAND
e-mail: nawrocki@et.put.poznan.pl

Abstract: We have built a measuring system for investigating electrical conductance in nanowires. Our measurements concern nanowires formed in both magnetic and nonmagnetic metals. The statistical results (histograms) of the quantization are compared for nonmagnetic and magnetic nanowires. The results of conductance quantization in cobalt nanowires have not been presented before. We have observed conductance quantization of macroscopic metallic contacts in a circuit with a mechanical contact and with a LC circuit. The investigation has been performed outdoors at room temperature.

1.   INTRODUCTION

We have measured the quantization of electrical conductance in macroscopic metallic contacts using the method proposed by Costa-Krämer et al. [1]. The quantization phenomenon occurs because of the formation of a nanometer-sized wire (nanowire) between macroscopic metallic contacts (according to the theory proposed by Landauer [2]). The measurements concern nanowires formed in non-magnetic metals (Au, Cu, W) and in magnetic metals (Ni, Co). Our measurements focused on nanowires formed by pairs of different metals, e.g. Au-Co, Cu-W.

B. Aktaş et al. (eds.), Nanostructured Magnetic Materials and their Applications, 383–392.

2.   MEASURING SYSTEM

The measuring system consists of a digital oscilloscope and a function generator connected to a PC computer by IEEE-488 interface. For setting macrowires in motion we used a piezoelectric device supplied by DC amplifier. The nanowire is formed when macrowires are very close to each other. It means the distance in nanometer size. The block diagram of the measuring system is presented in Fig. 1.

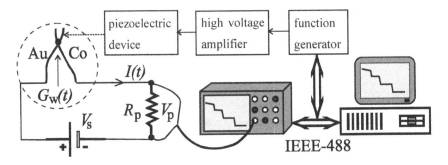

Figure 1. A system for measurements of conductance quantization

The measuring circuit consists of a bias voltage supply V_s, a pair of macroscopic wires, which make up the QPC contact under investigation and a serial resistor R_p of 1000 Ω. V_p voltage on the resistor R_p is a function of the measured conductance G_w of a nanowire:

$$V_p = \frac{V_s}{1 + \frac{1}{G_w R_p}}, \quad \text{for } G_w \gg 1/R_p: \quad V_p \cong V_s R_p \times G_w. \tag{1}$$

To record V_p, we use a digital oscilloscope LeCroy 93100M with a 8-bit analog-to-digital converter. The oscilloscope is equipped with IEEE-488 interface. The piezoelectric device is used to control the backward and forward movement of the macroscopic wires between which the nanowires occur. A high voltage amplifier controlled by a digital function generator supplies the piezoelectric device. Both electrodes (macroscopic wires) are made of wire 0.5 mm in diameter. The investigations have been carried in air at room temperature. The conductance was measured between two pieces of metal moved to contact by the piezoelectric tube actuator.

3. THEORY OF CONDUCTANCE QUANTIZATION

According to the Büttiker-Landauer theory, electrical conductance is a transmission between two reservoirs of electrons. In a constriction of a nonmagnetic metal with nanometer dimensions, called a nanowire, the ballistic transport of electrons occurs in conductive channels. The number of channels is proportional to the width of the nanowire. The conductance of such nanowire is described by the Landauer's formula [2]:

$$G = \frac{2e^2}{h} \sum_{n=1}^{N} T_n ,$$ (2)

where: e – electron charge, h – Planck constant, T_n – electron transmission in the channel number n.

Electron spin degeneracy in nonmagnetic nanowires causes the quantum of conductance equal to $G_0 = 2e^2/h \cong 12.9$ kΩ. Removing this degeneracy by a strong magnetic field would make the quantum of electrical conductance equal to $G_M = e^2/h$. The effect of the removal of spin degeneracy in nonmagnetic material was experimentally confirmed when investigating the quantization of conductance in GaAs/AlGaAs semiconductor, subjected to a magnetic field of 2.5 T at low temperature of 0.6 K [5]. For magnetic metals the conductance formula (3) contains the spin effect [3]:

$$G = \frac{e^2}{h} \left[\sum_{n=1}^{N_1} T_{n\uparrow} + \sum_{n=1}^{N_2} T_{n\downarrow} \right],$$ (3)

where: $T_{n\uparrow}$ – transmission of electrons with spin up, $T_{n\downarrow}$ – transmission of electrons with spin down.

In ferromagnetic nanowire a spin dependent density of states at both sides of the nanowire can be different. So, the number of occupied states with spin up N_1 is mostly not equal to the number of states with spin down N_2. Costa-Krämer suggested [3] that value of transmission in ferromagnetic nanowire is a random number between 0 and 1.

4.   CONDUCTANCE QUANTIZATION IN NANOWIRES

The quantization of electric conductance depends neither on the kind of metal nor on temperature. However, the purpose of studying the quantization for different metals was to observe how the metal properties affect the contacts between wires. We have investigated the conductance quantization of nanowires for three nonmagnetic metals (gold, copper and tungsten) and for magnetic metals (cobalt and nickel). To our knowledge, the results of quantization in cobalt nanowires and in nanowires formed by pairs of different metals have not been reported before. All measurements were carried out at room temperature. For nonmagnetic metals, the conductance quantization in units of $G_0 = 2e^2/h = (12.9 \text{ k}\Omega)^{-1}$ was previously observed for the following nanowires: Au-Au, Cu-Cu, W-W, W-Au, W-Cu. The quantization of conductance in our experiment was evident. All characteristics showed the same steps equal to $2e^2/h$. We observed two phenomena: quantization occurred when breaking the contact between two wires, and quantization occurred when establishing the contact between the wires [4]. The characteristics are only partially reproducible; they differ in number and height of steps, and in the time length. The steps can correspond to 1, 2, 3 or 4 quanta. It should be emphasised that quantum effects were observed only for some of the characteristics recorded. The conductance quantization has been so far more pronouncedly observable for gold contacts. Fig. 2a shows an example plot of conductance vs. time during the process of drawing a gold nanowire (non-magnetic). Fig. 2b shows the conductance histogram obtained from 5000 consecutive characteristics in the conductance range from $0.5G_0$ to $4\ G_0$, for the bias voltage $V_{bias} = 0,420$ V.

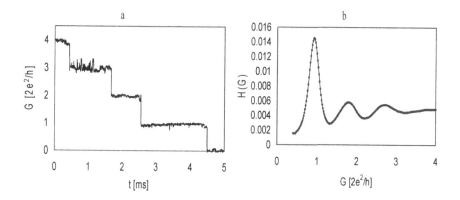

Figure 2. Conductance quantization in gold nanowires: a) G_w – conductance vs. time for V_{bias} = 0,420 V at room temperature; b) conductance histogram from 5000 conductance characteristics

Fig. 3a shows an example plot of conductance vs. time during the process of drawing a cobalt nanowire (magnetic material). Fig. 3b shows the conductance histogram obtained also from 6000 consecutive characteristics, for the bias voltage V_{bias} = 0,420 V. This histogram looks quite differently from the histogram for gold nanowires.

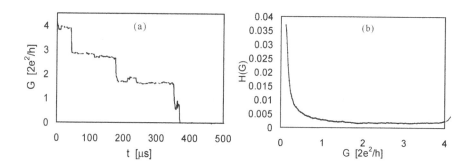

Figure 3. Conductance quantization in cobalt nanowires: a) G_w – conductance vs. time for V_{bias} = 0,420 V at room temperature; b) conductance histogram from 6000 conductance characteristics

From 6000 consecutive conductance characteristics G_w = f(t) for one pair of metals we calculated a histogram of quantization in the conductance range from 0.6 to $1.3G_0$. We have measured the conductance of nanowires formed by the following pairs of magnetic and nonmagnetic metals: Au-Au, Cu-Cu, Co-Co, Ni-Ni, W-W, Co-Au, Au-W, Au-Ni, Cu-Co, Cu-Ni, and Co-Ni. For magnetic metals (Ni and Co) we can observe some steps with a height of $2e^2/h$ on the conductance characteristics but no peaks on the histograms – Fig. 4.

The histograms presented indicate that the quantization process takes a different course in nonmagnetic and magnetic nanowires. In Fig. 4 three histograms for the following nanowires: Au-Au, Au-Co and Co-Co are shown. From Fig. 4 one can conclude that from two different metals forming a nanowire, it is the softer metal (gold or copper) that determines the properties of the nanowire.

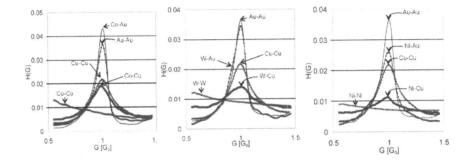

Figure 4. Conductance histograms from conductance characteristics from nanowires: a) nanowire formed by gold and a magnetic metal; b) nanowire formed by copper and a magnetic metal; c) nanowire formed by one metal

We also found that the sharpness of the histogram curve depends on the speed of moving electrodes (macroscopic wires) during the forming of a nanowire. Each histogram presented in Fig. 2, Fig. 3 and Fig. 4 in the paper was obtained at a speed of 24 μm/s. In Figure 5 the histograms for two values of speed (2.4 μm.s and 24 μm/s) are shown.

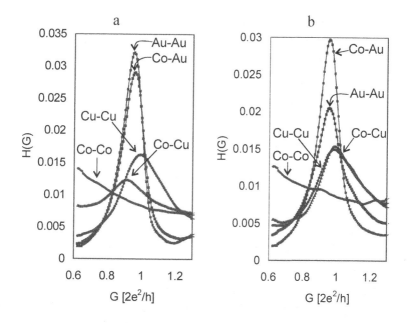

Figure 5. Conductance histograms from conductance characteristics measured at a different speed of electrods: a) at speed of 2.4 μm/s; b) at speed of 24 μm/s

5. QUANTIZATION OF CONDUCTIVITY IN *LCG* CIRCUIT

Transient states in a *LCG* circuit depend on the parameters *L, C* and *G* (or resistance *R*). Transients in the mechanical switch circuit may have oscillatory or non-oscillatory property, depending on the current value of conductance *G(t)* in this circuit – Fig. 6. The change of the property of the *LCG* circuit, occurring during a transient, follows a mathematical description. In a transient state the current in the circuit is described with (4) using the Laplace transform,

$$I(s) = \frac{V_s}{s} \frac{(s^2 LC + 1)G}{s^2 LC + sLG + 1)}.$$ (4)

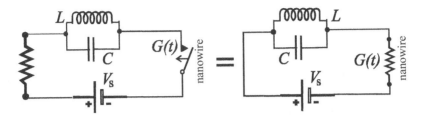

Figure 6. Transient states of $I(t)$ in the LCG circuit with a quantized conductance

From circuit theory it follows that the form of the transient signal in a circuit is characterized by the determinant Δ of the quadratic equation $M(s)$ in the denominator of the formula (5):

$$M(s) = s^2 LC + sLG + 1, \quad \Delta = (LG)^2 - 4LC \tag{5}$$

Output function (current in the circuit) has oscillatory character when $\Delta < 0$. Consequently, for the oscillating circuit, the limit value of conductance $G = G_{\lim}$ when $\Delta = 0$ is given by the relation

$$G_{\lim} = 2\sqrt{C/L} . \tag{6}$$

In a transient state, for the conductance less than the limiting value $G_{\lim}$ the current in the circuit exhibits an oscillatory character. For example, for the circuit parameters $L = 4$ mH and $C = 30$ pF, the limit conductivity $G_{\lim}$ amounts to 10^{-4} A/V. It means that in a transient process of quantization the electric current in the circuit is of oscillating form for the first quantization level $G_0 \cong 77.5 \times 10^{-6}$ A/V. For the second it is $G(2) = 2G_0$, the third $G(3) = 3G_0$, and the remaining quantization levels has non-oscillatory property. The transient form in *LCG* circuit during the phase of contact breaking is shown in Fig. 7 (simulation of the circuit state) [6].

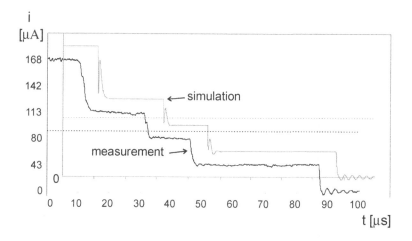

Figure 7. Transient states of *I(t)* in the circuit with a quantized conductance (measured curve and shefted simulation curve)

6. DISCUSSION

The conductance quantization has proved to be observable in an experimental set-up, giving opportunity to investigate the quantum effects in electrical conductivity. The quantization in metallic nanowires occurs in a way different for magnetic and nonmagnetic metals respectively. The quantization steps are equal to $2e^2/h$ in nonmagnetic nanowires and should be equal to e^2/h in magnetic nanowires. The histograms of metallic nanowires obtained after statistical processing of conductance characteristics for nonmagnetic samples are quite different from those obtained for magnetic nanowires. The curves of histograms from magnetic nanowires do not contain peaks. The latter phenomenon is explained by Costa-Krämer as caused by different quanta: $2e^2/h$ and e^2/h and the backscattering effect [3]. In our opinion, a curve of histograms without peaks for magnetic nanowires can be caused rather by the hardness of magnetic metals. We have obtained histograms without peaks for a nonmagnetic hard metal – tungsten. In Table 1 we present Brinell hardness for the metals under investigation.

Table 1. Hardness of investigated metals (in Brinell scale)

Metal	Sn	Au	Cu	Ni	Co	W
Brinell hardness [MPa]	50	180	400	850	1250	2500

The quantization in nonmagnetic metals occurs more frequently and its histograms have peaks because of easier formation of nanowires from soft metal than from hard metal. For a nanowire formed by a pair of metals, the magnetic metal and the nonmagnetic one, the histogram is like the histogram of nonmagnetic nanowire. Propably the nanowire formed from such pair of metals contains mostly atoms of nonmagnetic metal. The actual reasons for the flat form of histograms for magnetic nanowires (magnetic properties or hardness of material) can be proved in an experiment where the samples used are solid constrictions, instead of dynamically formed nanowires.

The above presented analysis of signals in RLC circuit shows that it is possible to determine the circuit parameters, especially capacity C, so that the oscillations appear only at the lowest quantum levels, where the contacts resistance is greater than 12 kΩ. This case proves that the quantum resistance for non-ideal ballistic electron transport is lossy and the energy accumulated in LC circuit is lost just in the quantum resistance. When designing an electronic circuit with a mechanical contact, one should take into account the fact that the resistance of the contact may take the value of 13 kΩ, which together with the capacity and inductance of the whole system (e.g. a residual one), may lead to the appearance of oscillations disadvantageous to systems with high sensitivity. It can be important for low drift amplifiers with a chopper in an input circuits.

REFERENCES

[1] Costa-Krämer J.L. et al., Nanowire formation in macroscopic metallic contacts, *Surface Sci.* **342** (1995) L1144.

[2] Landauer R., Conductance determined by transmission: probes and quantised constriction resitance , *J. Phys.: Cond. Matter* **1** (1989) 8099.

[3] Costa-Krämer J.L., Conductance quantization at room temperature in magnetic and non-magnetic metallic nanowires, *Physical Review B* **55** (1997) 4875.

[4] Martinek J., Nawrocki W., Wawrzyniak M., Stankowski J., Quantized conductance of the nanowires spontaneously formed between macroscopic metallic contacts, *Molecular Physics Reports* **20** (1997) 157.

[5] van Wees B.J. et al., Quantum ballistic and adiabatic electron transport studied with quantum point contacts, *Physical Review B* **43** (1991) 12431.

[6] Nawrocki W., Pajakowski J. „Transient stated in electrical circuits with quantized conductance", Intern. Conference on Nanoscience NANO-7, Malmoe (Sweden) 2002

HUGE MAGNETORESISTANCE IN QUANTUM MAGNETIC NANOCONTACTS

L.R. Tagirov [1,2], K.B. Efetov [3,4]

[1]*Kazan State University, Kazan 420008, Russia*
[2]*Kazan Physico-Technical Institute of RAS, Kazan 420029, Russia*
[3]*Theoretische Physik III, Ruhr-Universität Bochum, 44780 Bochum, Germany*
[4]*L.D. Landau Institute for Theoretical Physics, Moscow 117940, Russia*

Abstract: The quasiclassical theory of a nanosize point contacts (PC) between two ferromagnets is developed. The maximum available magnetoresistance in PC is calculated for ballistic and diffusive transport at the area of a contact. In the ballistic regime, the magnetoresistance in excess of few hundreds percents is obtained for the iron-group ferromagnets. The regime of quantized conductance through the magnétic nanocontact is considered. It is shown that magnetoresistance is tremendously enhanced at small number of open conductance channels. The quantum spin valve realization is discussed in detail, and recent observations of huge (up to 100'000%) magnetoresistance in the electrodeposited nickel nanocontacts are discussed in the framework of the developed theory.

Key words: conductance quantization, giant magnetoresistance, magnetic point contacts

1. INTRODUCTION

In recent experimental studies of Ni-Ni and Co-Co point contacts (PC) a surprisingly high negative magnetoresistance in excess of 200% has been discovered [1,2]. Somewhat smaller ($\sim 30\%$), but also very large for a single interface, magnetoresistance was observed in Fe-Fe point contacts [3]. The set up of the experiments was typical for observation of giant magnetoresistance (GMR), the effect observed earlier in hybrid systems involving ferromagnetic and normal metals [4,5]. However, for the multilayer structures the typical change of the resistance reaches $10\% - 50\%$, which is considerably lower than the corresponding values in

B. Aktaş et al. (eds.), Nanostructured Magnetic Materials and their Applications, 393–417.
© 2004 *Kluwer Academic Publishers. Printed in the Netherlands.*

Refs. [1,2]. Further development of the nanocontacts fabrication techniques raised the above values till 70000%-100000% [6-10]. The authors claim that they observed ballistic magnetoresistance (BMR), i.e. the magnetoresistance at ballistic (collisionless) electronic transport at the area and vicinity of a nanocontact.

A negative magnetoresistance can be due to scattering of conduction carriers on domain walls (DW). According to the general quantum-mechanical prescription, any inhomogeneity in the potential landscape results in reflection of quasiparticle wave function, which evokes an additional electric resistance. This effect has been considered for a free-standing domain wall in a number of works [11-14] and provided typical values of MR in a range of few percents. The low values of MR were obtained assuming that the widths of the DW is large, typically 150-1000 Å, which resulted in a low scattering amplitudes. Considering sharp DW in the ballistic regime van Hoof et al. estimated the magnetoresistance to get values ~ 70% [14]. The fact that the enhancement of impurity scattering in a sharp DW may give large MR was utilized in Ref. [2] to explain the anomalously large values of MR in the experiments on the point contacts [1,2]. However, the theory in Ref. [2] is essentially the perturbation theory. That is why the use of this theory to explain 300% or even larger effect seems to be unjustifiable extrapolation. One needs to develop the nonperturbative theory of electron scattering on the sharp DW at the area of a nanocontact. In this paper we summarize our studies of electron scattering by the constrained domain wall [15-17] aiming to demonstrate that DW scattering enables to provide huge magnitudes of negative magnetoresistance observed in the recent point contact experiments.

2.   EFFECTIVELY SHARP DOMAIN WALL

The diminishing of the width of DW when decreasing the size of the constriction between two oppositely magnetized domains was proposed by Bruno [18]. In his model the DW width becomes comparable with PC length, and the magnetization rotates almost entirely inside the constriction. This conclusion holds until the diameter of PC is smaller than its actual length. With further increase of the constriction size (diameter) one may expect that the wall will bend outside of PC. This behavior has been clearly demonstrated in recent micromagnetic simulations of domain walls in magnetic nanocontacts [19,20]. In their calculations for two bulk ferromagnetic rods connected by the nanosize thread Savchenko et al. [19] demonstrated, that the domain wall is bulged out the constriction on the distance of its size (Figs. 3a-5a of [19]) for parameters of the ferromagnet close to permalloy. They also checked that with increasing the material anisotropy the domain wall shrinks towards the connecting thread (Fig. 5 in Ref. [19]). Molyneux *et al.* [20] analyzed in detail nanocontacts between large-area thin films. They concluded, that the width of the constrained DW is about $2a + d$, where d is the length and a is the width of the connecting channel, respectively. In their calculations the magnetization relaxes almost isotropically outside the constriction. Finally, the authors of Ref. [20] conclude that 3D domain wall is more localized compared with the 2D (thin-film) one.

The micromagnetic calculation results can be easily understood using simple energy considerations. From the symmetry of the problem it is obvious that in a free-standing, infinite area DW the exchange energy relaxes into the chain of magnetic moments till the total anisotropy energy of the chain equals the loss of the exchange energy (classic 1D Landau-Lifshitz solution). In the 2D, thin-film case, the portion of the exchange energy, that did not relax inside of constriction, relaxes into the 2D plane outside of the neck. In this 2D-case, two half-circles (we use the conclusion of Ref. [20] about the isotropic relaxation of the magnetization) at mouths of the constriction accommodate the number of magnetic moments in the 1D domain wall chain, minus the number of the moments inside the constriction. In the case of 3D nanocontact, approximately the same amount of magnetic moments has to be accommodated by two semi-spheres at the mouths of the neck. It is clear, that the spatial extent of the domain wall will eventually decrease upon increasing the dimensionality of the magnetization relaxation space (chain -> area -> volume).

Coey *et al.* [21,22] have drawn attention to the fact, that in nanosize constrictions the continuum approximation, used in micromagnetic simulations, is no longer valid. They have analyzed DW in nanocontacts calculating explicitly lattice sums over the magnetic moments in the constriction and the adjacent space. The main conclusions are as follows: discrepancy between results of the continuum and the discrete approaches become marked as the characteristic dimensions fall below 10 inter-atomic spacings; it is possible to have very narrow domain walls with a width determined by the effective length of the constriction, the latter one can be as little as a few inter-atomic spacings.

Another necessary condition for realization of the sharp DW is conservation of the electron spin orientation when crossing the domain wall. The orientation conserves if the DW width d_w is shorter than the length d_s, at which the electron spin quantization axis adjusts varying direction of the local exchange field. For the ballistic transmission through PC $d_s = \min(\frac{2\pi v_F}{\omega_z}, v_F T_1)$, where T_1 is the longitudinal relaxation rate time of the carriers magnetization [23], and ω_z is the Zeeman precession frequency

[24]. If we assume the DW width 5 nm and the Fermi velocity $v_F \sim 10^5$ m/s, then the time-of-flight is about 5×10^{-14} s - too short compared with Zeeman or spin-relaxation times. At this condition the transmission process looks like transmission through the abrupt DW, and the description of the electron transport through PC with boundary conditions at PC interface is valid.

3. THE POINT CONTACT MODEL AND ITS SOLUTION

We believe that extremely large magnetoresistance can be obtained because of the strong spin-dependent reflection of carriers from the effectively sharp DW in the PC area. It is realized in ferromagnetic metals where there is big exchange splitting of conduction band. Mapped onto the parabolic conduction band structure the exchange splitting results in non-equivalent values of the spin-subband Fermi momenta, $k_{F\uparrow}$ and $k_{F\downarrow}$ (Ref. [25] gives $k_{F\uparrow} = 1.1\,\text{Å}^{-1}$ and $k_{F\downarrow} = 0.42\,\text{Å}^{-1}$ for iron).

At the *ferromagnetic* (F) alignment of magnetizations in the contacting domains there is no domain wall in the constriction, and the current flows through PC independently in each spin-subband. Then, the resistance of PC is actually the Sharvin resistance [26] of spin-channels connected in parallel. At the *antiferromagnetic* (AF) alignment of magnetizations the additional resistance appears, which is associated with reflection of electrons from the potential barrier created by the domain wall. In fact, at AF-alignment the spin-subband assignment in one of the magnetic domains is reversed with respect to the another one, and the current flowing from, say, majority (larger Fermi momentum) subband of one bank of the contact has to be accommodated by the minority (smaller Fermi momentum) subband of the another bank. Then, in terms of quantum mechanics [27], the incident electron waves will be partially reflected because of the Fermi momenta mismatch of majority and minority subbands ($k_{F\uparrow} \neq k_{F\downarrow}$). However, the partial reflection of electrons is not the sole reason for the enhanced resistance at the AF-alignment. When the angle of incidence becomes large enough (it depends on the ratio of spin-subband Fermi-momenta, $k_{F\uparrow}$ and $k_{F\downarrow}$) the minority subband can not further accept the momentum transferred from the opposite side of the PC, which is majority subband with the same spin projection. As a result, only a narrow incidence angles cone (for $k_{F\downarrow} \ll k_{F\uparrow}$ as in the example given above) around the normal direction to the interface is responsible for the charge transport across the PC. Electrons with more inclined trajectories are completely reflected. Thus, the partial transmission at the steep incidence, and the total reflection at slanted incidence provide high boundary resistance of PC at AF alignment of magnetizations.

3.1 Formalization of the model and its solution [15]

The PC model we consider to realize the physics described above is the circular hole of the radius a made in a membrane. The membrane divides the space on two half-spaces, occupied by the single-domain ferromagnetic metals. It is impenetrable for the quasiparticles carrying a current, and the connecting channel is assumed to be ballistic (shorter than the mean free path). The $z-$axis of a coordinate system is chosen perpendicular to the membrane plane. The electron motion on both sides of the contact can be described by the equations for quasiclassical (QC) Green functions derived by Zaitsev [28]. They are, in fact, equivalent to the Boltzmann equations in the $\tau-$approximation:

$$v_z \frac{\partial g_a}{\partial z} + \mathbf{v}_\| \frac{\partial g_s}{\partial \vec{\rho}} + \frac{1}{\tau}\left(g_s - \overline{g}_s\right) = 0,$$

$$v_z \frac{\partial g_s}{\partial z} + \mathbf{v}_\| \frac{\partial g_a}{\partial \vec{\rho}} + \frac{g_a}{\tau} = 0. \tag{1}$$

g_s and g_a are symmetric and antisymmetric with respect to $z-$ projection of quasiparticle momentum QC GF (Green functions integrated over the energy variable), $\mathbf{v}$ is the vector of the Fermi velocity, $v_z = v_F \cos\theta$, $v_\|^2 = v_F^2 - v_z^2$, angle θ is measured from the $z-$axis, v_F is the modulus of $\mathbf{v}$, the bar over $\overline{g}_s$ means the averaging over the solid angle. We assume that the spin-mixing process is weak, therefore we consider spin channels as independent and omit the spin-channel indices in (1) and in expressions below.

The boundary conditions to equation (1) for the specular scattering ($p_{F1\alpha}\sin\theta_1 = p_{F2\alpha}\sin\theta_2 \equiv p_\|$) at the interface $z = 0$ are [28]:

$$g_{a1}(0) = g_{a2}(0) = \begin{cases} g_a(0), & p_\| < p_{F1}, p_{F2} \\ 0, & p_\| > \min(p_{F1}, p_{F2}) \end{cases},$$

$$2Rg_a(0) = -D\left(g_{s2} - g_{s1}\right), \tag{2}$$

where subscript 1 or 2 labels left- or right-hand side of the contact, respectively, p_{Fi} is the Fermi momentum of i-th side, $p_{\parallel}$ is the projection of the Fermi momentum vector on the PC plane. D and $R = 1 - D$ are the exact quantum-mechanical transmission and reflection coefficients that can be considered either as phenomenological parameters or calculated for models of interest. For the F-alignment of the domain magnetizations, carriers move in a constant potential, and the transmission probabilities in the each spin-channel are equal to unity. For the AF-aligned domains the carriers move in a potential created by the magnetization profile in the geometrically constrained domain wall. The second line in the first boundary condition in Eq. (2) quantifies explicitly the case of the total reflection for inclined trajectories, described qualitatively above.

The density of a current through the contact per spin projection is expressed via the antisymmetric GF,

$$j^z(z,\vec{\rho},t) = -\frac{ep_{F\min}^2}{2\pi} \int\limits_0^{\pi/2} d\Omega_\theta \cos\theta g_a\left(z,\vec{\rho},t\right). \tag{3}$$

The total current through the area of the contact per spin projection is

$$I^z(z \to 0,t) = \int\limits_{area} d\vec{\rho} j^z(z = 0,\vec{\rho},t) = a\int\limits_0^\infty dk J_1(ka) j^z(0,k,t). \tag{4}$$

In the above equations $p_{F\min} = \min(p_{F1},p_{F2})$, $J_1(x)$ is the Bessel function, $j^z(0,k,t)$ is the Fourier-transform of current density, Eq. (3), over the in-plane coordinate ρ. The cylindrical symmetry of the problem has been used upon derivation of Eq. (4).

We search a solution for g_s using the standard Ansatz in kinetic theories ($k_B = \hbar = 1$):

$$g_s(\varepsilon) = 2\tanh\frac{\varepsilon}{2T} + f_s(\varepsilon), \tag{5}$$

where the first term is the equilibrium value of g_s in the leads far away of PC, and the second term is deviation from the equilibrium. Substitution of (5) into Eqs. (1) and Fourier transformation over the variable ρ lead to the second order differential equation, which has an exact solution [15]. After some lengthy algebra which involves matching of the solutions by the boundary conditions, Eq. (2), we obtain the final expression for the current per spin projection through PC at small applied voltage [15]:

$$I^z = \frac{e^2 k_{F\,\mathrm{min}}^2 a^2 V}{2\pi} \int\limits_0^\infty \frac{dk}{k} J_1^2(ka) \langle D\,F(k,\theta)\cos\theta \rangle,$$ (6)

where

$$F(k,\theta) = 1 - \left[\frac{1}{2(1-\lambda_1)\kappa_1 l_{z1}} + \frac{1}{2(1-\lambda_2)\kappa_2 l_{z2}} \right]$$

$$\times \frac{\overline{D}}{1 + \frac{\tilde{\lambda}_1}{2(1-\lambda_1)} + \frac{\tilde{\lambda}_2}{2(1-\lambda_2)}},$$ (7)

$$\lambda_i(k) = \frac{1}{kl_i}\arctan kl_i, \quad \tilde{\lambda}_i = \left\langle \frac{D}{\kappa_i l_{zi}} \right\rangle = \int\limits_0^1 dx \frac{D(x)}{\sqrt{1+k^2 l_i^2(1-x^2)}},$$ (8)

where $<\dots>$ means averaging over the solid angle, $l_i = \tau_i v_{Fi}$ is the mean free path, $l_z = l\cos\theta$, $x = \cos\theta$. Eqs. (6) and (7) express the current through the orifice in terms of parameters D_α, l_α, a, $k_{F\alpha}$ characterizing the system.

3.2   Magnetoresistance of a point contact

Now we calculate the magnetoresistance of PC between two identical ferromagnets. It can be expressed via the conductance $\sigma = I/V$ as follows:

$$MR = \frac{R^{AF} - R^F}{R^F} = \frac{\sigma^F - \sigma^{AF}}{\sigma^{AF}},$$ (9)

where R^F (σ^F) stands for the resistance (conductance) at F alignment of magnetizations of contacting ferromagnets, and R^{AF} (σ^{AF}) is for the AF alignment of magnetizations. For the F-alignment the net current is the sum of currents for the both (independent) spin channels, $D_\alpha = 1$, $\tilde{\lambda}_i = \lambda_i$. Labeling the quantities by arrow-up/down notations we write down

$$\sigma^F = \sigma_{\uparrow\uparrow}^z + \sigma_{\downarrow\downarrow}^z = \frac{e^2 \left(k_{F\uparrow}^2 + k_{F\downarrow}^2 \right)\left(\pi a^2\right)}{4\pi^2} \int\limits_0^\infty \frac{dk}{k} J_1^2(ka)$$

$$\times \left\{ \frac{k_{F\uparrow}^2}{k_{F\uparrow}^2 + k_{F\downarrow}^2} \frac{k^2 l_\uparrow^2}{\left(1 + \sqrt{1 + k^2 l_\uparrow^2}\right)^2} + (\uparrow\downarrow) \right\}. \tag{10}$$

The prefactor in Eq. (10) is nothing but the sum of Sharvin [26] conductances for the spin channels. For the AF alignment of magnetizations the conductance is:

$$\sigma^{AF} = \frac{e^2 k_{F\downarrow}^2 \left(\pi a^2\right)}{\pi^2} \int_0^\infty \frac{dk}{k} J_1^2(ka) \int_0^1 dx x \left(D(x)\right)_{\uparrow\downarrow}$$

$$\times \left\{ 1 - \left[\frac{1 - \lambda^\uparrow}{\sqrt{1 + k^2 l_\uparrow^2 \left(1 - x^2\right)}} + \frac{1 - \lambda^\downarrow}{\sqrt{1 + k^2 l_\downarrow^2 \left(1 - x^2\right)}} \right] \right.$$

$$\left. \times \frac{\left(\overline{D}\right)_{\uparrow\downarrow}}{2\left(1 - \lambda^\uparrow\right)\left(1 - \lambda^\downarrow\right) + \tilde{\lambda}_{\uparrow\downarrow}^\uparrow (1 - \lambda^\downarrow) + \tilde{\lambda}_{\uparrow\downarrow}^\downarrow (1 - \lambda^\uparrow)} \right\}, \tag{11}$$

where $\left(D(x)\right)_{\uparrow\downarrow}$ stands for the transmission coefficient of the interface at AF-alignment, $x = \cos(\theta_\downarrow)$. The equations (9)-(11) solve the problem of finding the magnetoresistance of PC at arbitrary ratio of the contact size to the mean free paths of electrons in the ferromagnetic metal. Starting with the equations (6)-(8) one can derive magnetoresistance of the point contact made of different ferromagnetic metals (heterogeneous magnetic contacts).

4.   MAGNETORESISTANCE AT CLASSICAL (NON-QUANTIZED) CONDUCTION

To simplify the general analysis of the solution we use the step-like shape for the potential barrier created by DW. The approximation of DW profile by the abrupt potential gives maximum available magnetoresistance for a particular choice of other physical parameters. The transmission probability, $\left(D(x)\right)_{\uparrow\downarrow}$, can be found from the solution of Schrödinger equation for the particle moving in the step-like potential landscape [27]:

$$\left(D(x)\right)_{\uparrow\downarrow} = \frac{4(v_{z1}^{\uparrow})_{\uparrow}(v_{z2}^{\uparrow})_{\downarrow}}{\left[(v_{z1}^{\uparrow})_{\uparrow} + (v_{z2}^{\uparrow})_{\downarrow}\right]^2} = \left(D(x)\right)_{\downarrow\uparrow}, \tag{12}$$

with $v_{z2}^{\uparrow} = v_{z1}^{\downarrow}$ for the AF alignment. With the transmission coefficient Eq. (12) the dependence of MR on the mean free paths of electrons in the bulk leads and on the size of the constriction can be analyzed. Our expressions (6) to (11) cover the regimes of conduction from ballistic $\left(l_\alpha \gg a\right)$ to diffusive $\left(l_\alpha \ll a\right)$ in the vicinity of PC. However, the PC size is beyond the quantization regime, $2a \sim 10 - 15$ Å, which will be considered later.

4.1   Ballistic regime of conductance in the vicinity of PC

For the purely ballistic transport $[a/l_\uparrow \to 0$, where $l_\uparrow$ ($l_\downarrow$) is the majority (minority) electrons mean free path] all integrals in Eqs. (10), (11) can be evaluated analytically, and the magnetoresistance reads (the difference in the effective masses in the conduction spin-subbands has been neglected):

$$MR = \frac{(1-\delta)\left\{5\delta^3 + 15\delta^2 + 9\delta + 3\right\}}{8\delta^3(\delta+2)}, \tag{13}$$

where

$$\delta = \frac{k_{F\downarrow}}{k_{F\uparrow}} = \frac{v_{F\downarrow}}{v_{F\uparrow}} \le 1. \tag{14}$$

If $\delta = 1$, then $MR = 0$, *i.e.* the magnetoresistance vanishes in the contact of non-magnetic metals. For the set of δ values we obtain from (1): $\delta = 0.5$, $MR = 238\%$; $\delta = 0.4$, $MR = 455\%$; $\delta = 0.33$, $MR = 780\%$; $\delta = 0.3$, $MR = 1012\%$. Note here, that being expressed via the density of states (DOS) polarization,

$$P \equiv P_{DOS} = \frac{1-\delta}{1+\delta}, \tag{15}$$

the magnetoresistance (1) reads:

$$MR = \frac{P\left(8 - 3P + P^3\right)}{3 - 8P + 6P^2 - P^4}. \tag{16}$$

It predicts more steep growth of magnetoresistance upon increasing the polarization P_{DOS} compared with the Julliere's formula [29],

$$MR_{Jull} = \frac{2P^2}{\left(1 - P^2\right)}. \tag{17}$$

The reason is in the dependence of the transmission coefficients on energy, incidence angles and the cutoff of the slanted trajectories in our approach.

Let us recall here the experimental data on magnetoresistance of magnetic PC by García *et al.* Ni-Ni PC showed maximal $MR \simeq 280\%$ [1], Co-Co PC showed maximal $MR \simeq 230\%$ [2]. To obtain the MR values 280% (Ni) and 230% (Co) we have to use the values $\delta(\text{Ni}) \simeq 0.47$ and $\delta(\text{Co}) \simeq 0.5$. These numbers are in the range of the values, obtained experimentally from the single photon threshold photoemission, $\delta(\text{Co}) \simeq 0.4$ [30], and from ferromagnet/superconductor point contact spectroscopy: $\delta(\text{Ni}) \simeq 0.59 - 0.65$ [31], $\delta(\text{Ni}) \simeq 0.71$ [32]; $\delta(\text{Co}) \simeq 0.62 - 0.65$ [31]; $\delta(\text{Co}) \simeq 0.68$ [32].

If we use the experimental data of Ref. Soulen for iron, $\delta(\text{Fe}) \simeq 0.59 - 0.65$, then in our theory we obtain $MR(\text{Fe}) = (100 - 140)\%$, which is larger than the experimentally measured $MR(\text{expt}) = 33\%$ [3]. The justification of our model suggests that observed MR does not solely confined to a value of polarization δ (P). We believe that the basic condition for observation of upper MR limit, $d_w \ll d_s$, is not fulfilled in the Fe-Fe PC experiment [3].

4.2   **Arbitrary mean free path. Numerical results.**

In general case the angular integrals in (10) and (11) can be still evaluated analytically, whereas the integrations over k can be done only numerically. The results for the magnetoresistance (9) as a function of the contact radius are shown on Fig. 1. The point sequences show maximum available MR, that could be realized in PC with the physical parameters displayed in the figure. MR exponentially drops till the size of the contact approaches the mean free path of a material. Then, it shows a smooth crossover from ballistic to diffusive regimes of conduction. When the radius of the nanohole is much larger than the mean free path ($a \gg l_\uparrow, l_\downarrow$) the giant MR values can be obtained, if the condition of validity of our model,

$d_w \ll d_s$, is realized in an experiment. In the opposite limit, $d_w > d_s$, when PC size is so large that DW becomes smooth and wide, the electron spin

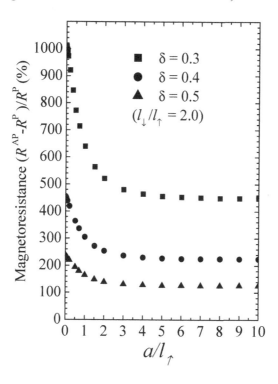

Figure 1. The dependence of the magnetoresistance on the PC radius.

spin will track the local exchange field in the domain wall, and MR will level off at Levy-Zhang [13] impurity scattering enhancement mechanism, which can give 2-11% magnetoresistance. The requirement of abrupt DW with constant width, irrespective of the PC size, can be technologically controlled, if very thin (2-4 monolayers of the thickness $\sim \lambda_F$) nonmagnetic interlayer is deposited on the PC plane before depositing the second electrode. Then, just like in CPP transport in multilayers [4,5], the contacting domains will be exchange decoupled, so the magnetization will acquire sudden reversal within the spacer thickness $\sim \lambda_F$. In this case our analysis is valid for an *arbitrary* size of PC [15]. Another realization is very thin, magnetically "dead" layer inside the nanocontact [33,34], which appears either accidentally, or is created technologically during the deposition process.

5.   CONDUCTANCE QUANTIZATION AND MAGNETORESISTANCE IN MAGNETIC POINT CONTACTS [16,17]

Since experiments with two-dimensional electron gas in a semiconductor [35,36] it is demonstrated that electric conduction is quantized, and elementary conductance quantum is equal to $2e^2/h$. When measured on tiny contacts of nonmagnetic semiconductors and metals the conductance quantization is limited to low temperatures by thermal fluctuations, and the factor 2 is attributed to the two-fold spin degeneracy of conduction electron states. Recently, sharp conductance quantization steps have been observed in nanosize point contacts of ferromagnetic metals at room temperature [37-40]. It is possible, because phonon and magnon assisted relaxation processes are quenched due to a large, $\sim 1eV$, exchange splitting of the conduction band. In addition, Oshima and Miyano [38] found a clear indication of the odd-valued number N of open conductance channels ($\sigma = N(e^2/h)$) in nickel point contacts from room temperature up to 770K. Ono *et al* [40] presented the evidence of switching from $2e^2/h$ conductance quantum to e^2/h quantum at room temperature in the nickel nanocontacts of another morphology. Obviously, the change of conductance quantum from $2e^2/h$ to e^2/h is a result of lifting-off the spin degeneracy of the conduction band. Recent calculations [41,42] confirmed the e^2/h conductance quantization in ferromagnetic metals, which is due to non-synchronous opening of "up" and "down" spin-channels of conduction. Experimental observations of conductance quantization steps in point contacts of ferromagnetic metals at room temperature give anticipation that conduction quantization may be responsible for the giant magnitude and the giant fluctuations of magnetoresistance in tiny magnetic contacts.

5.1   Basic formulas for conductance and magnetoresistance

Here we apply our model described above to the case, when conductance of the constriction is quantized. The generalization on the case of conductance quantization means proper re-definition of the transmission coefficient D in the formulas for the conductance. We assume that the connecting channel has the cylindrical shape of arbitrary (but shorter than the mean free path) length d. The channel plays the role of a filter, which selects from the continuous domain of quasiparticle incidence angles only those, which satisfy the energy and momentum conservation laws, and conditions for quantization of the transverse motion of an electron in the channel. As the diameter of the channel is assumed to be very small, we may use the ballistic-limit versions of Eqs. (10), (11) to calculate the conductance of the channel:

$$\sigma^F = \sigma_{\uparrow\uparrow} + \sigma_{\downarrow\downarrow}$$

$$= \frac{e^2}{h} \sum_{m,n} \{ D_{\uparrow\uparrow}(x_{mn}) + D_{\downarrow\downarrow}(x_{mn}) \}, \tag{18}$$

$$\sigma^{AF} = \frac{2e^2}{h} \widetilde{\sum_{m,n}} D_{\uparrow\downarrow}(x_{mn}). \tag{19}$$

Similar formulas may be also obtained within the Landauer-Büttiker scattering formalism [43]. In the above expressions $\sigma_{\alpha\alpha}$ is the conductance for the α-th spin-channel, and $x_{mn} = \cos\theta$ is the cosine of the quasiparticle incidence angle, θ, measured from the cylinder axis direction. $D_{\alpha\beta}(x)$ is the quantum mechanical transmission coefficient for the connecting channel, evaluated in Ref. [16].

Quantization of the transverse motion in the channel imposes the condition that the projection of the incident quasiparticle momentum, parallel to the interface,

$$p_{\parallel} = p_{F\alpha} \sin\theta = p_{mn} \equiv \hbar a^{-1} Z_{mn}, \tag{20}$$

where $p_{F\alpha}$ is the Fermi momentum for the α-th spin-channel, Z_{mn} is the n-th zero of the Bessel function $\mathbf{J}_m(x)$ [16]. This is the *first basic* selection rule. Tilde in Eqs. (6) and (7) means that the summations should be done over the open conduction channels, satisfying the condition:

$$x_{mn} \equiv \cos\theta = \sqrt{1 - (\hbar Z_{mn}/p_{F\alpha}a)^2} \leq 1. \tag{21}$$

When the magnetizations alignment is ferromagnetic, the Fermi momenta on both sides of the contact are equal in each spin-channel. The energy and momentum conservation is already taken into account in Eq. (6) (both the ingoing and outgoing quasiparticles have the same Fermi energy, and the specular character of the scattering is satisfied automatically).

At the antiferromagnetic alignment, conservation of the parallel to the interface momentum ($p_\parallel \equiv p_{F1\alpha}\sin\theta_1 = p_{F2\alpha}\sin\theta_2$, where the subscript 1 or 2 labels left- or right-hand side of the contact, respectively) introduces an *additional* selection rule into Eq. (8):

$$p_{F\alpha} = \min(p_{Fj\uparrow}, p_{Fj\downarrow}). \tag{22}$$

This selection rule strictly holds, when the electron spin conserves upon transmission through the DW. Such a conservation is realized in the model of effectively abrupt DW [15,16] discussed above and in the model of quantum DW [41].

The magnetoresistance is defined by Eq. (9). For the numerical calculations we used the step-like potential barrier, as before, and the sloping potential landscape to approximate the constrained domain wall profile [18]. The former one matches the limiting case of the infinitely large slope and gives maximum available magnetoresistance. The results of calculations for the potential step can be directly compared with our previous quasiclassical calculations, Eq. (1).

5.2   Results of magnetoresistance calculations for the contact of cylindrical cross-section [16]

The numerical routine consists of the summation over the consecutive values of zeros Z_{mn} of the Bessel function, $\mathbf{J}_m(Z_{mn}) = 0$, satisfying the constraint, Eq. (21). At the ferromagnetic alignment of magnetizations $p_{F\alpha} \equiv p_{F\uparrow}$ for the $\sigma_{\uparrow\uparrow}$ contribution to the conductance σ^F, Eq. (18), and $p_{F\alpha} \equiv p_{F\downarrow}$ for the $\sigma_{\downarrow\downarrow}$ contribution. At the antiferromagnetic alignment the minority Fermi momentum of the either spin projection should be used instead of $p_{F\alpha}$ in Eqs. (20) and (21) to calculate the conductance σ^{AF}, Eq. (19). The results are displayed in Figures 2 and 3. The parameter $\delta = p_{F\downarrow}/p_{F\uparrow} \le 1$, Eq. (13), characterizes the conduction band spin-polarization. Calculations revealed that the results depend on the absolute value of $p_{F\uparrow}$, and we have chosen $p_{F\uparrow} = 1\,\text{Å}^{-1}$ for the presentation.

Fig. 2 displays the results of the calculations for $\delta = 0.68$. The panel (a) shows the dependence of F- and AF-conductance on the channel radius. The parameters d and $\lambda = dp_{F\uparrow}\hbar^{-1}$ are the length and dimensionless length of the connecting channel, respectively. The chosen value, $\lambda = 10.0$, corresponds to the connecting channel length 10Å (1 nm). The panel (b) shows the dependence of magnetoresistance on the channel radius. The panels (c) and (d) display the magnetoresistance against F-conductance for the sloping (c) and the step-like (d) potential landscapes in the channel.

Physically, Fig. 2 corresponds to the case, when the AF-alignment conduction opens in the interior part of the first F-conductance plateau. It allows us to make the following conclusions: (1) the F-alignment

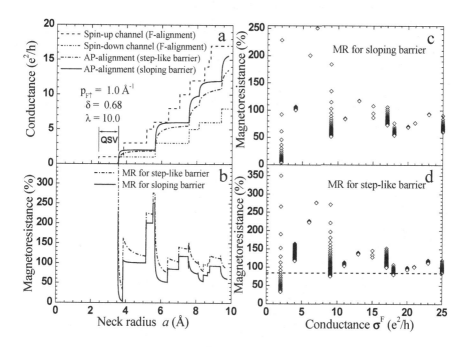

Figure 2. The dependence of conductance (a), and MR (b) on the radius of the neck a. Panels (c) and (d) show dependencies of MR on the number of the open conductance channels at the F-alignment of the magnetizations. The maximal MR=582% for the step-like potential at $\sigma^F = 2e^2/h$ is not shown.

conductance is spin-dependent, the conductions of spin-channels open non-synchronously (see panel (a)), thus resulting in e^2/h quantization of the conductance [41,42]; (2) the magnetoresistance has very sharp and high peak, when the first spin-down (minority) conductance step appears. Simultaneously, the AF-conductance starts growing (panel (b) in correlation with panel(a)); (3) there are no MR points at some numbers of open F-conductance channels (panels (c) and (d)). This is due to degeneracy of quantized motion in the circular area of the contact cross-section; (4) when increasing the nanohole radius (panel (b)), or the number of open channels (panels (c) and (d)), the amplitude of oscillations and sub-steps of magnetoresistance decreases, and the asymptotic value (panel (d)) is given by our quasiclassical theory, Eq. (13); (5) the magnetoresistance, drawn against the F-alignment conductance, is a multivalued function of F-conductance, σ^F (panels (c) and (d)).

Let us discuss some of the above findings in detail. The finding 2) indicates, that the finite magnetoresistance appears when the first conduction channel opens for the spin-down electrons at F-alignment. If the spin-polarization of the conduction band is such, that the spin-down conductance channel appears at the first spin-up conductance plateau, the MR peak should descent to $N^F = 2$ open channels of F-conductance, $\sigma^F = (e^2/h)N^F$. This is the case in our Fig. 2, the experimental results by García *et al.* [1-3] for Ni-Ni and Fe-Fe point contacts clearly reveal the similar tendency.

The finding 4) shows that above 10-15 open conductance channels (in e^2/h quantization), or 8-10Å of nanohole radius, the magnetoresistance fluctuations become relatively small, and the mean value of magnetoresistance converges well to that obtained in quasiclassics (Eq. (13) and Fig. 1).

The finding 5) is crucial for the interpretation of the experimental data. Panel (b) shows a very sharp peak at $a \sim 3.5$ Å. This peak is followed by decay up to $a \sim 3.85$ Å until the new spin-down conduction channel opens at F-alignment. Drawn against the number N^F of open channels (panels (c) and (d)) all points which belong to the peak described above correspond to the single abscissa point, $N^F = 2$. This means that magnetoresistance is a *multivalued* function of the number of open conduction channels at F-alignment, at least when the temperature effects and quenched disorder may be neglected. The magnetoresistance does not oscillate as a function of conductance σ^F, but there are distributions of MR at fixed values of F-alignment conductance, σ^F. The origin of these distributions is clarified by the panel (a) of Fig. 2 in correlation with the panel (b): in spite of the actual radius (in fact, it maybe more correct to speak about the area of a contact) at a trial attempt may vary in the range $3.5 - 3.9$ Å, it gives identical values of F-conductance because of the quantization. At the same time, the AF-conductance depends on the contact radius (area), and this results in the different values of magnetoresistance. The multivalued behavior leads to extremely large fluctuations in the measured magnetoresistance data at the same F-conductance values. Because of the peak shape, the density of points decreases drastically from the small values of magnetoresistance towards the larger ones. As a consequence of decreasing density of points, large, or giant magnetoresistance values appear much less probable, than the low magnetoresistance values. This MR distribution should not be interpreted as poor reliability and reproducibility of measurements.

Increasing the degree of conduction-band spin polarization (decreasing the parameter δ) we observe, that opening of the spin-down conductance channel moves towards next steps in the F-conductance. In Fig. 3 we moved over the threshold. The figure is drawn with the parameter $\delta = 0.57$, so that the spin-down conductance appears now at the second plateau of spin-up conductance at F-alignment (panel (a)). The MR points appear now at $N^F = 4$ open conductance channels at F-alignment, because of the two-fold degeneracy for the particle motion in the cross-section plane. The figure reveals new feature: the lowest number of F-alignment conduction channels, at which the finite magnetoresistance data appear, is confined to the degree of conduction band polarization. For the given set of parameters the threshold is located approximately at $\delta \approx 0.63$. Thus, the minimal number of open channels N^F, at which the finite magnetoresistance points appear, gives us the possibility to estimate lower bound for the conduction band spin-polarization parameter δ.

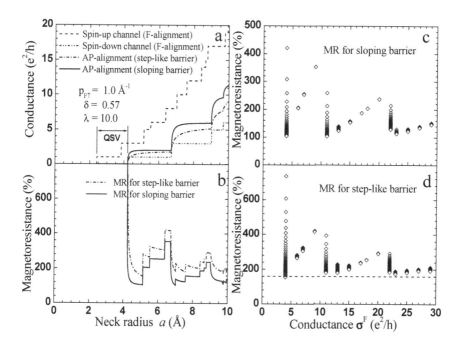

Figure 3. The same as in Fig. 2, but for $\delta = 0.57$. The maximal MR=3953% and MR=1017% for the step-like potential and MR=1612% for the sloping potential at $\sigma^F = 4e^2/h$ are not shown.

Our calculations, panels (c) of Figs. 2 and 3, show that the finite domain wall width does not influence qualitatively the results, which can be deduced from the calculations for the step-like potential barrier due to DW. All conclusions hold, but only the magnitude and the overall width of MR distributions decrease as compared with the results for the abrupt potential of DW in the constriction (panels (d) of Figs. 2 and 3).

5.3   Quantum spin valve regime

Let us emphasize here, that if the antiferromagnetic alignment conductance, σ^{AF}, is equal to zero, then, according to the definition Eq. (9), magnetoresistance diverges. It is just the case if conductance of the constriction is quantized. Figures 2 and 3 clearly show, that there is a range of the contact size (cross-section area) at which the F-alignment conductance is finite, but the AF-alignment conductance is zero. Then magnetoresistance is infinitely large in our idealized model with no reversal of the carriers spin upon transmission through the neck. We called this regime of the point contact operation as quantum spin valve (QSV - indicated in Figs. 2-5) [16,17]. In a more realistic treatment, which allows the electron-spin flip, the AF conductance may be very small but finite at either area of the contact. Then, the magnitude of MR may reach huge values of tens of thousand percents if the spin-flip process is slow. It is likely that the QSV regime has been realized in the experiments by Chopra and Hua [9], which observed up to 100000% MR in electrodeposited Ni nanocontacts.

Our calculations show (see panels (a) of Figs. 2 and 3), that higher the polarization of the conduction band, wider the range of neck size, at which the regime of quantum spin-valve can be realized. Magnetic half-metals with 100% polarization of conduction band would be almost always quantum spin valves at nanometer range of the contact size.

5.4   Influence of the cross-section asymmetry [17]

As we already mentioned above, every experimental point on MR comes from a unique realization of the point contact, the shape and size of which are unknown. One may naturally imagin, that the contact cross-section shape is far from the cylindrical one. The easiest way to introduce deviation from the cylindrical symmetry is to assume the rectangular cross section. It is trivially quantized, and brings principal alteration - lifting-off of the degeneracy in the contact cross-section introducing two independent quantum numbers for the transverse motion. In an actual calculation routine we have to substitute into Eqs. (20) and (21)

$$\lambda_{mn} = \pi \sqrt{\left(\frac{m}{a}\right)^2 + \left(\frac{n}{b}\right)^2} \qquad (23)$$

instead of Z_{mn}/a [17]. In Eq. (23) a and b are the width and height of the neck, m and n are positive integer. The results are displayed in Figures 5 and 6. Below we summarize briefly main changes compared to the case of the cylindrical neck.

Figure 4 displays the results of calculations for the neck of square cross-section ($b = a$) and $\delta = 0.68$. Panel (a) shows the dependence of F- and AF-conductance on the channel size a. As before, d and $\lambda = dp_{F\uparrow}\hbar^{-1}$ are the length in Å and dimensionless length of the connecting channel, respectively. Panel (b) shows the dependence of the magnetoresistance on the channel size a. The panels (c) and (d) display the magnetoresistance against F-conductance for the sloping (c) and the step-like (d) DW potential landscapes in the channel. The magnitude of MR beyond the quantum spin valve regime is well above 200% for very moderate polarization of the conduction band ($\delta = 0.68$).

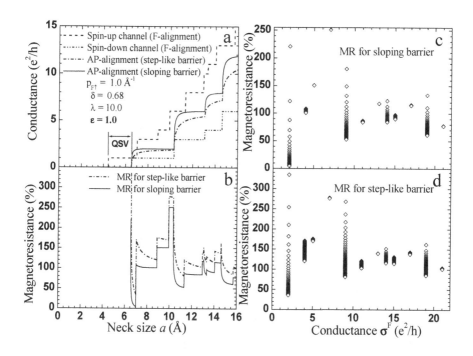

Figure 4. The dependence of conductance (a), and MR (b) on the cross-sectional size of the neck a. Panels (c) and (d) show dependencies of MR on the number of the open conductance channels at the F-alignment of the magnetizations. The maximal MR=563% for the step-like potential at $\sigma^F = 2e^2/h$ is not shown.

In Fig. 5 we change the aspect ratio ε of the sides of the rectangular, $\varepsilon = b/a$, it is drawn with $\varepsilon = 1.5$. Main changes can be summarized as follows: (1) the AF-conductance opens now at three open channels of F-conductance ($\sigma^F = 3e^2/h$; panels (c) and (d)); (2) the range of the neck sizes with zero AF-conductance (quantum spin valve regime) becomes wider; (3) the overall magnitude of MR increases (panels (b)–(d)); (4) magnetoresistance points appear at almost every number of open F-conductance channels [compare panels (c) and (d) in Figs. 4 and 5] reflecting the lifting-off of the degeneracy in the contact cross-section plane (recall that $a \neq b$). We emphasize an issue which has important implication to point contact GMR experiments: number of open F-conductance channels, at which the AF conductance opens ($\sigma^F = 2e^2/h$ in Fig. 4 and $\sigma^F = 3e^2/h$ in Fig. 5), depends not only on the polarization of the conduction band, but also on the asymmetry of the point contact cross-section.

5.5   Discussion of the experiments

Our model calculations show that in the quantized conductance regime the minimal number of open F-conductance channels, at which magnetoresistance points appear, is determined by the conduction band

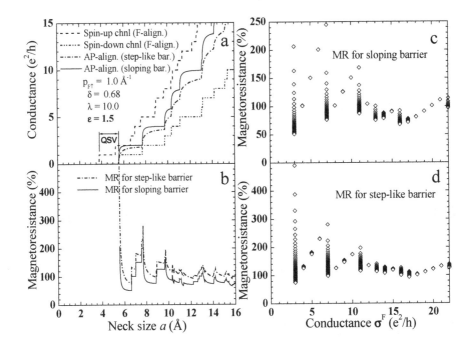

Figure 5. The same as in Fig. 4, but for $\varepsilon = 1.5$. The maximal MR=758% for the step-like potential and MR=322% for the sloping potential at $\sigma^F = 3e^2/h$ are not shown.

polarization δ. In Fig. 2 the AF-alignment conductance channel opens at the first plateau of spin-up conductance. The magnetoresistance appears at $N^F = 2$ open F-conductance channels. Analysis shows, that the threshold of magnetoresistance rise moves from $N^F = 2$ to $N^F = 3$ open F-conductance channels at $\delta \simeq 0.63$. The experiments by García *et al* on Ni-Ni contacts (Fig. 2b of Ref. [1], Fig. 1a of Ref. [2] and Fig. 2 of Ref. [3]) and on Fe-Fe contacts (Fig. 1a of Ref. [3]) clearly indicate, that MR points appear close to $N^F = 2$ for both materials. This means that δ for both, Ni and Fe, is *larger* than 0.63 for our choice of the parameter $p_{F\uparrow} = 1\,\text{Å}^{-1}$.

In contrast, the experimental data for Co-Co contacts, Fig. 2a of Ref. [2], and Fig. 2 of Ref. [3], indicate, that MR points appear at $N^F \sim 3$ open channels, that is at the second plateau of spin-up F-conductance (see panel (a) in correlation with panel (b) of Fig. 5). This suggests that the polarization of the conduction band in Co is higher (and δ is smaller), as compared with Ni and Fe, and allows us to estimate the lower bound for $\delta(\text{Co}) \simeq 0.47 - 0.63$.

Next information may come from the scatter of MR points at small numbers of open F-conductance channels, N^F. We interpret these distributions of experimentally measured MR values as manifestation of the multivalued behavior of magnetoresistance as a function of F-conductance.

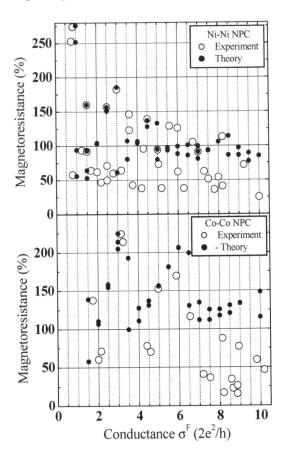

Figure 6. Experimental data and calculated points (δ (Ni)=0.64 and δ (Co)=0.57) for MR put on the same graph. The experimental data are taken from Ref. [3].

In Fig. 6 the calculations of MR are displayed together with the results of experiments on Ni and Co point contacts. Besides the conduction band polarization δ, the contact size a and the length of the channel d determine the values of magnetoresistance in our model. In actual calculations we varied the contact length in the range $\lambda = 4.0 - 10.0$ and the aspect ratio in the range $\varepsilon = 1.0 - 2.0$. Maximal theoretically available MR values for the length of the channel $\lambda = 10.0$ are indicated in the Figure captions. As far as the contact size and shape are not controlled in the break-junction-type experiments by García *et al.* the comparison with the experiment has only qualitative, illustrative meaning. Observing the tunneling and Andreev-reflection data on δ and our figures we may conclude, that using realistic values of δ and the aspect ratios in the range

$\sim 1.0 - 2.0$ we may easily reproduce maximal values as well as giant fluctuations of MR data obtained by García *et al.* [1-3].

The discrepancy of calculated MR with experiments at $N^F \geq 7-8$ (in e^2/h quantization) can be referred to various reasons: (1) when the diameter of the constriction becomes large the domain wall is no longer effectively abrupt (independent of the actual shape), and the magnetoresistance begins to drops very fast converging to the values 2-11% given by the Levy-Zhang mechanism of scattering enhancement in the domain wall [13]; (2) one or several impurities or lattice defects may be located just at the constriction causing additional random deviations of conductance values from integer numbers of e^2/h (see, for example, [44-46] and references therein); (3) the (longitudinal) shape of the constriction may deviate substantially from the uniform one. According to calculations by Torres *et al.* [47] for variable cross-section constrictions with the hyperbolic geometry, the conductance quantization steps survive at the opening solid angles up to $90°$, at least for the small number of open conductance channels. This also leads to deviations of conductance values from integer numbers of e^2/h; (4) independent on the actual shape of the neck, when its length is comparable or longer than the cross-section size, the dipole-dipole anisotropy energy may cause fluctuations between Bloch, Neel or more complicated vortex types of domain walls. Coey *et al* concluded [21] that giant MR of a nanocontact may be reduced somewhat by fluctuations, but not eliminated. We do not expect any effects, which come from the weak localization [48], because the measurements have been made on pure metals at room temperature.

6.   CONCLUSION

In conclusion, we have investigated theoretically the giant magnetoresistance in nanosize magnetic point contacts made of ferromagnetic metal. Our calculations show that the magnitude of magnetoresistance is dramatically enhanced when the ballistic regime of conductance is realized. The ballistic magnetoresistance (BMR) in the quasiclassical regime of conductance can easily reach few hundred percents at experimentally approved polarizations of the ferromagnet conduction band. Next, the regime of quantized conductance through the point contact is considered, and the conductance is calculated for the ferromagnetic (F) and antiferromagnetic (AF) alignments of magnetizations in contacting ferromagnets. Calculations show that BMR of the quantum point contact experiences huge enhancement at first few open conduction channels for the F-alignment of magnetizations. At the same time giant fluctuations of BMR are predicted as a result of conduction quantization. The influence of asymmetry of the contact cross section on the quantum BMR is also analyzed. At certain range of the contact area the F-alignment conductance is finite, but the AF-alignment conductance is zero because of conductance quantization. In this case BMR is infinitely large as far as the electron-spin is conserved upon transmission through the point contact. We called this regime of the magnetic point contact operation as quantum spin-valve (QSV). In a more realistic model BMR has to be limited from above by the conduction-electron spin-reversal process, and can reach tens of thousand percents. It is very likely, that recent observations of huge, 3 000% to 100'000% BMR in nickel point contacts have origin in conductance quantization and realization of the QSV regime. This huge magnetoresistance property survives for every shape of the nanocontact and disorder, provided that: (1) conductance at the ferromagnetic alignment is quantized (steps are not destroyed); (2) the domain wall in the constriction is effectively sharp (the conduction electron spin flip rate is slow).

ACKNOWLEDGMENTS

This work has been supported by Deutsche SFB 491. The authors acknowledge the fruitful cooperation with Dr. B. Vodopyanov. LRT gratefully acknowledges extensive discussion of conductance quantization in nanocontacts with Prof. Thomas Schimmel and support by the EU grant BMR-505282-1 and the RFBR grant No 03-02-17656.

REFERENCES

[1]   N. García, M. Muñoz, and Y.-W. Zhao, Phys. Rev. Lett. **82**, 2923 (1999).
[2]   G. Tatara, Y.-W. Zhao, M. Muñoz, and N. García, Phys. Rev. Lett. **83**, 2030 (1999).
[3]   N. García, M. Muñoz, and Y.-W. Zhao, Appl. Phys. Lett. **76**, 2586 (2000).
[4]   M. A. M. Gijs, G. E. W. Bauer, Adv. Phys. **46**, 285 (1997).
[5]   J.-Ph. Ansermet, J. Phys.: Cond. Matt. **10**, 6027 (1998).
[6]   N. Garca, M. Munoz, V.V. Osipov *et al.*, J. Magn. Magn. Mater. **240**, 92 (2002).
[7]   H.D. Chopra, S.Z. Hua, Phys. Rev. B **66**, 020403(R) (2002).
[8]   H. Wang, H. Cheng, N. Garcia, cond-mat/0207516 (22 July 2002).
[9]   S.Z. Hua, H.D. Chopra, Phys. Rev. B **67**, 060401(R) (2003).
[10]   J.-E. Wegrowe, T. Wade, X. Hoffer *et al.*, Phys. Rev. B **67**, 104418 (2003).
[11]   G.G. Cabrera and L.M. Falicov, Phys. Stat. Solidi (b) **61**, 539 (1974); **62**, 217 (1974).
[12]   L. Berger, J. Appl. Phys. **49**, 2156 (1978); **69**, 1550 (1991).
[13]   P.M. Levy, Sh. Zhang, Phys. Rev. Lett. **79**, 5110 (1997).
[14]   J.B.A.N. van Hoof *et al.*, Phys. Rev. B **59**, 138 (1999).
[15]   L.R. Tagirov, B.P. Vodopyanov, K.B. Efetov, Phys. Rev. B **63**, 104468 (2001).

[16] L.R. Tagirov, B.P. Vodopyanov, K.B. Efetov, Phys. Rev. B **65**, 214419 (2002).

[17] L.R. Tagirov, B.P. Vodopyanov, B.M. Garipov, J. Magn. Magn. Mater. **258-259**, 61 (2003).

[18] P. Bruno, Phys. Rev. Lett. **83**, 2425 (1999).

[19] L.L. Savchenko, A.K. Zvezdin, A.F. Popkov, K.A. Zvezdin, Fiz. Tverd. Tela (St. Petersburg) **43**, 1449 (2001) [Phys. Solid State **43**, 1509 (2001)].

[20] V.A. Molyneux, V.V. Osipov, E.V. Ponizovskaya, Phys. Rev. B **65**, 184425 (2002).

[21] J.M.D. Coey, L. Berger, Y. Labaye, Phys. Rev. B **64**, 020407 (2001).

[22] Y. Labaye, L. Berger, J.M.D. Coey, Journ. Appl. Phys. **91**, 5341 (2002).

[23] A. Overhauser, Phys. Rev. **89**, 689 (1953).

[24] J.F. Gregg, W. Allen, K. Ounadjela *et al.*, Phys. Rev. Lett. **77**, 1580 (1996).

[25] M.B. Stearns, J. Appl. Phys. **73**, 6396 (1993).

[26] Yu.V. Sharvin, Zh. Exp. Teor. Fiz. **48**, 984 (1965) [Sov. Phys. - JETP **21**, 655 (1965)].

[27] L.D. Landau and E.M. Lifshitz, *Quantum Mechanics*, §25, Butterworth-Heinemann, Oxford, 1995.

[28] A.V. Zaitsev, ZhETF **86**, 1742 (1984) [Sov. Phys. - JETP **59**, 1015 (1984)].

[29] M. Julliere, Phys. Lett. A **54**, 225 (1975).

[30] J.C. Gröbli *et al.*, Physica B **204**, 359 (1995).

[31] R.J. Soulen *et al*, Science **282**, 85 (1998); J. Appl. Phys. **85**, 4589 (1999).

[32] S.K. Upadhyay, A. Palanisami, R.N. Louie and R.A. Buhrman, Phys. Rev. Lett. **81**, 3247 (1998).

[33] N. García, H. Rohrer, I.G. Saveliev, Y.-W. Zhao, Phys. Rev. Lett. **85**, 3053 (2000).

[34] N. García, M. Muños, G.G. Qian, H. Rohrer, I.G. Saveliev, Y.-W. Zhao, Appl. Phys. Lett. **79**, 4550 (2001).

[35] B.J. v. Wees, H. v. Houten, C.W.J. Beenakker *et al*, Phys. Rev. Lett. **60**, 848 (1988).

[36] D.A. Wharam, T.J. Thornton, R. Newbury *et al*, J. Phys. C **21**, L209 (1988).

[37] J.L. Costa-Krämer, Phys. Rev. B **55**, 4875 (1997).

[38] H. Oshima, K. Miyano, Appl. Phys. Lett. **73**, 1103 (1998).

[39] F. Ott, S. Barberan, J.G. Lunney *et al*, Phys. Rev. B **58**, 4656 (1998).

[40] T. Ono, Y. Ooka, H. Miyajima, Appl. Phys. Lett. **75**, 1622 (1999).

[41] H. Imamura, N. Kobayashi, S. Takahashi and S. Maekawa, Phys. Rev. Lett. **84**, 1003 (2000).

[42] A.K. Zvezdin, A.F. Popkov, JETP Lett. **71**, 209 (2000).

[43] R. Landauer, IBM J. Res. Dev. **32**, 306 (1988); M. Büttiker, IBM J. Res. Dev. **32**, 317 (1988).

[44] C.S. Chu, R.S. Sorbello, Phys. Rev. B **40**, 5941 (1989).

[45] P.F. Bagwell, Phys. Rev. B **46**, 12573 (1992).

[46] P. García-Mochales, P.A. Serena, N. García, J.L. Costa-Krämer, Phys. Rev. B **53**, 10268 (1996).

[47] J.A. Torres, J.I. Pascual, J.J. Sáenz, Phys. Rev. B **49**, 16581 (1994).

[48] G. Tatara and H. Fukuyama, Phys. Rev. Lett. **78**, 3773 (1997).

8

PART 8: APPLICATIONS OF NANOSTRUCTURED MAGNETIC MATERIALS

MAGNETOIMPEDANCE OF THE NOVEL MAGNETIC NANOSTRUCTURES IN EXTRA HIGH FREQUENCY BAND: PHYSICS, TECHNIQUE AND APPLICATION

Sergey Tarapov

Institute of Radiophysics and Electronics NAS of Ukraine,12 Ac Proskura St., 61085, Kharkov, Ukraine; e-mail: tarapov@ire.kharkov.ua

Abstract: One of the promising areas for application of nanostructures is high-frequency electronic industry. Thus, the problem of study of nanostructures in the frequency band which coincides with the band of their application attracts today attention of experts. In given report the results of extra-high frequency (EHF) electromagnetic wave interaction with various layered nanostructures are discussed. The behavior of Giant Magnetoresistance (GMR) effect depending on the applied static magnetic field is compared with similar EHF-phenomenon - Giant Magnetoimpedance (GMI). We used the Electron Spin Resonance (ESR) technique at frequencies 10-40 GHz, 65-80 GHz and the non-resonance EHF technique at 40-140 GHz as main experimental methods of research. Examples of applications of GMI nanostructures under study for design of electronically controlled EHF-devices are discussed.

Key words: nanostructures, multilayers, Electron Spin Resonance, extra-high-frequency band

1.   INTRODUCTION

Nanostructures appear to be promising materials for application in the Extra High Frequency (EHF) technology and design because their magnetic properties are rather sensitive to the influence of alternative current of this frequency band [1-3]. On this reason we studied the nanostructures (such as layered magnetic metallic systems) in the frequency band which coincides with the band of their prospective application by the Electron Spin Resonance (ESR) technique[5, 6]. In this paper the resumptive review of

B. Aktaş et al. (eds.), Nanostructured Magnetic Materials and their Applications, 421–432.

researches performed in IRE NASU (Kharkov, Ukraine) and KhPI (Kharkov, Ukraine) is presented and results of the studies are analyzed.

2. SPECIMENS

We worked mainly with GMR multilayered (ML) structures Fe(6 nm)/[Co(t nm)/Cu(d nm)]$_8$ and Fe(6 nm)/[Co(t nm)/Cu(d nm)]$_{30}$ with magnetoresistance ratio of 5-10% [4,5]. The Co-layers thickness varied from t=1 nm to t=4 nm. The Cu-layers thickness varied in the range d of 1.7 nm-7.4nm. The films have been produced using a combination of two methods: triode ion-plasma sputtering for cobalt and iron deposition and magnetron sputtering for copper deposition. The films have been deposited on non-cooled glass-ceramic substrates with specially prepared surfaces covered by amorphous oxide to diminish surface roughness. Multilayer samples under study were polycrystalline and exhibited rather well defined laminated structure according to data of X-ray diffraction and electron transmission microscopy of oblique cross-sections. Also it was proved that any detectable interfacial mixing in ML structures under study is absent. In more detail the procedure of manufacturing and physical properties of such multilayers is described in Ref. [7].

3. EXPERIMENTAL AND ANALYSIS

We used the ESR method to study the magnetic characteristics of the films [4]. ESR method allows to separate the contribution of both Fe - buffer layer and Co - multilayers to the magnetic structure of the system. A typical line shape [8], where two peaks ("sharp" - for Fe and "weak" - for Co) are presented is shown in Fig. 1. As the experimental data was detected in rather wide frequency range 20-75 GHz, we were able to get resonance field - resonance frequency dependence with high enough accuracy (Fig. 2). More detailed procedure is described in [2, 6, 8]. Here we note only that using known Kittel formulas for these dependencies we determined a dynamic magnetization for each component of the multilayered structure (for Co and Fe) separately.

The data on dynamic magnetization for all set of specimens allow to make general conclusion that the specimens with largest Cu-layer thickness (t=7.4 nm) have the largest magnetization values as well. At the same time, the magnetization is almost independent from Co-layer thickness.

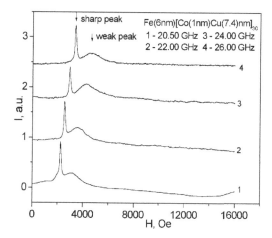

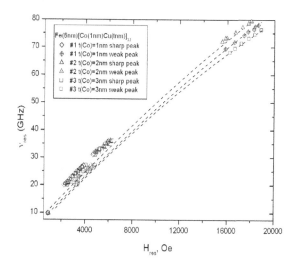

Figure 1. ESR data for Co/Cu multilayers: a) Typical ESR-line shapes for Co (weak peaks) and Fe (sharp peaks) in the Fe(Co/Cu) multilayers in millimeter waveband; b) Resonance frequency-field dependence.

Another, more important conclusion is that the multilayered nanostructures with large magnetization magnitudes are more perspective (suitable) for the application. Because of this, with increase of the magnetization magnitude the external magnetic field magnitude, at which the GMI effect occurs, decreases. So we can suppose that multilayers with highest distances between Co layers are most suitable for EHF application. Of course, this distance should not be larger than that at which interlayer exchange interaction is noticeable.

In order to investigate in detail the DC-resistance behavior in presence of the external magnetic field, the dependence $\Delta R/R = f(H)$ (Fig. 2a) was detected during the static experiments. Besides, the same characteristics have been detected for AC currents, i.e. for the frequencies f~30-140 GHz. An interesting result was revealed, that the behavior of the transmission coefficient (T) of the EHF wave, propagating through the ML structure (i.e. $\Delta T/T = f(H)$) in the entire frequency band 30-140 GHz, completely repeats the same dependence for DC- resistance: $\Delta R/R = f(H)$ [8]. The dependencies of such type are shown in Fig. 2b. Let us note that T coefficient is in direct proportion to the EHF impedance $\dot{Z}$ within the accuracy of the experiment. Characteristic points on these graphs are the static field magnitudes H_m (the field of maximum magnetic disordering of the ML structure ≈200 Oe) and H_j (the field at which the domain walls start to rotate ~70 Oe).

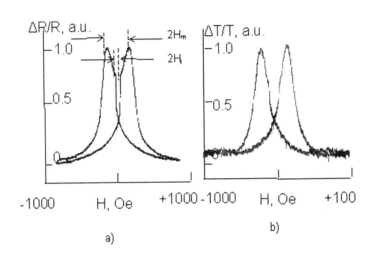

Figure 2. GMR (a) and GMI (b) responses for Fe(Co/Cu) nanostructures: a) DC response b) AC response at 120 GHz.

The dependence presented in Fig. 2 is of hysteresis type. The curve that describes the magnetoresistance variation when the static magnetic field magnitude changes from the negative magnitude to positive one is bilaterally symmetrical to the curve that describes the reverse variation of the static field. The magnitude of this hysteresis equals 200-300 Oe. As it is known this phenomena is defined by the domain type of the magnetic structure under study.

We studied this hysteresis in small area of variation of external static magnetic fields in order to determine the optimal conditions for application of ML nanostructures in high-frequency technologies. As a result the so-called "partial" hysteresis (Fig. 3. position 1) on the EHF dependence $\Delta T/T = f(H)$ has been detected. In Fig. 3 the segment of $\Delta T/T = f(H)$ dependence is presented for the case when the reverse trace of the curve is started from the field magnitude close to $+H_m$. In such case one can observe typical hysteresis dependence with a hysteresis magnitude of about 200 Oe.

The remarkable fact is that the reverse lines on the partial hysteresis (1) for the field, which is scanned in small band ~50 Oe, are located closely to direct lines, so the loops of "partial" hysteresis are rather small and have approximately linear type of the field dependence. At the same time the type of the impedance variation on the external H-field have the similar character as for the main impedance.

Another interesting result is that thorough study of the curves presented in Fig. 3 allows to state that linear segments of $\Delta T/T$ dependence take place. This result is rather valuable from the point of view of ML application as EHF amplifying-transforming devices.

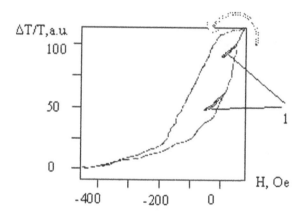

Figure 3. The general hysteresis and the "partial" (1) hysteresis of the magnetoimpedance for the multilayered nanostructures Fe(Co/Cu) for f=44 GHz.

On the basis of the results described above it is necessary to make certain conclusions which determine the possible areas of application nanostructures (particularly - multilayers):

1. $\Delta T/T$ = f(H) repeats completely the $\Delta R/R$ = f(H) dependence in the entire frequency band f ~ 30-140 GHz.

2. The relaxation times of the processes, which are responsible for the magnetoimpedance have values $\tau \approx f \leq 10^{-11}$s.

3. The linear-type dependence and the small width of partial hysteresis allows to assume that while applying the external field *H*, which varied by a certain time-law, it is possible to reveal the similar time dependence for the impedance.

4. Specimens with largest magnetization are most promising for the GMI applications in EHF band.

Thus the nanostructures under study are rather promising for the EHF technology applications. Let us note that main directions of the researches, which should overcome disadvantages of these structures are:

- to increase the $\Delta T/T$ = f(H) magnitude up to (30-50%)

- to increase the magnetization and anisotropy fields in order to decrease the external field of the electron spin resonance.

4. APPLICATIONS OF NANOSTURCTURES FOR EHF DESIGN

In spite of some imperfections mentioned in previous chapter ML nanostructures seem to be promising elements for the EHF-band application due to reasons listed above.

The remarkable outcome one can make from them is that the usage of these ML nanostructures allows to arrange electrical/electronical management of their EHF parameters in contrast to mechanical ones (as it takes place in the majority of contemporary devices). Let us note that this is a traditional way of technology: to replace the mechanical management of the device with the electronical one. This way leads to improvement of the reliability of the devices and to increasing the accuracy of processes control.

We would like to present below several directions conducted by our Laboratory in order to describe possible applications of nanostructures in some areas of EHF design.

4.1   Attenuator. Frequency Filter

An effective frequency filter of millimeter (mm) wavelength band can be created on the base of Disk Dielectric Resonator (DDR) [11] with flat surface covered with GMI multilayer.

A model of such filter on the cryogenic test-bench is presented in Fig.4. Several types of disk dielectric resonators (DDR) have been manufactured for the electrodynamical test experiments. We used high quality materials: isotropic polycrystalline quartz; single-crystal quartz and single-crystal sapphire. Taking into account the mentioned frequency band we calculated their geometrical parameters such that it will have in their spectra the whispering gallery modes. The radius of the DDR is about $R \approx (4\text{-}7) \cdot \lambda \cdot \varepsilon^{-1/2}$ (λ is the wavelength; ε is the DDR material permittivity). The thickness of DDR is about $t \approx 0.5 \cdot \lambda \cdot \varepsilon^{-1/2}$. The GMR multilayer is deposited on the both of flat surfaces of the disk. The disk dielectric resonator with ML placed between two coupling waveguides shown in Fig. 4a. An external static magnetic field $H_{st} \sim 300$ Oe is created by the coil located under the DDR.

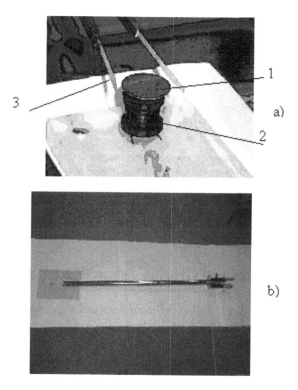

Figure 4. Modeling of the EHF filter based on the GMI multilayer structures: a) DDR covered by GMI multilayer with quartz waveguides; b) DDR placed on the cryogenic module for experiments at T<4K.

The principle of operation of such EHF filter is based on the fact that DDR has rather high quality factor for "whispering gallery" oscillation modes:

$Q{\sim}3\text{-}5{\times}10^4$ at the room temperature and $Q{\sim}(1\text{-}3){\times}10^5$ at liquid helium temperatures. This provides the width of DDR's resonance curve (Fig. 5) of order $\delta f{\approx}30$ MHz for $T{=}300$ K and $\delta f{\approx}2\text{-}7$ MHz for T<4.2 K. The transmission coefficient changes its magnitude $\delta T{\sim}70\%$ while the frequency of the input signal varies in the band of $f_0{\pm}\delta f$. In order to realize the electronical tuning/control of the filter it is necessary to vary the static magnetic field. As a result, the flat boundary of DDR changes its impedance and the eigen-frequency of the oscillation changes.

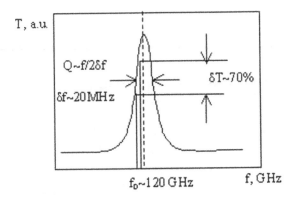

Figure 5. Resonance curve for the electronically controlled frequency filter on the GMI structure implemented in DDR.

Thus, one can shift the resonance frequency f_0 of the DDR in the range of about 5-7%. For example, in the 150 GHz band the frequency tuning range is approximately 5-9 GHz.

4.2   Antenna design (Griding - structure)

The multilayered structures with GMI can be used successfully in antenna design. An electrically managed phase-locked antenna device designed on the basis of GMI multilayared strips is shown in Fig. 6. The ML-strips are located in such a way that they form an antenna array [9]. When the incident wave with the wave vector $\vec{k}$ falls on the array at some angle θ_I, it excites the reflected wave propagating at the angle θ_R. The θ_R angle is defined by the geometrical parameters (the width) of the strips and by high-frequency currents $\overset{\bullet}{J}$ distribution in the strips as well. Strips are under the influence of the static magnetic field, produced by conventional wire conductors (2) with the DC currents located between ML-strips (1).

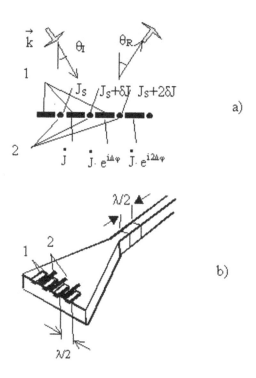

Figure 6. Phased-locked antenna array on the GMI elements.
a) Principal flow chart b) design of the horn antenna array.

The distribution of static currents is presented in Fig. 6a. Variation of the static current on dJ leads to the variation of the static magnetic field created by these wires by some δH value. In its turn, it calls a definite phase shift $(\Delta\varphi)$ in AC currents which are exited in ML strips. Thus, it is possible to manage the EHF-currents distribution in the ML-strips while changing the DC current in the wires. Due to this a certain phase shift between EHF-currents in the strips can be provided. Following the known rules [9] of the phase-locked antenna-arrays design it is possible to arrange an electronical spatial scan of the antenna pattern. It is possible to change the angle of directivity of such antenna array varying the distribution of DC - currents amplitude (δJ).

In Fig. 6.b the model of the horn antenna combined with the GMI multilayered array providing the scan of the angle of radiation is presented as well. Preliminary experiments demonstrated the possibility to vary the antennas angle of radiation by the magnitude of about $\alpha \sim 5^0$-9^0.

4.3   EHF amplifying devices

On the reason of very small relaxation times of GMI processes $\tau \approx 1/f \approx 10^{-11}$s, it is possible to employ ML structures in the area of amplifying (both linear and non-linear) radio-signals in the band 20-150 GHz [10].

4.3.1   Transistor. (The "linear" GMI-EHF-device)

The electromagnetic structure (resonator, waveguide, etc.) covered with the GMI nanostructure can be considered as a transistor-type device of the EHF band. The principle of such transistor operation is presented in Fig. 7. Here the first (increasing) segment belongs to the $\Delta T/T = f(H)$ dependence of GMI structure is shown.

Let us choice an EHF resonator included the GMI structure as an active element. For example the Disk Dielectric Resonator (Fig. 4) can be used.

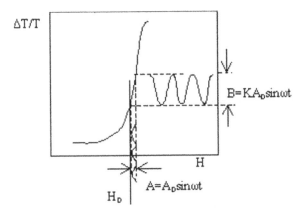

Figure 7. Principle of transistor on GMI phenomenon design.

Then let us choose the linear fragment of the $\Delta T/T = f(H)$ curve such as it presented in Fig. 7. In other words let us apply to the GMI - element located in electrodynamical structure the external static magnetic field of the definite value H_0. Then impose the "information" EHF signal ($A = A_0 \cdot \sin\omega t$) with some amplitude ($A_0$) on the external magnetic field. Here ω belongs to the GHz band. It is evident from Fig. 7 that the variation of T follows this signal under the same harmonic law ($B = K \cdot B_0 \cdot \sin\omega t$) with the accuracy restricted only by linearity of the segment choused. Here the amplification coefficient (K) is defined by the slope of $\Delta T/T = f(H)$ curve.

4.3.2   Harmonic generator. The "non-linear" GMI-EHF-device

The same principle is the basis of the design of device which uses non-linear properties of GMI nanostructures, i.e. - harmonic generator. Let us note that the principle of devices of such kind, based on ferrites properties, is used in the microwave radio science for a long time. But in the given case when the ML structure is the active element, the external magnetic fields values (500 Oe instead 5000 Oe), required for its operation, are noticeably lower. The ML structure shows much smaller high-frequency loss in comparison with the traditional ferrite system, due to the physical nature of the materials.

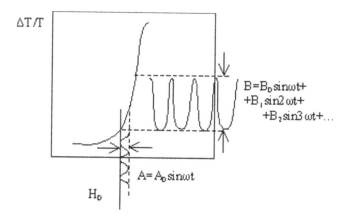

Figure 8. Principle of GMI EHF harmonic generator

Let us take the same DDR resonator covered with the GMI multilayer as it was described in the previous chapter. Then impose some external static magnetic field H_0. Now let us apply the harmonic signal not to the linear segment of $\Delta T/T = f(H)$ but to some essentially non-linear segment as it is shown in Fig. 8. In such a case notwithstanding the applied signal is a harmonic one with the frequency ω the resulting signal $\Delta T/T = B(t)$ acquiring essentially non-harmonic character. In the given case it can be presented by a Fourier series of signals ($B = B_0 \cdot \sin\omega t + B_1 \cdot \sin 2\omega t + B_2 \cdot \sin 3\omega t + \ldots$) with frequencies: ω; 2ω; 3ω; $4\omega \ldots$etc..

5.   CONCLUSIONS

Thus, using the multilayered nanostructures designed for application in the EHF, it is possible to affirm that the GMI nanostructures are promising elements which can be a basis for the electronically controlled Extra High Frequencies devices. Electrical (instead of the mechanical) type of control of the elements properties provides increasing of reliability, reproducibility and accuracy of EHF elements and devices. Short relaxation times of the tuning process, $\tau \sim 10^{-11}$s, provide operation of such devices at frequencies at least not smaller than $f \sim 140$ GHz.

The work is partially supported by STCU grant #1916.

REFERENCES

1. A.B.Rinkevich, L.N.Romashev, V.V.Ustinov, JETP (2000) **117**, N5, 960.
2. D.P.Belozorov, V.N.Derkach, S.I.Tarapov etc., Int. J Infrared Millimeter Waves, (2001) **22**, N11, 1669.
3. A.Ludwig, M. Tewes, S. Glasmachers, etc.M. Lohndorf, E. Quandt, J. Magn. Magn. Mater, (2002) **242-245**, 1126.
4. S.Tarapov, Basic of High-Frequency Electron Spin Resonance Experiment at Very Low Temperatures, Publ. Center of GIT, Gebze, Turkey, (2000) 93p
5. V.N.Derkach, S.V.Nedukh, A.G.Ravlik, I.G.Shipkova, S.T.Roschenko, S.I.Tarapov, Radiophysics and Electronics, (2002) **7**, N1, 115.
6. D.P.Belozorov, V.N.Derkach, S.I.Tarapov, Foreign Radio Electronics. Advances in Modern Radio Science (2002) **12**, 48.
7. S.T.Roschenko, A.G.Ravlik, I.G.Shipkova, Yu.A.Zolotarev, Phys. Met. Metallogr. (2000) **90** N3, 58.
8. D.P.Belozorov, V.N.Derkach, S.V.Nedukh, A.G.Ravlik, S.T.Roschenko, I.G.Shipkova, S.I.Tarapov, F.Yildiz, Intern. Journ. of Infrared and Millimeter Waves, (2001) **22**, N11, 1669.
9. C.Balanis, Antenna Theory, Wiley &Sons, 1997, 942p.
10.A.A.Vertiy, S.P.Gavrilov, S.I.Tarapov, V.P.Shestopalov, Reports of Acad. of Sci. USSR, (1992) **323**, N2, 270.
11.V.N.Derkach, R.V.Golovashchenko and A.S.Plevako, Proc. of 12th Int. Conf. "Microwave & Telecommunication Technology" (CriMiCo'2002). September 9-13, 2002, Sevastopol, Crimea, Ukraine. Sevastopol: "Weber" (2002) pp.548-549.

MAGNETOREFRACTIVE EFFECT IN MAGNETIC NANOCOMPOSITES IN REFLECTION: DEPENDENCIES ON INCIDENT ANGLE AND POLARIZATION OF LIGHT

A. Granovsky [a], A. Kozlov [a], A. Yurasov [a], M. Inoue [b], J.P. Clerc [c]

[a] Faculty of Physics, Lomonosov Moscow State University, 119992 Moscow, Russia
[b] Toyohashi University of Technology, Toyohashi 441-8580, Japan
[c] Ecole Polytechnique, Universitaire de Marseille, Technopole de Chateau Gombert, 13453 Marseille, France

Abstract: We calculate the magnetorefractive effect in ferromagnetic metal-insulator nanocomposites in reflection as a function of an incident angle for p- and s-polarization of light. The MRE is larger for p-polarized light and attains the maximum value for the incident angle ϕ_0 close to the Bruster angle. The difference between the MRE for p-and s-polarized light when $30^0 < \phi_0 < 70^0$ is noticeable, but not well pronounced for metallic samples, and by contrast, can be very large in insulating samples. The available experimental data on angular and polarization dependencies of the MRE in various nanocomposites are analyzed.

Key words: magnetorefractive effect, nanocomposites.

1.   INTRODUCTION

Recently, a new magneto-optical effect, called as the magnetorefractive effect (MRE), has attracted considerable attention (see [1-3] and references therein). The MRE consists in magnetization-induced changes in optical properties (reflection, transmission, absorption) of any kind of magnetic material with large magnetoresistance. Since the MRE is a frequency analogue of magnetoresistance, its main mechanism, and hence its magnitude, strongly depend on the magnetic system under consideration. The MRE is due to spin-dependent scattering [4,5] in all-metallic magnetic

B. Aktaş et al. (eds.), Nanostructured Magnetic Materials and their Applications, 433–440.

multilayers (for example, Fe/Cr [4,6]) and granular alloys (for example, Co_xAg_{1-x} [7]) exhibiting giant magnetoresistance. In manganites with colossal magnetoresistance the MRE arises because of the magnetization-induced phase transition [8]. At last, the MRE in magnetic nanocomposites with tunnel type magnetoresistance originates from high frequency spin-dependent tunneling [1-3]. The MRE theory for nanocomposites [1-3] is based on the assumption that the tunnel junction between adjacent granules is a capacitor. The same tunnel junction in nanocomposites close to the percolation threshold is assumed to be responsible for both the tunnel magnetoresistance at low frequencies and the MRE at high frequencies. In spite of some limitations of the theory discussed in details in [3], this simple approach is in a qualitative agreement with the most of experimental data for the MRE in nanocomposites in reflection at normal incidence.

The main goal of this communication is to extend this theory to the case of an arbitrary incident angle for s- and p-polarized light. For the best of our knowledge, the angular dependence of the MRE has not been theoretically studied yet in both metallic and non-metallic nanostructures.

2.   THEORY

Let us consider reflection of p-polarized from a thick magnetic sample magnetized in its plane. One can observe in this geometry three different magnetooptical effects: the Kerr effect, the orientational effect [9], and the MRE.

The Kerr effect is linear with magnetization M and therefore can be easily determined from experiment. Besides, its angular dependence is well known.

Both the orientational effect and the MRE are quadratic in magnetization. But since the orientational effect is due to the spin-orbit interaction squared, it is extremely small in comparison with the MRE in magnetic nanostructures with noticeable magnetoresistance [1]. Neglecting traditional magnetooptical Kerr and orientational effects, for p-polarized light incident from a transparent insulator (medium 1 with real refractive index n_1) at an angle ϕ_0 on the magnetic sample (medium 2 with a complex refractive index $\eta_2 = n_2 - ik_2$), the reflectance R can be written as

$$R = |r_{12}^p|^2, \tag{1}$$

$$r_{12}^p = \frac{g_1 n_2^2 - g_2 n_1^2}{g_1 n_2^2 + g_2 n_1^2}, \tag{2}$$

$$g_1 = \sqrt{n_1^2 - n_1^2 \sin^2 \phi_0}, \ \ g_2 = \sqrt{n_2^2 - n_1^2 \sin^2 \phi_0}. \tag{3}$$

By definition of the MRE in reflection, we have

$$\frac{\Delta R}{R} = \frac{R(M=0) - R(M)}{R(M=0)}, \tag{4}$$

$$\frac{n_2 - n_2^0}{n_2^0} = cM^2; \ \ \frac{k_2 - k_2^0}{k_2^0} = dM^2, \tag{5}$$

where n_2^0 and k_2^0 are the refractive and extinction coefficients in demagnetized state of the magnetic sample, correspondingly, c and d are the MRE parameters which depend on its mechanism. These simple expressions (1-5) determine completely the angular dependence of the MRE in refraction for p-polarized light in any kind of materials. Similar expressions can be easily obtained for the s-polarized light.

If medium 1 is vacuum ($n_1 = 1$) and the incidence of light is close to normal ($\phi_0 = 0$), it immediately follows from (1-5) that

$$\left(\frac{\Delta R}{R} \right)_{\phi_0 = 0} = (1 - R) M^2 \left[c \frac{1 - \left(n_2^0 \right)^2 + \left(k_2^0 \right)^2}{\left(1 - n_2^0 \right)^2 + \left(k_2^0 \right)^2} - 2d \frac{\left(k_2^0 \right)^2}{\left(1 - n_2^0 \right)^2 + \left(k_2^0 \right)^2} \right]. \tag{6}$$

If $cM^2 \ll 1$, $dM^2 \ll 1$ (that is true when magnetoresistance is not too large, see Eq. (10) below), expression for the angular dependence of the MRE can be written as follows

$$\frac{\Delta R}{R} = \frac{4M^2}{[(a_1\cos\varphi_0 - g_1)^2 + (a_2\cos\varphi_0 - g_2)^2][(a_1\cos\varphi_0 + g_1)^2 + (a_2\cos\varphi_0 + g_2)^2]}$$

$$[(a_1^2\cos^2\varphi_0 - g_1^2)(b_1 g_1\cos\varphi_0 - a_1\cos\varphi_0 \mathrm{Re}(\frac{b}{2g})) +$$

$$+(a_2^2\cos^2\varphi_0 - g_2^2)(b_2 g_2\cos\varphi_0 - a_2\cos\varphi_0 \mathrm{Im}(\frac{b}{2g})) +$$

$$+2b_1\cos\varphi_0(2a_1 a_2 g_2\cos^2\varphi_0 - g_1 g_2^2 - g_1 a_2^2\cos^2\varphi_0) +$$

$$+2b_2\cos\varphi_0(2a_1 a_2 g_1\cos^2\varphi_0 - g_2 g_1^2 - g_2 a_1^2\cos^2\varphi_0) -$$

$$-2\mathrm{Re}(\frac{b}{2g})(a_1 a_2^2\cos^3\varphi_0 + a_1 g_2^2\cos\varphi_0 - 2g_1 a_2 g_2\cos\varphi_0) -$$

$$-2\mathrm{Im}(\frac{b}{2g})(a_2 a_1^2\cos^3\varphi_0 + a_2 g_1^2\cos\varphi_0 - 2g_1 a_1 g_2\cos\varphi_0)]$$

(7)

where

$$a=a_1-ia_2,\ a_1=n^2-k^2,\ a_2=2nk,\ b=b_1-ib_2,\ b_1=2cn^2-2dk^2,\ b_2=2nk(c+d),$$ (8)

$$g=g_1-ig_2=\sqrt{(a_1-ia_2)^2-\sin^2\phi_0},$$ (9)

in which index 2 indicating the magnetic medium and upper index 0 indicating the demagnetized state have been omitted.

The expression (7) has a general character and can be applied to any material. In the case of magnetic nanocomposites in the framework of the approach developed in [1]

$$dM^2 = \frac{\Delta\rho}{\rho}\frac{1}{1+\left(\dfrac{k}{n}\right)^2};\qquad cM^2 = \frac{\Delta\rho}{\rho}\frac{\left(\dfrac{k}{n}\right)^2}{1+\left(\dfrac{k}{n}\right)^2},$$ (10)

where

$$\frac{\Delta\rho(H)}{\rho} = \frac{\rho(0) - \rho(H)}{\rho(0)} \tag{11}$$

is tunneling magnetoresistance in a magnetic field H.

3. RESULTS AND DISCUSSION.

Figs. 1,2 show possible angular dependences of the MRE in magnetic nanocomposites. The dependences were calculated using Eqs. (7-10). Optical parameters are taken to those corresponding to nanocomposites $Co_x(Al_2O_3)_{1-x}$ with x close to the percolation threshold x_c for the wavelength $\lambda = 9$ μm [10]. Fig. 1 corresponds to the insulator side $x<x_c$ of the metal-insulator transition. Fig. 2 shows the angular dependence for metallic samples $(x>x_c)$. For both cases we assume that magnetoresistance is of 3%. These parameters for optical constants, magnetoresistance, and the wavelength of incident light are typical for the MRE measurements in nanocomposites [1].

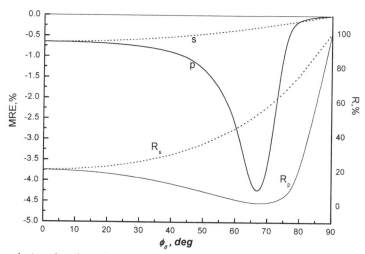

Figure 1 Angular dependencies of the magnetorefractive effect $\Delta R/R$ and the reflectance R in insulating nanocomposites (below the percolation threshold, $x<x_c$) for p-polarized (solid line) and s-polarized (dot line) light. $\Delta\rho/\rho=3\%$, n=2.5, k=0.5.

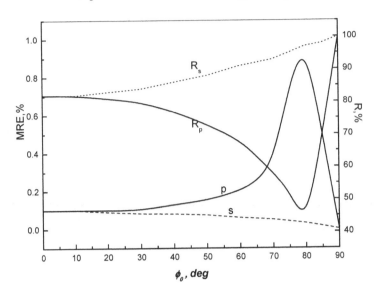

Figure 2 The same as on Fig. 1 for metallic nanocomposites (above the percolation threshold, $x > x_c$). $\Delta\rho/\rho = 3\%$, $n=4$, $k=8$.

Some of results shown on Figs. 1,2 can be summarized as follows:

1. There is a correlation between the MRE and the reflectance for both polarizations of light, as could be expected, but the correlation is not linear.

2. The MRE does not depend on magnetization orientation in the sample plane. It was confirmed in [11].

3. For small incident angles ($\phi_0 < 10^0$) the MRE does not depend on polarization of light. It is in agreement with experimental data shown on Fig. 4 of ref. [1] for CoFeZr- SiO$_{1.7}$ nanocomposites .

4. The MRE is larger for p-polarized light and attains the maximum value for ϕ_0 close to the Bruster angle ϕ_B. For p-polarized light the MRE increases with ϕ_0 at $\phi_0 < \phi_B$. The Bruster phenomenon is well pronounced for p-polarized light in the case of insulating samples below the percolation threshold ($x < x_c$, Fig. 1) but does not exist for s-polarized light and in the case of metallic samples ($x > x_c$, Fig. 2). Therefore, the difference between the MRE for p-and s-polarized light when $30^0 < \phi_0 < 70^0$ is noticeable but not well pronounced for metallic samples, and by contrast, can be very large in insulating samples. It allows us to explain experimental data on the MRE polarization dependence in CoFe-Al$_2$O$_3$ nanocomposites [12].

These results are in qualitative agreement with available experimental data [1,11, 12]. In some sense, it can be viewed as a confirmation of the theoretical approach to the MRE in nanocomposites developed in [1].

Besides, these results show how to further increase the MRE by tuning a granular alloy composition and an incident angle.

It should be pointed out that angular dependence of the MRE might be more complicated than shown on Figs. 1, 2 if the interference of light takes place. It is rather straightforward to take it into account by considering three-layer model, namely, transparent insulator- magnetic sample of finite thickness –semi-infinite substrate.

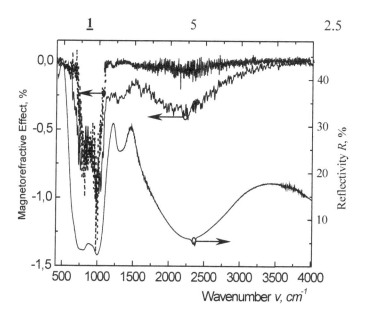

Figure 3. Frequency spectra of MRE in a magnetic field of 1700 Oe for a $(Co_{0.4}Fe_{0.6})_{48}(Mg_{52}F)$ film at $\phi_0=10^0$ (solid line) and $\phi_0=45^0$ (dot line).

Let us now discuss the puzzling behavior of the MRE in nanocomposites CoFe-MgF shown on Fig. 3 for the two incident angles $\phi_0=10^0$ and $\phi_0=45^0$. One can see on Fig. 3 that if the MRE increases with ϕ_0 for p-polarized light at $\lambda=2.5 \div 7.5$ μm, as could be expected from the above calculations, by contrast, the MRE is *less* for $\phi_0=45^0$ than that for $\phi_0=10^0$ in the vicinity of $\lambda=10$ μm. From the analysis of experimental data on the MRE at $\phi_0=10^0$ it was concluded in [1-3] that the theory fails to explain the frequency dependence of the MRE in nanocomposites CoFe-MgF. It was proposed that the matrix MgF or magnetic clusters embedded in matrix MgF are responsible for the disagreement between the theory and experiment. The data on the MRE for $\phi_0=45^0$ (Fig. 3) are also in a contradiction with the theory based on the Fresnel formula (1). To explain this behavior one should suppose that spin-dependent tunneling probability depends on ϕ_0, that seems completely impossible, or to ascribe it to an unknown mechanism of the MRE in CoFe-MgF. So, the angular dependence of the MRE in CoFe-MgF also indicates on an additional mechanism of the MRE, associated with magnetic excitations in MgF matrix.

In conclusion, the calculated angular and polarization dependencies of the MRE in magnetic nanocomposites provide a qualitative explanation for a number of experimental data and promise the attainment of higher magnitude of the MRE. However, the developed theory is unable to describe the specific features of the angular and frequency dependencies in nanocomposites CoFe-MgF.

ACKNOWLEDGMENTS

This study was supported in part by the Russian Foundation for Basic Research (grant no. 03-02-16127), and   by the Scientific Program "Universities of Russia" (grant no. 01.03.003).

A.G. is grateful to Toyohashi University of Technology (Japan), and to Ecole Polytechnique, Universitaire de Marseille (France) for hospitality.

REFERENCES

1. A. Granovsky, I. Bykov, E. Gan'shina, et al. *JETP 96* (2003) 1104.
2. A. Granovsky and M. Inoue, *J. Magn. Soc. Korea* 8 (2002) 45.
3. A. Granovsky and M. Inoue, *Proceedings of the International Conference on Magnetism, ICM-2003*, Rome, 26 July –1 August 2003, Italy. (to be published in *J. Magn. Magn. Mat.*)
4. J.C. Jacquet and T. Valet, *in Magnetic Ultrathin Films, Multilayers and Surfaces, MRS Symposium Proc.* **384** (1995) 477.
5. A.B. Granovsky, M.V. Kuzmichev, and J.P. Clerc, *JETP* **89** (1999) 955.
6. S. Uran, M. Grimsditch, E. Fullerton, and S.D. Bader, *Phys.Rev.B* **57** (1998) 2705.
7. V.G. Kravets, D. Bosec, J.A.D. Matthew et al., *Phys. Rev. B* **65** (2002) 054415-1.
8. Yu.P. Sukhorukov, N.N. Loshkareva, E.A. Gan'shina et. al., Techn. Phys. Lett. **25** (1999) 551.
9. G.S. Krinchik and V.S. Gushchin, *JETP Lett.* 10 (1969)24
10. G.A. Niklasson and C.G. Granqvist. J. Appl. Phys. **55** (1984) 338211 I. Bykov, E. Ganshina, A. Granovsky, and V. Guschin, *Phys. Sol. State* **42** (2000) 498.
12. D. Bozec, V.G. Kravets, J.A.D. Matthew et al., *J. Appl. Phys.* **91** (2002) 8795.

Index